Free Radicals in Chemistry and Biology

Authors

Milan Lazár, Ph.D., D.Sc.
Polymer Institute
Slovak Academy of Sciences
Bratislava, Czechoslovakia

Jozef Rychlý, Ph.D.
Polymer Institute
Slovak Academy of Sciences
Bratislava, Czechoslovakia

Viliam Klimo, Ph.D.
Polymer Institute
Slovak Academy of Sciences
Bratislava, Czechoslovakia

Peter Pelikán, Ph.D., D.Sc.
Department of Physical Chemistry
Slovak Technical University
Bratislava, Czechoslovakia

Ladislav Valko, Ph.D., D.Sc.
Department of Physical Chemistry
Slovak Technical University
Bratislava, Czechoslovakia

CRC Press, Inc.
Boca Raton, Florida

Library of Congress Cataloging-in-Publication Data

Free radicals in chemistry and biology.

Bibliography: p.
Includes index.
1. Free radicals (Chemistry) 2. Biochemistry.
I. Lazár, Milan.
QD471.F735   1989   541.3'94   88-10399
ISBN 0-8493-5387-4

Direct all inquiries to CRC Press, Inc., 2000 Corporate Blvd., N.W., Boca Raton, Florida, 33431.

© 1989 by CRC Press, Inc.

International Standard Book Number 0-8493-5387-4

Library of Congress Card Number 88-10399
Printed in the United States

PREFACE

Although the modern history of radical chemistry is only some decades old, the results of research in this field are of enormous value. The contemporary study of reactions and properties of free radicals is required not only by theoretical chemistry but also by technological applications of radical reactions, such as combustion, production of plastics and rubber, thermal processing of crude oil, halogenation of organic compounds, oxidation of hydrocarbons, stabilization of polymers, motor oils, edible oils, and medications against aging, etc. Numerous reactions in cosmic space on one hand and in living cells on the second hand also proceed by radical mechanisms. Radical reactions have thus significant impact on many processes in the chemical industry and energetics and on the clarification of chemical changes in both living and nonliving nature.

In this text an attempt has been made to outline particular features of free radical chemistry as they are displayed in each respective field. We hope that such an approach may contribute to the search for a common language among specialists of diverse branches of science and industry. On the other hand, however, we were aware that not all subjects could be included or equal weight given to individual topics.

In preparing the manuscript of the book we have followed the classical division of contents involving generation of free radicals, their elementary reactions, and their reactivity. Successively we have built up the more complex systems such as nonbranched and branched chain reactions, oscillatory reactions, etc. In our view, the examples throughout the book cover the new and most illustrative cases of radical occurrence in chemistry and biochemistry. The latter are, of course, considerably less numerous since there still prevails an idea that free radicals are some "enfant terrible" of well-functioning living organisms. One thing is sure now, that they can be neither omitted nor their role be overestimated in living systems. This may be ensured only by the correct understanding of their potential transformation pathways.

It was not our goal to quote all available literature on free radicals in chemistry and biology. We have considered it more desirable to put forward a synthesis of a limited number of overviews and important original articles, where the given topic was already critically compiled or brought some new ideas and conceptions which attracted our attention.

We wish to express our sincere gratitude to the staff of CRC Press, particularly to Ms. Marsha Baker and Ms. R. Taub who improved the final version of the book, to Mrs. T. Lazar and L. Rychla for valuable assistance in typing the manuscript, to Ms. D. Farkas for her help in drawing the figures, and finally to all colleagues of the Polymer Institute of Slovak Academy of Sciences and Technical University of Bratislava who in some way contributed to the accomplishment of the manuscript.

Bratislava, May, 1987.

M. Lazár
J. Rychlý
V. Klimo
P. Pelikán
L. Valko

THE AUTHORS

Milan Lazár, Ph.D., D.Sc., was born in 1927 and was graduated from Slovak Technical University in 1951. He received his Ph.D. in 1957 and his D.Sc. in 1967. Between the years 1951 and 1960, Dr. Lazár was a research worker at the Institute of Cables and Insulators. Since 1960 he has been conducting research at the Polymer Institute of the Slovak Academy of Sciences, of which he was head from 1960 to 1970. Dr. Lazár has coauthored over 120 articles primarily on the subject of free radicals in polymer systems. He has also coauthored several books on the subject.

Josef Rychlý, Ph.D., was born in 1944 and was graduated from Technical University in Pardubice in 1966. He obtained his Ph.D. in 1971. Since 1967, Dr. Rychlý has been a research worker at the Polymer Institute of the Slovak Academy of Sciences. His area of research was stabilization of free radicals. Since 1977, he has been dealing with thermo-oxidation and flammability of polymeric materials. In 1981, he spent 6 months in the Laboratory of Combustion of Professor M. Lucquin at University Lille in France. Dr. Rychlý has coauthored more than 80 scientific papers and two books.

Viliam Klimo, Ph.D., was born in 1947 and was graduated from Slovak Technical University in 1971. He obtained his Ph.D. in 1977. Since 1971, Dr. Klimo has been a research worker at the Polymer Institute of the Slovak Academy of Sciences. His area of expertise is in quantum chemistry and the theory of elementary chemical processes. Dr. Klimo has coauthored more than 30 scientific papers and two books.

Peter Pelikán, Ph.D., D.Sc., was born in 1939 and studied physical chemistry at Slovak Technical University where he received his M.S. degree in 1969. He obtained his Ph.D. in 1971. Since 1971 he has been a reading lecturer at the university. For more than 15 years Dr. Pelikán has devoted his research to quantum chemical calculations of electronic structure and the reactivity of molecules, especially of transition metal compounds. He has published over 150 articles in the area of quantum chemistry covering theoretical problems of coordination chemistry, stereochemistry, catalytic activity, and radical chemistry. He is also the coauthor of several books on these subjects.

Ladislav Valko, Ph.D., D.Sc., was born in 1930 and was graduated from Slovak Technical University in 1954. He received his Ph.D. in 1961 and D.Sc. in 1979 from Slovak Technical University and the Commenius University, respectively. Dr. Valko's research is in the area of reaction mechanisms, with special emphasis on thermal elimination and isomerization reactions. He spent one year as a Research Fellow at Dr. Marchal's laboratory at C.N.R.S. in Strasbourg. In 1981 he was awarded the state prize. Currently, Dr. Valko is professor of Physical Chemistry at Slovak Technical University in Bratislava and corresponding member of Slovak Academy of Sciences. He has coauthored numerous articles and books.

TABLE OF CONTENTS

Chapter 1

INTRODUCTION

The first experimentally founded opinions on free radicals date to the beginning of our century when the existence of triphenylmethyl radicals was disclosed incidentally upon the preparation of hexaphenylethane.[1] In interpreting this unexpected result, Gomberg abandoned Kekule's commonly accepted ideas of four-valent carbon atoms in organic compounds and advanced a new concept of the existence of free radicals as a reactive by-product in some reactions. The properties of "a compound with three-valent carbon" stimulated chemists to synthesize analogous compounds. The molar weight and paramagnetic susceptibility of aryl ethanes were measured somewhat later and the dissociation degree of ethane into free radicals varying with the type of substituents was determined, accordingly.

Generalization and application of free-radical theory in interpreting chemical reactions, however, proceeded into the body of chemical knowledge very slowly. Even in the consciousness of prominent chemists, these ideas were viewed more as rarities than as a new way of interpreting some known but misunderstood, apparently mystical reactions (explosive decomposition of oxidized diethyl ether, combustion, etc.). The growth of free-radical theory was hindered particularly by the rapidly developing ionic theory, which successfully explained the course of numerous substitution reactions on aromatic compounds. If we realize that the chemistry of that time was focused on reactions just of these compounds, then it is not to be wondered at, that the idea of the important role of free radicals in chemical reactions became plausible only after the failure of ionic mechanisms in some reactions.[2,3] Especially important in this respect was the implication of advantages of free-radical theory in the kinetics of chain reactions.[4]

The tumultuous period of reevaluation and formulation of new ideas in chemistry may well be documented in the case of the paper on synthesis and properties of the stable α,γ-bis(di-)phenylene-β-phenylpropenyl radical. The manuscript of the paper was originally rejected and published only after 25 years in its original version.[5] The emergence of free radical theory into biochemistry was even longer overdue. In many textbooks of a biochemical nature the term radical does not appear until the 1980s. Only in recent times has the knowledge of free radical chemistry gradually become applied to living matter.

The proof of the existence and the search for new free radicals encouraged the development and construction of ingenious instruments. As a result of these new experimental methods, more profound information about the properties of reactive free radicals was obtained. Indeed, the most influential contribution to the present state of knowledge about free radicals was the discovery of electron spin resonance (Zavoiskii, 1944); this discovery led to the determination of the structure of many radicals and to the invention of chemically induced spin polarization (H. Fischer, H. R. Ward, 1967), which made it possible to differentiate the products formed by fast radical or nonradical mechanisms.

The concentration of free radicals as intermediates of chemical transformations of molecules in reaction medium is usually very low. When compared to neutral molecules, the number of free radicals in the volume unit is often 8 to 10 orders lower. The fact that free radicals play an important role in reaction mechanisms even at such extreme dilution documents their high reactivity. The high reactivity is usually attributed to free radicals as the principal feature of particles with unpaired electrons and with the tendency to form new covalent bonds.

The common feature of all radical-reaction mechanisms is that they may be constructed from several kinds of elementary reactions. The variety of radical reactions then follows from the different structure of radicals and from the large spectrum of mutual interrelations of respective elementary reactions.

A newly formed radical may participate in successive elementary reactions with surrounding particles. By repetition of such subsequent steps the chain character of the reaction may be asserted. A small number of free radicals are thus capable of evoking the chemical transformation of a large assembly of molecules. This kinetic viewpoint may be illustrated by the stoichiometric coefficient x of radical regeneration (Scheme 1).

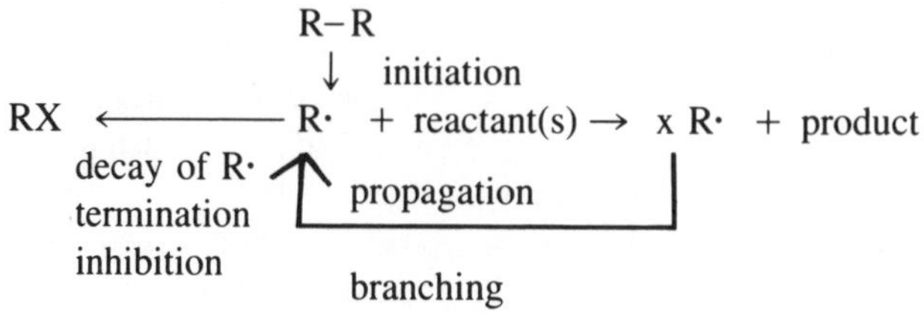

$$R-R$$
$$\downarrow \quad \text{initiation}$$

RX $\longleftarrow$ R$\cdot$ + reactant(s) $\rightarrow$ x R$\cdot$ + product

decay of R$\cdot$ $\uparrow$
termination   propagation
inhibition

branching

SCHEME 1

Depending on its value, we distinguish chain (x > 0) and nonchain (x = 0) complex reactions of radicals. If one radical gives only one new radical (x = 1), the chain reaction is nonbranched. In branched chain reactions (x > 1), two radicals are formed from one radical in some propagation step.

The requirement of the presence of suitable thermodynamic driving forces is a fundamental concept here. It involves the insertion of a relatively weak bond into the substrate and the formation of a relatively stable strong bond in the product.

The generation of free radicals in chain reactions is called initiation. The transformation of reactants to products occurs in some elementary reaction of a more or less complex sequence of the propagation stage. For example, a poly-addition chain reaction has a relatively simple pattern, in which primary radicals R$\cdot$ are linked together with molecule M. The forming radical RM$\cdot$ adds another molecule M and the same kind of addition repeats many times until the macroradical R(M)$\cdot_n$ terminates. The kinetic chain of addition reactions ends by combination of two radicals into one molecule, or by disproportionation, which gives two molecules. Somewhat more complicated is a chain reaction composed of transfer reactions which are mutually interrelated. One kind of transfer reaction initiates the formation of new radicals which make possible the course of the first reaction.

The cycle of the propagation reactions may be, of course, composed of more than only addition or several transfer reactions, but also from other types of propagation elementary steps, such as fragmentation or isomerization of radicals, etc.

The specific feature of chain reactions is large sensitivity of reaction rate to all reactions which either increase or decrease the radical concentration in the system. The resulting effect of changes of radical concentration is inhibition or retardation and the appearance of distinct limits at which a branched chain reaction acquires an explosive course.[6]

The chain reactions as a whole are exothermic processes where the released reaction heat may also contribute to their propagation. In the time lag before the occurrence of respective elementary reaction, the reaction heat may be nonhomogeneously distributed over the system of reacting species. This may cause further acceleration.

The complex reactions may occur, in fact, in each system where free radicals are present. According to the reactivity of the substrate, the respective type of complex reaction is realized. The presence of an unpaired electron in the active intermediate is, however, always essential at such kind of reactions. A deeper insight into the mechanisms of radical reactions may be achieved from the detailed study of respective elementary reactions. Analyzing their course, the conception of more subtle intermediates, such as activated transient complexes which should appear on the reaction pathway from radicals to products, seems to be of importance, too.[7] The structure of the transient state of the elementary reaction may be deduced from experimental data on the structure and properties of radicals, reactants, and reaction products appearing in a given reaction step.

The unpaired electron in atoms may be located either on its s-orbital (hydrogen, atoms of alkaline metals) or on the p-orbital (atoms of halogens, oxygen, nitrogen, etc.). The unpaired p-electron may also be located on the hybridized s-p orbital of the carbon atom which is the integral part of a larger polyatomic particle. Some compounds of transition elements have unpaired electrons on d-orbitals of corresponding atoms.

Reactivity of atoms and radicals depends on the type of orbital and atom on which the unpaired electron is localized, and on neighboring atoms forming radicals. As a consequence, there exists a large group of free radicals which are stable due to either steric hindrance of a radical site or to extensive delocalization of a radical center which does not allow it to enter some elementary reaction.

REFERENCES

1. **Ihde, A. J.,** The history of free radicals and Moses Gomberg's contributions, *Pure Appl. Chem.,* 15, 1, 1967.
2. **Mayo, F. R.,** The evolution of free radical chemistry at Chicago, *J. Chem. Educ.,* 63, 97, 1986.
3. **Walling, Ch.,** Forty years of free radicals, in *Organic Free Radicals,* Pryor, W. A., Ed., ACS Symp. Ser. 69, Washington, D.C., 1978, 1.
4. **Semenov, N. N.,** *Chemical Kinetics and Chain Reactions,* Clarendon Press, Oxford, 1935.
5. **Koelsch, C. F.,** Synthesis with triarylvinyl magnesium bromide α, γ-bisdiphenylene-β-phenylallyl, a stable radical, *J. Am. Chem. Soc.,* 79, 4439, 1957.
6. **Bamford, C. H. and Tipper, C. F., Eds.,** *Comprehensive Chemical Kinetics, Gas Phase Combustion,* Elsevier, Amsterdam, 1977.
7. **Benson, S. W.,** The range of chemical forces and the rates of chemical reactions, *Acc. Chem. Res.,* 19, 335, 1986.

Chapter 2

GENERATION OF FREE RADICALS

I. INTRODUCTION

Free radicals as particles with unpaired electrons are usually formed at homolytic fission of some chemical bond from the molecule of a proper precursor. The selectivity of the respective bond splitting depends on the bond dissociation energy and on the manner of energy supply.

Mechanical rupture of the macromolecule when each of the two newly formed fragments retains one electron from the bond electron pair, linking the atoms of the macromolecular chain, is the most illustrative example. Energy necessary for bond cleavage may be supplied, furthermore, as thermal or electromagnetic radiation having the form of ultraviolet light, $\gamma -$ and roentgen radiation. Free radicals also appear directly in the chemical reactions of molecules where the rearrangement of reactants may provide energy sufficient for splitting of the weakest bonds. In the process of subsequent decay and a new formation of the chemical bond, free radicals then play the role of reaction intermediates. Sometimes, the visualization of unpaired electrons from the electron structure of transition metals ions, by their redox reactions with a suitable substrate, may be considered as the manner of free radical generation, too.

Most, if not all, methods of generation are usually complex, and formation of free radicals proceeds as a result of several complementary processes. In electrical discharge, for example, the molecules release the radicals and ions not only by their collisions with electrons and ions but also as a result of the high temperature of corresponding plasma.

Because of their high reactivity, primarily formed free radicals cannot be easily detected and only the more stable radicals, which are produced in the sequence of reactions, are observed very often.

By the generation of free radicals from molecular precursors free radicals are formed in pairs. In each pair, radicals are in close proximity and may react reversibly to form the original molecule. The escape of free radicals from the initial pair occurs predominantly in the gaseous phase. In the condensed system, the diffusion of particles is markedly slower and only a part of the free radicals from pairs gets out of the "cage" of surrounding molecules. From this aspect, conditions, are more favorable in electrolytical oxidation of some anions at electrodes. Free radicals do not appear there in pairs, but their spatial distribution is heterogeneous, the concentration of free radicals in the system being lower than that at the electrode.

II. THE FORMATION OF FREE RADICALS BY MECHANICAL FORCES

The interaction energy in the assemblies of low molecular compounds (100 atoms at maximum) related to one molecule is one or two orders lower than the dissociation energy of the arbitrary chemical bond in the molecule. The molecules of such compounds can, therefore, only exceptionally be homolytically decomposed by the effect of mechanical forces, and the yield of free radicals will be consequently very low. On the other hand, in the case of macromolecules, intermolecular interactions may easily prevail over the strength of chemical bonds and forced deformation of the polymer will always be accompanied by the splitting of a main chain of the macromolecule.

Mechanical degradation of macromolecules is the result of the cooperative effect of mechanical forces and statistically fluctuating oscillations of particles of the surrounding medium. The potential energy barrier of the strained bond is decreased by mechanical deformation of the macromolecular system so that the heat energy is sufficient to break the most weakened and strained bonds. Hence, the fracture of the chemical bond by the effect of mechanical forces may be understood as the process of mechanically activated thermal decomposition.[1]

Estimating the efficiency of scissions of chemical bonds at mechanical degradation of macromolecular compound, the value 2.10^{-15} cm^2 of the effective cross-section of the molecule was considered; the length of molecule compared to this value may be taken as infinitely high. In such a case the cut across the packed array of macromolecules should yield $5 \cdot 10^{14}$ free radicals at maximum per 1 cm^2 of newly formed surface of polymer compounds.[2] Such an amount of free radicals was observed experimentally when a polymer sample was cut by a sharp instrument.*

A somewhat different situation was found in a polyamide fiber which instead of being cut was stretched until the fiber was broken down. In this type of procedure, about 100 times more free radicals were formed than when the fibrous sample was cut with sharp scissors. The reason for such a large difference consists in the fact that stretched macromolecules always break down at their weakest bonds, not only at one chosen cross section. For the appearance of one macrocrack in polyamide fiber, one strained macromolecule should fracture, e.g., at 100 weak bonds, approximately along its backbone. The breaking of macromolecules in the stretched fiber does not occur all at once but gradually.[3]

This may explain why the actual strength of polymer samples is considerably lower than the theoretical values corresponding to the strength of individual bonds in macromolecules. To achieve the theoretical strength the fracture of all macromolecules should occur simultaneously and not as a stepwise impairment of the weakest links.

In mechanically strained macromolecular systems, free radicals are formed, of course, even before the macroscopic fracture of the sample. This is the case of the initial phases of a polymer tearing off and pressing at the extrusion and mixing of polymer melts and solutions. In the presence of solvents, the yield of free radicals decreases to a larger extent than corresponds to the decrease of concentration of macromolecules in the system. This may be interpreted by the increased possibilities of macromolecule motion in the system and thus by easier adaptability to the effect of external deformation forces.

Similar to the effect of a solvent, is the effect of increasing temperature up to the critical value of thermal cleavage of chemical bonds. At higher temperatures, macromolecules become more mobile and the yield of primary radicals generated by mechanical destruction decreases.

Rupture of chemical bonds in macromolecules occurring during their mechanical degradation may well be observed on records of the electron spin resonance spectra which make it possible to estimate the concentration of free radicals in the sample as well as their structure. For the registration of ESR spectra of highly reactive carbon-centered free radicals in organic polymers, it is necessary to perform the measurements at low temperature of, at least, liquid nitrogen (77 K) and at perfectly inert atmosphere free of oxygen.[4,5]

Free radicals in polymer systems containing incorporated sulfur may be observed considerably more easily. In the crosslinked natural rubber, free radicals are formed which, at room temperature and in the air, decay only after several tens of hours.[6]

Primarily formed radicals are usually transformed to more stable ones. The scission of polyoxymethylene macromolecules at first gives two kinds of radicals, namely, $RCH_2O \cdot$ and

---

*   The determination of the amount of formed free radicals was based upon the presence of stable low molecular free radicals scavenging reactive radicals arising at the cut surface of the polymer. The estimation of the difference in concentration of these stable radicals in the initial sample and after cutting is relatively simple and may be performed by some of the available methods (ESR spectrometry, colorimetry, etc.).

ROĊH$_2$. These primary particles subsequently abstract hydrogen atom from unbroken parts of macromolecular chains and change thus to one type of secondary radicals, ROCHOR. The radical formation is mainly caused here by ruptures of the tie-molecules linking crystalline regions.[7]

As was already pointed out, the formation of free radicals depends on the extent of deformation, and the high concentration of radicals may be recorded even before the macroscopic rupture of the sample. This fact explains the chemical reasons for fatigue and aging of organic polymers during their exposure to mechanical forces. The alkyl radicals formed there are prone to react with atmospheric oxygen and the original structure and the properties of the polymer system are changed accordingly.

Mechanical rupture of macromolecules and production of free radicals may also occur in biological systems. It may be assumed that surgical operations, ripping of muscles, smashes, and bruises are accompanied by the formation of free radicals in macromolecular tissue. Relatively very high levels of free radicals observed in the hearts of animals may reflect either mechanical rupture of some protein macromolecules or radical intermediates of reactions in intense metabolism. Since no experimental studies are available to date, speculations about the role of bioradicals formed by mechanical degradation of macromolecules in living systems seem to be premature.

To complement the known facts about mechano-generation of free radicals, it is necessary to point out some results obtained with mechanical destruction of hydrocarbon polymers using the glass ball mill method, where products of an ionic nature were observed along with free radicals.[8]

The assumption of heterogeneous scission of polyolefin macromolecules as the first reaction step does not appear to be very likely. Ions apparently may be formed in subsequent heterogeneous reactions of free radicals with reactive centers on the glass surface. Such an idea is supported by observation of the emission of electrons and cations accompanying the fracture of polybutadiene[9] and epoxy composites[10] which is larger in the case of filled than unfilled polymer.

The effect of shock waves on the condensed medium is another example of mechanochemical generation of free radicals, complicated, moreover, by the effect of increased temperature. The method consists of applying a rapid increase of pressure (up to 100 GPa) over a short time interval (<1 nsec) which leads to the appearance of a strong shear in the sample and to an increase in temperature.

For solid materials (anthracene, acrylamide, etc.) exposed to the shock wave there may be registered an ESR spectrum which is always a narrow single line having g = 2.002, which corresponds to organic free radicals.[11]

We remind the reader here that similar spectra may be received after mechanical grinding of solids such as anthracene, acridine, and calophony in achate mortar.[12] Shock waves do not only generate radicals but also influence their propagation reactions which may sometimes lead to quite unexpected reactions which are difficult to be realized in a nonshocked system.

Of interest is also the effect of shock waves on gaseous hydrocarbons. Methane, for example, under such conditions may be pyrolyzed to ethylene, acetylene, hydrogen, and carbon.[13] The temperature generated in the shock tubes where methane is adiabatically compressed to 2 MPa at a reaction time of 2.5 msec is between 1750 and 2700 K. In such a case, the course of the process is, however, almost identical with the pure pyrolysis.

III. THERMAL HOMOLYSIS OF CHEMICAL BONDS

Thermal heating of the system of molecules increases the velocity of their motion and mutual collisions bringing about more and more intense reversible deformations. At certain temperatures some of the weakest bonds in polyatomic molecules break down and each molecule cleaves into two free radicals. From the overall assembly of molecules only those which have acquired certain critical energy in several incidental collisions decompose.

The process of transformation of kinetic molecular energy to the energy which causes fission of chemical bonds proceeds through several stages. The first step, the accumulation of internal energy of a molecule by one strong collision or several collisions, can be the determining rate in the overall process. This is usually true for fissions of atoms from molecules containing less than five atoms. Also, at a sufficiently high temperature, the decomposition of all types of molecules becomes collision controlled and hence bimolecular.[14]

More complex molecules usually decompose by the first order, a unimolecular mechanism with a rate controlled by bond fission. The ideas about the character of such unimolecular reactions led to the development of the RRKM (Rice-Ramsperger-Kassel-Marcus) theory[15] which is able to predict successfully the rate constants of thermolytic bond fission.

According to this theory, the redistribution of collision energy has a statistical character. The excess of internal energy can be divided between the fixed and nonfixed parts. The first one is not effective in the unimolecular decomposition. Only the second part, which must be larger than the critical amount of energy necessary for bond scission is of importance here. Mainly the vibrational modes are effective during the decomposition step and the energization process is considered as one of translational-vibrational energy transfer. The time interval of energy redistribution among all vibration degrees of freedom is not greater than 10^{-11} sec.

The lifetime of energized molecules and their decomposition to the products can be described by statistical considerations and by ideas of transition state theory. The rate constant for conversion of energized molecules to products is proportional to the statistical probability that the critical energy will be found in the respective vibrational mode leading to the scission of a given bond. The main parameters which play an important role here are the number of rotational-vibrational states and their density for the given excitation energy in the energized molecule and in the corresponding transition state configuration.

The probability of decomposition of a compound at a certain temperature depends above all on the strength of the weakest bond. The data in Table 1 on thermolysis of bonds give, however, only a very approximate orientation in their actual reactivity since standard bond energy may be markedly different from the dissociation energy, which is the only decisive factor of the facility of bond splitting.*

Free radicals coming into existence in homolysis of simple biatomic molecules are in dynamic equilibrium (Equation 1) with parent molecules

$$A - A \rightleftharpoons A\cdot + A\cdot \tag{1}$$

The equilibrium shifts towards the higher concentration of free radicals with increasing temperature and depends on dissociation energy of bonds linking the respective atoms of the molecule. At 2000 K and at atmospheric pressure, iodine [D(I-I) = 147 kJ·mol^{-1}] is almost entirely dissociated to atoms (~95%) while hydrogen [D(H-H) = 436 kJ·mol^{-1}] only imperceptibly (~0.1%).

---

* For diatomic molecules the dissociation energy is equal to the standard energy of the bond. This is not, however, the case of three or polyatomic molecules where the difference between energy calculated from overall atomization energy and dissociation energy exists. As an example we may give the three-atomic molecule A-B-A consisting of two kinds of atoms A and B. The average (standard) energy of the A-B bond may be calculated from the atomization energy of the molecule divided by two. As a rule, energy necessary for dissociation of one bond A-B in the molecule A-B-A (dissociation energy) differs from the standard energy of bond A-B. A relatively high difference may exist also for one kind of chemical bond in different compounds.[18]

Table 1
STANDARD BOND ENERGIES[16,17] AND
APPROXIMATE TEMPERATURE OF
THERMOLYSIS AT WHICH HALF OF ORIGINAL
CHEMICAL BONDS DISSOCIATES PER 1 HR

X – Y	D kJ/mol	T °C	X – Y	D kJ/mol	T °C
H – O	464	980	H – H	436	1090
H – C	414	850	C – C	357	670
H – N	389	780	S – S	229	440
C – O	351	680	Si – Si	226	430
H – S	343	650	F – F	169	260
C – Cl	334	630	N – N	162	230
C – Br	280	480	Se – Se	160	230
C – S	272	460	O – O	140	160
C – I	230	350	Li – Li	103	50
O – Cl	205	280	Na – Na	69	– 60
O – F	188	235	K – K	55	– 100

A. The Wall Effect

Adsorption of forming radicals on the walls of the reaction vessel and especially chemisorption of the precursor of free radicals can play an important role in the overall course of the thermal decomposition of molecules. The wall effect manifests itself at thermolysis of gases at lower temperatures and at low concentrations of radicals in the system. It may accelerate the decomposition reaction as much as by several orders on one hand, and influence the rate of radical decay on the other hand. The effect depends on the wall quality and composition as well as on the designed surface treatment, its history, and on its degree of aging. The wall effect of the reaction vessel consisting in the decrease of dissociation energy of the breaking bond is partially brought about by the contribution of the heat Q released during adsorption (Equation 2) of two radicals A on the surface of the reactor walls.

$$AA + surface \rightarrow \cdot A_{ads} + \cdot A_{ads} + Q \qquad (2)$$

Since desorption of atoms or low molecular radicals from the reactor walls usually require lower energy than corresponds to the dissociation energy of the weakest bonds in the nonadsorbed molecule, the process of thermolysis is accelerated. The wall effect at chemisorption leads to the replacement of the original A-B bond by two new bonds A-M and B-M with metal M, of which at least one is weaker than the original A-B bond.

At chemisorption of hydrogen on polycrystalline platinum, the dissociation energy of Pt-H bond is 226 kJ·mol^{-1}, which is approximately one half of the dissociation energy of the H-H bond.[19]

Information about thermolysis of bond O-H in chemisorbed deuteroformic acid on the surface of a single crystal of Cu (110) (face-centered cubic crystal of copper with the plane view — Scheme 1)

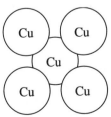

Scheme 1

Table 2

**TEMPERATURE OF THE
THERMOLYSIS OF
HYDROCARBONS R^{1}R^{2}R^{3}C–CR^{1}R^{2}R^3
AT WHICH HALF OF THE
ORIGINAL MOLECULES IS
DECOMPOSED DURING 1 HR.[23-25]**

R^1	R^2	R^3	°C ($\tau/2 = 1$ hr)
H	H	H	695
CH$_3$	H	H	590
CH$_3$	CH$_3$	H	565
CH$_3$	CH$_3$	CH$_3$	490
C$_2$H$_5$	CH$_3$	CH$_3$	420
C$_2$H$_5$	C$_2$H$_5$	C$_2$H$_5$	285
t–C$_4$H$_9$	CH$_3$	CH$_3$	195
t–C$_4$H$_9$	t–C$_4$H$_9$	H	141

may be obtained from the temperature dependence of desorption of reaction products.[20]

Hydrogen appears at 273 K, indicating that the O-H bond in formic acid is already broken below this temperature, which is lower by at least 600 or 700 K than the temperature of homogeneous thermolysis in the gaseous phase. The evolution of D$_2$ and CO$_2$ at 475 K, both appearing simultaneously, with identical rates, indicates that only single-surface species should decompose. The D$_2$/CO$_2$ ratio equalled 0.5 and simultaneously performed spectroscopic measurements confirmed the stoichiometry of the intermediate to be ·DCO$_2$. Such particles may be formed by the metallation of the O-H bond of the formic acid by copper. The proposed scheme (Equations 3 to 5) of dehydrogenation of deuteroformic acid on the crystalline surface of copper (the data above the arrow indicate the temperature of the beginning of the apparent course of the reaction)

$$(Cu)_n(DCOOH)_2 \xrightarrow{180\ K} (Cu)_n(DCOO) + (Cu)_n(H) + DCOOH \tag{3}$$

$$2(Cu)_n(H) \xrightarrow{273\ K} H_2 + 2(Cu)_n \tag{4}$$

$$(Cu)_n(DCOO) \xrightarrow{475\ K} D_2 + CO_2 + (Cu)_n \tag{5}$$

demonstrates the wall effect of the reactor more clearly. Even though it involves the generation of free radicals by chemical reaction, we cannot speak about pure process of chemical bond thermolysis. The different transfer of excitation energy taking place between the surface of the reactor and the precursor of radicals, particularly at various physical methods of free radical generation, does not seem to be neglected in the interpretation of wall effect, as well.[21,22] The increased delivery of this energy from reactor walls shifts the dissociation equilibrium (Equation 2) to the left.

B. Thermolysis of Polyatomic Molecules

Free radicals primarily formed in decomposition of polyatomic molecules undergo not only reverse combination to parent molecules but also further subsequent reactions. At thermolysis of polyatomic molecules, there is thus formed the mixture of different free radicals.

Table 3
ARRHENIUS PARAMETERS OF PEROXIDE THERMOLYSIS IN HYDROCARBON SOLVENTS AND THE TEMPERATURE T FOR DECOMPOSITION OF HALF THE ORIGINAL PEROXIDE DURING 10 HR

Type of peroxide	log A s	E kJ/mol	T (K)	Ref.
Dialkyl peroxides				
$[CH_3O-]_2$	16.1	159	399	27
$[CH_3CH_2O-]_2$	16.1	156	392	28
$[(CH_3)_3CO-]_2$	15.9	159	403	28
Diacyl peroxides				
$[CH_3C(O)O-]_2$	15.8	134	341	28
$[C_2H_5C(O)O-]_2$	15.4	130	338	28
$[C_6H_5C(O)O-]_2$	13.9	125	351	28
$[CH_3CF_2C(O)O-]_2$	13.0	91	254	29
Peroxy esters				
$CH_3C(O)O_2C(CH_3)_3$	17.6	166	389	28
$(CH_3)_3CC(O)O_2C(CH_3)_3$	15.7	128	328	28
$C_6H_5C(O)O_2C(CH_3)_3$	15.5	143	370	28
$C_6H_5CH_2C(O)O_2C(CH_3)_3$	13.7	119	338	28
$(C_6H_5)_3CC(O)O_2C(CH_3)_3$	14.3	103	283	28
Tetroxides				
$[(CH_3)_3CO_2-]_2$	16.6	73	370	30
$[C_6H_5(CH_3)_2CO_2-]_2$	17.1	69	165	30
$[(CH_3)_3C(CH_3)_2CO_2-]_2$	17.5	68	160	30
$[(C_3H_7)(CH_3)_2CO_2-]_2$	19.6	76	163	30

A certain type of radical is received more easily when some of the bonds in the parent molecule have lower dissociation energy than the others. This may be achieved by the proper choice of a free-radical precursor. The dissociation energy of the respective bond is first determined by the kind of linked atoms, by the repulsion effect of neighboring and mutually nonbonded atoms, and by the degree of delocalization of the unpaired electron of the forming free radical (Table 2).

The rate of decomposition of the free radical precursor is usually expressed by the corresponding rate constant k or by $\tau/2$, at which one half of the initial amount of a given compound is decomposed at a certain temperature. From temperature dependence of values of k, Arrhenius parameters A and E from the known Equation 6

$$k = Ae^{-E/RT} \qquad (6)$$

can be easily determined.

For selective generation of free radicals by thermolysis we use compounds with some unstable bonds. Such compounds are widely used as initiators of free radical chain reactions. The best known thermal initiators — peroxides — have, e.g., bonds O-O in the molecule. The average O-O single bond energy is the smallest when compared with values for homonuclear single bonds X-X where X is a neighboring atom of the same period or of the same group of the Periodic Table. The destabilization of the O-O bond results from lone pair electron repulsion. Lone pair-lone pair repulsion causes also a weakening of S-S or Se-Se bonds, but due to the larger covalent radii of these atoms the effect is much smaller than for the O-O linkage.[26] If destabilization resulting from lone pair repulsion is lower for some reason, a strengthening of the O-O linkage occurs and vice versa (Table 3).

Table 4
APPROXIMATE TEMPERATURES T OF DECOMPOSITION OF SOME PEROXIDIC INITIATORS AT WHICH THEIR HALF-TIME $\tau/2$ IN INERT SOLVENT IS 1 AND 10 HR

Peroxide	T/°C	
	$\tau/2 = 1$ hr	$\tau/2 = 10$ hr
HO–OH[a]	380	345
CH$_3$O–OH[a]	360	320
(CH$_3$)$_3$SiO–OC(CH$_3$)$_3$	250	225
(C$_2$H$_5$)$_3$GeO–OC(CH$_3$)$_3$	240	215
CF$_3$O–OCF$_3$[a]	240	215
(CH$_3$)$_3$COOH	195	170
C$_6$H$_5$(CH$_3$)$_2$CO–OH	170	145
(CH$_3$)$_3$CO–OC(CH$_3$)$_3$	150	125
[C$_6$H$_5$(CH$_3$)$_2$CO–]$_2$	135	115
(CH$_3$)$_3$CO–O(OC)C$_6$H$_5$	125	105
F$_5$SO–OSF$_5$[a]	120	105
C$_6$H$_5$(CO)O–O(OC)C$_6$H$_5$	95	75
$^-$O$_3$ SO–OSO$_3$$^{-}$[b]	85	65
[CH$_3$(CH$_2$)$_3$(CO)O–]$_2$	85	65
[(CH$_3$)$_2$CHO(CO)O–]$_2$	70	50
O$_3$[a]	30	15
HO–ONO$_2$[a]	15	−5
[C$_6$H$_5$ (CH$_3$)$_2$CO]$_2$	−16	−40

[a]   Decomposition in the gaseous phase.
[b]   Decomposition in water.

The effect of increased or decreased electron density on the peroxide oxygens, and thereby stabilization or destabilization of the peroxide, may be demonstrated on phenyl ring substitutions of dibenzoylperoxide. Electron withdrawing substituents in the meta and para positions retard decomposition while the reverse is true for electron donating substituents. Ortho substituents accelerate the decomposition by a steric effect. Certain ortho substituents (iodo, vinyl etc.) increase the rate of peroxide decomposition by anchimerically assisted reactions.[31]

The formation of free radicals by anchimerically accelerated bond homolysis of the generalized type (Equation 7)

$$R^1–A–B–R^2 \rightarrow ·A–B–R^1 + ·R^2 \qquad (7)$$

is possible if the new bond formed between B and R^1 is considerably stronger than the original bond between A and R^1. The rate-determining step is a combination of the neighboring group having empty orbitals and a heteroatom with a lone pair of electrons. Today, a number of unimolecular homolytic reactions with an anchimeric effect is known.[32]

Comparison of the decomposition rates[33,34] of structurally different peroxides indicates many possibilities in the choice of free radical initiators (Table 4). It should be borne in mind that data in the above two tables are valid only for diluted solutions of peroxides in an inert solvent. The concentrated solutions of peroxides have a lower temperature of decomposition, and their decomposition is of the chain and often explosive character.

In the first step of decomposition, the molecule of peroxide is broken down into two oxyradicals. The pair of oxyradicals remains in the place of its formation in the cage of surrounded molecules of liquid medium only about 10^{-11} sec. Before both radicals diffuse away, they may either reversibly dimerize or enter other elementary reactions. Back dimerization within a radical pair decreases only the rate of free radical production; any other reaction in the cage decreases the initiator efficiency.

At the decomposition of diacetyl peroxide, primary radicals partially decarboxylate to methyl radicals (Equation 8) which recombine with acetyloxy radicals to produce thermally stable methyl acetate (Equation 9)

$$CH_3COO\cdot \rightarrow \cdot CH_3 + CO_2 \tag{8}$$

$$\cdot CH_3 + CH_3COO\cdot \rightarrow CH_3COOCH_3 \tag{9}$$

From the viewpoint of the generation of free radicals, inefficient decomposition of the free radical initiator is a chain reaction of free radicals with the parent peroxide. The process is called induced decomposition. The peroxide decomposes thus into an inactive compound and only one free radical is formed per one free radical consumed. The overall number of free radicals potentially available for initiation of a given reaction is reduced. The free radical which induces the decomposition can be derived from the original peroxide molecule or, after the transfer reaction of this radical, from the reaction medium. The overall rate of peroxide decomposition, therefore, often depends on the character of the reaction medium.

The induced decomposition of dialkyl peroxides may occur as a hydrogen atom abstraction (Equation 10).

$$R_2CH-OO-CHR_2 + R_2CHO\cdot \rightarrow R_2\overset{\cdot}{C}OOCHR_2 + R_2CHOH$$
$$\downarrow \tag{10}$$
$$R_2CO + R_2CHO\cdot$$

As may be expected from the absence of reactive α-hydrogens in di-*tert*-alkyl peroxides and the presence of secondary and more reactive tertiary hydrogens in primary and secondary dialkyl peroxides, the susceptibility of dialkyl peroxides to induced decomposition increases in the order, *tert*-alkyl > alkyl > *sec*-alkyl peroxide.

The radical attack at the peroxidic O-O bond (Equation 11)

$$
\begin{array}{c}
\phantom{R-\overset{\cdot}{C}H-X +}\ O-R' \\
R-\overset{\cdot}{C}H-X + \ \Big| \qquad \rightarrow R-CO-R' + R'O\cdot \qquad (11) \\
\phantom{R-\overset{\cdot}{C}H-X +}\ O-R' \quad OX
\end{array}
$$

is particularly significant in reactions of nucleophile radicals with hydroxy or alkoxy groups in the α-position. As a matter of fact, rates of peroxide decomposition are appreciably higher in ethers or in primary and secondary alcohols (also amines) than in hydrocarbons.[33] Induced decomposition of dibenzoylperoxide is further complicated by radical addition to the para position of the phenyl group. When compared to dialkyl peroxide the stability of hydroperoxides is much more influenced by the reaction medium since spontaneous and induced decomposition[35] is usually accompanied by bimolecular reaction such as in Equation 12.

$$2\ R-O-O-H \rightarrow RO\cdot + RO_2^{\cdot} + H_2O \tag{12}$$

From the variety of known peroxide initiators, dibenzoylperoxide is the most often used; the temperature of its application is in the range of 60 to 90°C.

Initiation of radical reactions at zero degrees centigrade requires the usage of thermally unstable peroxides. Since the treatment of such compounds may already be dangerous, unstable peroxides are usually prepared ''in situ'', i.e., directly in the reaction medium from more stable compounds.[36] One example of several possibilities may be given on the synthesis of diethylperoxy dicarbonate from ethyl chloroformate and hydrogen peroxide in the presence of $NaHCO_3$ or $NaOH$[37] (Equation 13).

$$2\ C_2H_5OCOCl + H_2O_2 + 2\ NaOH \rightarrow C_2H_5OCO-OO-COOC_2H_5$$

$$+\ 2\ NaCl + 2\ H_2O \qquad (13)$$

Peroxides found wide applications as initiators of polymerization of vinyl monomers. They are the key intermediates at low temperature oxidation in atmospheric and stratospheric conditions, combustion, and in biochemical oxidative processes.

In addition to peroxides, azo-compounds and particularly azobisisobutyronitrile are also used. The decomposition of the latter is faster than that of dibenzoylperoxide at the same temperature. The equal rates of decomposition may be achieved at the shift of temperature of benzoylperoxide decomposition to about 10 K higher. The weakest linkages at the thermal decomposition of azobisisobutyronitrile are C-N bonds.[38,39]

$$(CH_3)_2C-N=N-C(CH_3)_2 \rightarrow 2(CH_3)_2C\cdot + N_2 \qquad (14)$$
$$\begin{array}{ccc} | & | & | \\ CN & CN & CN \end{array}$$

In the primary reaction step, two cyanoisopropyl radicals and one molecule of nitrogen (Equation 14) are formed. The question is whether the decomposition proceeds as a stepwise or concerted reaction.

The unsymmetrical phenyl-azo-triphenylmethanes are examples of azo-initiators which decompose unequivocally by stepwise scission (Equations 15 and 16) of C-N bonds.[40]

$$X(C_6H_4)-N=N-C(C_6H_5)_3 \rightarrow X(C_6H_4)-N=N\cdot + \cdot C(C_6H_5)_3 \qquad (15)$$

$$X(C_6H_4)-N=N\cdot \rightarrow X(C_6H_4)\dot{} + N_2 \qquad (16)$$

The stepwise reaction is now preferred to the concerted two-bond cleavage also in the case of symmetric azoalkanes.[41] It is of interest that *trans*-azoalkanes isomerize to *cis*-isomers before their dissociation.

As expected, the structure of azo-initiators has great influence on their thermal stability. Provided that methyl groups are bound to nitrogen instead of cyanoisopropyl groups, the temperature of decomposition of the azo-compound increases by about 200°C. As is demonstrated in Table 5, the thermal stability of some azoalkanes decreases with increasing bulkiness of alkyl substituents and with increasing delocalization of the unpaired electron on the radical produced in thermolysis of the azo-compound.

The effect of solvents on the course of thermolysis of azoalkanes is, however, considerably lower than that for peroxides. It consists mainly in different solvation of activated complex and parent molecules.

To the thermal initiators belong also organometallic compounds with dissociation energy of the metal-carbon δ-bond between 50 and 300 kJ/mol. For structurally similar compounds, the values of this dissociation energy depends on the position of the central atom in the

Table 5

ARRHENIUS PARAMETERS OF THERMOLYSIS OF
AZOALKANE[42-46] $(R^1R^2R^3C-N=)_2$ AND TEMPERATURE T OF
DECOMPOSITION AT HALF-LIFE 10 HR

R^1	R^2	R^3	A (sec^{-1})	E (kJ/mol)	T °C ($\tau/2$ = 10 hr)
H	H	H	$5.0\ 10^{15}$	214	275
CH_3	CH_3	CH_3	$7.6\ 10^{16}$	184	172
$C(CH_3)_3$	CH_3	CH_3	$3.8\ 10^{16}$	175	157
$CH_2C(CH_3)_3$	CH_3	CH_3	$4.1\ 10^{15}$	153	120
$CH_2C(CH_3)_3$	$CH_2C(CH_3)_3$	CH_3	$1.3\ 10^{14}$	129	85
CN	CH_3	CH_3	$2.6\ 10^{15}$	131	67
C_6H_5	C_6H_5	H	$3.0\ 10^{13}$	111	46

periodic system increasing in a given transition period from left to right and within a group on going from top to bottom.

Organometallic compounds such as dimethyl zinc [(H$_3$C-Zn) = 176 kJ/mol], diethyl zinc [D(H$_3$CH$_2$C-Zn) = 146 kJ/mol], tetraethyl lead [D(H$_3$CH$_2$C-Pb) = 130 kJ/mol], or tetramethyl lead [D(CH$_3$-Pb) = 146 kJ/mol] are relatively stable.[47] In the past, the organometallic compounds played an important role in the first-known procedures of the generation of alkyl radicals and in their proof by the method of metallic mirrors. Tetramethyl lead was thermolyzed in such a procedure at about 450°C, and generated methyl radicals reacted with metallic zinc mirror coated on the glass surface of the apparatus. Under such conditions the zinc mirror was gradually removed by its transformation to dimethyl zinc, which is volatile at a given temperature.

The thermal decomposition of tetraethyl lead has found wide application as the anti-knock additive in the combustion of some fuels in spark engines.

Alkyl transition metal compounds decompose at low temperatures. Intramolecular β-hydrogen abstraction can occur when the alkyl ligand has hydrogen atoms in the β-position, with respect to the metal which has an empty coordination site. The thermal stability of alkyl transition metal compounds can be markedly increased by suitable ligands, such as carbon monoxide, cyclopentadienyl, phosphites, and the cyano group which stabilizes the metal-carbon σ bond. Alkyl metal complexes of practically all transition metals are known.[48] Some of them are the key intermediates of many industrially important processes (e.g., of Ziegler-Natta low pressure polymerization of alkenes, hydroformylation, Fischer-Tropsch synthesis, alkene isomerization, and ethylene oxidation in the Wacker-Hoechst process).

Outside the laboratory or chemical plant, nature gives us important representatives of such compounds, namely, vitamin B$_{12}$ or coenzyme B$_{12}$ (Scheme 2), which are substantial components of mammalian metabolism.[49]

The cobalt atom in vitamin B$_{12}$ lies in the center of the nucleus of four five-membered rings (the corrin ring system) and is linked to the benzimidazole group on one side of the plane and to the R group on the second. The plane of the benzimidazole ring is nearly perpendicular to that of the nucleus. The cobalt atom is in close contact with five nitrogen and one carbon atoms. The benzimidazole substituent has some freedom in motion and, by its approaching or taking away the strength of the binding of opposite ligand R on the central atom of cobalt, will be gradually changed. This may be the principle of the mechanism of turning on and turning off the catalytic effect of vitamin B$_{12}$.

The cobalt-carbon bond dissociation energies[50] in coenzyme B$_{12}$ and related organocobalt compounds (denoted as $L_5Co^{3+}R^-$) lie in the range 70 to 105 kJ/mol. The weakness of this bond permits homolytic dissociation under mild conditions — at ambient temperature and at physiological pH in aqueous solution in the dark — which was confirmed also by model experiments.[51]

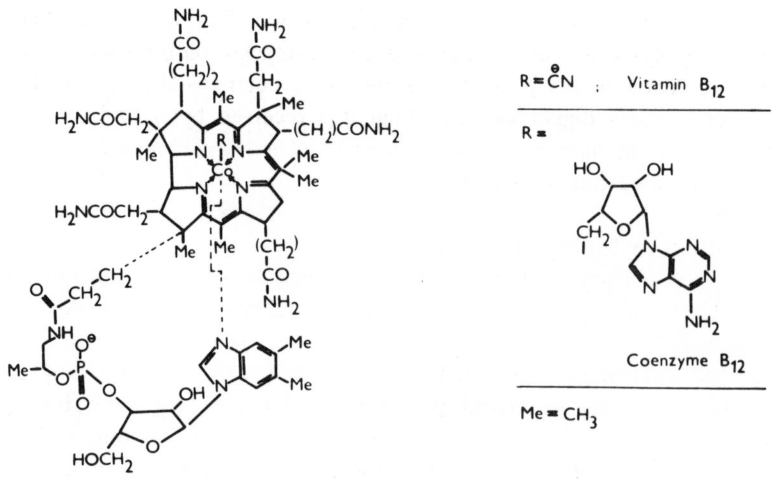

------ bonds were extended purposefully for the illustrative reasons

SCHEME 2

The dissociation of the Co-C bond can be described as the inner sphere redox change (Equation 17) which involves a decrease in the formal oxidation state of cobalt.

$$L_5Co^{3+}R^- \leftrightarrow L_5Co^{2+} + R\cdot \tag{17}$$

The cobalt(II) species and alkyl may coexist as a cage radical pair and can be detected by ESR spectroscopy. In light of these facts, it appears that coenzyme B_{12} fulfills its biochemical role by functioning as a free radical reservoir from which 5-deoxyadenosyl radicals are reversibly released.[50]

C. Sonolysis

The effect of sonic and especially of ultrasonic waves consists predominantly in high temperature thermolysis of molecules in microdomains of a liquid medium.[52-54] The formation of free radicals initiated by sonolysis is connected with transformation of elastic vibrations of sonic waves into the other forms of energy in microscopic cavities of the liquid phase. Cavities or, if you prefer, bubbles arise in the liquid as a consequence of thermal fluctuations in the motion of molecules. In the statistically formed microhole in the liquid the molecules of diluted mono or biatomic gas then assemble. It was estimated that 1 cm^3 of liquid contains about 10^{14} bubbles of average diameter 8.10^{-5} cm. Larger bubbles rise to the surface; smaller ones are relatively rapidly dissolved in the liquid and thus disappear. The concentration of the mechanical energy of environmental molecules of liquid medium proceeds by a fast compression of a microbubble which is thus warmed up to a high temperature of several thousands of degrees. On the cavity surface and at its proximity molecules decompose into radicals. The role of compression may be envisaged from the more than one thousand times higher velocity of propagation of ultrasonic waves when compared with the speed of the statistical motion of bubbles. High local accumulation of energy accompanying the compression of gaseous bubbles is indicated by the formation of about 10^8 radical pairs per one cavity, which cannot be achieved by any other method of radical generation. The ratio of radicals escaping to the reaction volume is, however, small. On the compressed cavity, free radicals immediately react and reaction products are rapidly cooled down by the environmental medium and by the subsequent expansion of the bubble following the passage of an ultrasonic wave. At an ultrasonic frequency of 15 kHz the average lifetime of one bubble corresponds to 10 cycles of compression and expansion. Taking into account the above mechanism of sonolysis, it is obvious why its course depends on the kind of dissolved gas.

The power of contemporary generators of ultrasonic energy is about 10^5 to 10^6 W cm^{-3}, which is several orders higher in value than that received by γ or UV radiation. The comparable local accumulation of energy in material may be obtained only by powerful lasers or by a pulsating source of roentgen rays.

D. Decomposition by Multiphoton Absorption

The thermolysis of chemical bonds in molecules may be performed also by means of an intense flow of infrared photons. The method is considerably more selective than conventional thermal sources.[55-58]

The absorbed quantum of infrared radiation increases the vibration amplitude of the chemical bond. Energy of one infrared photon is not, however, high enough to bring about the dissociation of a ground state chemical bond; its absorption will only shift the state of vibrational excitation by one step higher on the energy scale. For bond dissociation, it is necessary that a higher amount of infrared photons be absorbed all at once. At slow absorption of infrared radiation by a molecular system as it is, e.g., in the case of conventional heat sources (vapor heat exchanger, gas burner, electric resistance oven, etc.), higher vibrationally excited states of one bond are dissipated throughout the molecules and deactivated by collisions with other molecules to rotational and translational levels of ground vibrational state. (Conventional heat sources excite all vibrational states of molecules approximately to the same extent.)

On the other hand, in the case of laser excitation, the flow of infrared quanta to the heated molecules is 10^6 times faster. The difference consists also in the fact that equal vibrational quanta are supplied by laser.

The selective stimulation of chemical reactions by infrared laser is made possible by vibrational excitation of a specific chemical bond in the molecule. At resonance of characteristic vibrations of the molecule with laser frequency, only a certain local part of the molecule is heated, which is enough for bond rupture. The vibrational temperature of this local part of the molecule may achieve even 2000 K without the marked change of mean temperature of surrounding medium. The selected infrared frequency may be focused on particular bonds and cause their primary fission. It should be recalled that specific scission of bonds in the molecule may be used in laser isotope separation of deuterium by decomposition of fluoroform at natural deuterium abundance.[59]

The specificity of small molecule excitation may be demonstrated on dissociation of SF_6 containing two sulfur isotopes.[55] Sulfur hexafluoride with sulfur-32 decomposes with infrared photons of wavelengths 10,610 nm, whereas[34] SF_6 splits the fluorine out already by photons of lower energy (10,820 nm). The overall amount of absorbed photons necessary for splitting of the S-F bond is 35.

The process of multiphoton absorption and of subsequent homolysis of bonds is not, however, so unambiguous. Provided that one of the six S-F bonds in the SF_6 molecule is replaced by the S-Cl bond and thus the substituted molecule is excited by photons absorbed by the S-F bond selectively, we find that the weaker S-Cl bond is breaking preferentially. This indicates that the competition of the transfer of vibrational energy inside the molecule with accumulation of energy on the primarily excited bond may lead to the earlier excitation of vibrational levels pertaining to vibrations of different wavelength. The laser-induced decomposition proceeds, therefore, very frequently as a concerted reaction with parallel scission of several bonds.[60] The increase in selectivity of bond rupture may probably be achieved by shortening the duration time of individual pulses at constant intensity of the flow of photons.

Interaction of laser radiation with solid materials is not so specific as it is in the case of

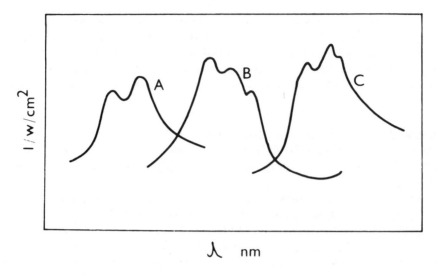

FIGURE 1.   The dependence of light intensity I on wave length (λ) of absorbed (A) or emitted (B, C) light at fluorescence (B) and phosphorence (C) of compounds having absorption spectrum A.

less condensed media, but various additional effects occur.[61] Laser irradiation of organic glass (polymethyl methacrylate) induces the thermolysis of polymer with formation of soot. Absorption of stimulated infrared radiation in polymer defects contributes to the increase of thermoelastic stresses which results in the rupture of chemical bonds and in the formation of free radicals. The absorption of radiation is increased further by the radicals themselves and by the transfer of energy between excited vibrational sublevels of their electronic states. The concentration of free radicals increases during the laser pulse simultaneously with the absorption coefficient of irradiated material. Such positive feedback leads to the mechano-chemical absorption wave, to the loss of thermal stability of polymer, and to its abrupt decomposition. If the intensity of incident laser radiation is not sufficient to produce such instability during one pulse, the gradual accumulation of residual mechanical forces and of free radicals may bring about the same effect in the series of pulses.

IV. PHOTOLYSIS

The incident light may refract, dissipate, or absorb in the molecular systems. When irradiated molecules are capable of absorbing the light, free radicals will arise also in this kind of initiation. At the same time, the flow of photons should have a higher energy than corresponds to the respective bond of the free radical precursor.

The compounds decomposing by the effect of light into radicals are called photoinitiators. The most frequent group of photoinitiators[62] involves carbonyl compounds, namely, ketones (benzophenone, acetophenone, benzil, diacetyl), quinones (benzoquinone, anthraquinone), and benzoinethers. The light sensitive compounds are, moreover, diazocompounds, organic halogen compounds, peroxides, organometallic compounds, thiols, disulfides, xanthoesters, etc. The mechanism of photolysis depends on the photoinitiator structure, but some general laws are valid for it, too.

The photoinitiator molecule absorbs light only when the energy of photons corresponds to the transition of electrons of the molecule to some of its excited states. Because of several possibilities for such transitions, each kind of molecule gives the characteristic absorption spectrum (Figure 1). Having absorbed the quantum of light, the electron in the molecule is displaced and the interatomic distances and bond angles subsequently change. The energy of the photons also determines which of the electrons of the molecular systems will be displaced. The transition of one electron from bond orbital (two electrons with opposite spins between linked atoms) or nonbond orbital (two electrons with opposite spins on one atom) to antibond unoccupied orbital may occur. The electrons from the original electron pair are separated by electronic excitation and in the molecule two reactive centers of radical character are formed.

Provided that at excitation of the electron to antibond orbital the orientation of spins remains unchanged, the singlet excited state is obtained; if the spin orientation in the excited state changes, the triplet excited state arises. The transition from the singlet to the triplet state is often connected with the energy decrease. The back transition from singlet (fluorescence) or from triplet (phosphorescence) to the ground state may be accompanied by the emission of photons. Radiation processes are not fully reversible since the emitted photons have lower energy (longer wavelength). The higher the lifetime of the electronically excited state, the higher the shift to longer wavelengths of light emission. The lifetime of the molecule in the excited singlet state is about 10^{-9} to 10^{-7} sec while that in the triplet state is 10^{-6} to 10^2 sec. The slower transition from the triplet state to the ground state of the molecule is caused by a necessary change of spin symmetry. The difference in the shape of the absorption and emission spectrum is brought about by the different probability of excitation of chromophores in the molecule. The relatively large width of absorption (or emission) bands in spectra may be explained by the excitation of chemically identical molecules to the different vibrational levels.

The excited molecule has another alternative, namely, to transmit the absorbed electronic energy to other molecules. Such a process is then called the transfer of energy or quenching of the excited state. Subsequent photolysis of the secondary excited molecule is photosensibilization, the energy donor being called the photosensitizer. The energy transfer of electronically excited molecules is implemented especially on the level of triplets. The molecules of quenchers have approximately the same levels of excited states as deactivated particles and they are capable of efficiently transforming the electronic energy into vibrations of atoms in chemical bonds.

The photolysis of a molecule into two radicals proceeds at that time when the energy of the excited singlet state exceeds the dissociation energy of some chemical bond in the molecule. The rupture of some bond may occur even after back restoration of the ground electronic state of the molecule which is still vibrationally excited. If such a "hot" molecule is capable of splitting, the split should proceed very quickly since the process of dissipation of the energy excess along the neighboring bonds and molecules takes place in a very short time interval of 10^{-12} to 10^{-8} sec.

It is recalled that ultraviolet light may also split out to a small extent electrons from some molecules. The ionization of a molecule by ultraviolet light was ascertained by identification of solvated electrons in irradiated ($\lambda < 200$ nm) low temperature hydrocarbon glasses.

A. Direct Photoscission of Bonds

The direct photolysis into free radicals may be most simply realized on biatomic molecules having a relatively weak chemical bond. Gaseous molecular iodine gives free radicals at irradiation by blue-green light of a wavelength of about 500 nm. Also, other halogens decompose similarly, except that the wavelength of the light used is shorter.

After sufficiently intense flash irradiation of gaseous chlorine, the absorption spectrum of its molecule disappears and the pressure in the system increases almost two times. This may be explained by dissociation of chlorine-chlorine bonds and by doubling the number of particles in the system. Worth noting in this experiment is that the temperature does not increase markedly after the irradiation. By flash photolysis with flashes of duration from 10^{-6} to 10^{-12} sec, the extremely fast photochemical reactions may well be investigated.

Table 6
VALUES OF ENERGY OF EXCITED STATES (E) OF
MOLECULES, DISSOCIATION ENERGY D OF THE
WEAKEST BOND C-H, AND QUANTUM YIELD (QA)
OF HYDROGEN ABSTRACTION DURING
PHOTOLYSIS

Compound	E (eV)	D (C-H) (eV)	QA
Aromatic hydrocarbons	2.5 — 3.5	4.5	10^{-5}
Alkylaromatic hydrocarbons	2.5 — 3.5	3.6	10^{-5}
Ketones	3.7	4.0	10^{-1}
Aliphatic hydrocarbons	6-7	4.0	10^{-1}
Aldehydes	3.4	3.0 — 3.7	$10^{-2} - 1$

In addition, polyatomic molecules are decomposed into radicals by ultraviolet photolysis. At the cleavage of an equivalent bond, the increasing number of atoms in the molecules, however, causes the decrease of the radicals yielded. It is reasonable to assume that in an excited molecule all the absorbed energy is randomized over a large number of degrees of freedom immediately after rapid conversion. For this reason the dissociation reaction of a large molecule is expected to be quenched more effectively by molecules of diluting gases. At the photolysis of 1-alkenes (CH_2=CH–CH_2–R) the relative yields of allyl radical decrease, indeed, as the size of the molecule becomes larger.[63]

The higher complexity of molecules, however, manifests itself in nonuniformity of photochemical excitation and of subsequent reactions. Comparing one kind of reaction we may verify the rule that photochemical splitting out of a hydrogen atom from hydrocarbons may take place only when the energy of the excited state will be higher than the strength of the C-H bond[64] (Table 6). This may be seen in the example of aldehydes and aliphatic hydrocarbons which have the highest quantum yield of scission of C-H bonds among the compounds of this Table. (Quantum yield is the number of reacted or formed molecules per one absorbed light quantum.) At closer inspection, we may find that formaldehyde has a lower dissociation energy of the C-H bond than acetaldehyde and it, therefore, eliminates hydrogen with a higher quantum yield.

A lower quantum yield of hydrogen elimination from aliphatic hydrocarbons and from some aldehydes is partially connected with parallel dissociation of the C-C bond.

Carbonyl compounds (organic acids, anhydrides, ketones, esters) which have the fundamental absorption band at 310 to 360 nm are more easily split on their C-C bond than on the C-H bond. The light quantum absorbed by the carbonyl group (Figure 2) is energetically sufficient to bring about the rupture of the simple bond C-C but is not able to split the C-H bond. At photolysis of acetone (Equation 18) there are, therefore, formed methyl and acetyl radicals.

$$CH_3COCH_3 \rightarrow \cdot CH_3 + CH_3\overset{\cdot}{C}=O \tag{18}$$

Since the acetyl radical spontaneously decomposes in a subsequent step (Equation 19) to another methyl radical and CO, the photoproduction of a methyl radical by this type of photochemical reaction has many advantages and is frequently used.

$$CH_3\overset{\cdot}{C}=O \rightarrow \cdot CH_3 + CO \tag{19}$$

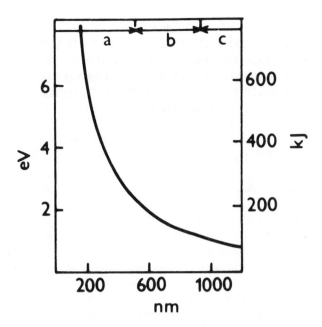

FIGURE 2. Plot of photon energy in dependence on the wavelength:
a — ultraviolet region, b — visible region, c — infrared region.

Quantum yield of ketone photolysis depends on their structure, the state of aggregation, and on the energy of the incident light beam. The quantum yield of CO at photolysis of acetone in a gaseous phase is, e.g., 1.0, at wave length 253.7 nm, and only 0.7 at 313 nm. In the condensed phase the quantum yield of photolysis is considerably lower.

The frequently used photoinitiator, benzoin, (Scheme 3)

$$\underset{\substack{\| \quad |}}{O \quad OH}$$
$$C_6H_5-C-CH-C_6H_5$$

SCHEME 3

and related ethers[65] are split into radicals efficiently even with longer wavelength of ultraviolet light ($\lambda = 366$ nm).

At photolysis of halogen hydrocarbons, the nonbond electron of halogen is excited to the antibonding orbital of the carbon-halogen bond. The transition occurs at an energy exceeding the bond strength and it undergoes efficient light-induced cleavage.

From extensive early studies conducted principally in the gas phase it was demonstrated that the absorption of light by an alkyl iodide leads to homolytic fission of the C-I bond, giving electronically excited alkyl radicals. The photolysis of gaseous iodoalkanes by the wavelength of range 250 to 270 nm taking place in the presence of mercury(I) halides gave radicals and mercury(II) halide.[66] The photobehavior of alkyl halides in solution is, however, more complex.[67] The experimental data are most consistent with a mechanism involving initial light-induced homolytic cleavage of the carbon-halogen bond (Equation 20)

$$R-X \leftrightarrow (R\cdot \ \cdot X) \rightarrow R\cdot + X\cdot \tag{20}$$

followed by diffusion of radical particles away from the reaction cage or by electron transfer (Equation 21) within the caged radical pair.

$$(R \cdot \ \cdot X) \rightarrow R^+ + X^- \tag{21}$$

The radical/ionic photo behavior is controlled by the nature of substituents and by the polarity of the liquid medium. For most alkyl iodides, ionic intermediates predominate. The photo behavior of alkyl bromides has substantially more radical character than ionic ones. A polar medium is the essential requirement for the generation of vinyl cationic intermediates. The quenching and sensitization studies on photolysis of vinyl halides reveal that the radical products are derived from the triplet excited states while the corresponding singlet excited states are responsible for generation of vinyl cations as intermediates.[68]

Photolytic cleavage of a weak bond is used for the production of free radicals, particularly from thermal initiators. For this purpose, there are synthesized aromatic photoinitiators capable of light absorption, which from the populated excited states convert the energy of the chromophore into a homolytic dissociation of the unstable bond (such as the O-O bond, etc.).[69] The homolytic dissociation of the O-O bond in peroxidic compounds is possible also by the electronic energy transfer from the photosensitizers.[70] By photolysis of diaroyl-peroxides aroyloxy radicals are formed similarly as in thermal decomposition. The considerable difference may, however, be seen in a higher degree of decarboxylation of photo-excited aroyloxy radicals.[71]

B. Indirect Photoproduction of Free Radicals

Excitation of electrons at electronic n-π* transition to different orbitals causes the oxygen of the carbonyl group to become electrophilic as in the alkoxy radical. This is a reason why ketones containing at least one alkyl group with more than three carbon atoms decompose to alkenes. In the excited intramolecular complex the hydrogen is transferred to reactive oxygen and the 1,4-biradical formed is split at the bond having a β-position with respect to the transitionally formed radical center on α carbons (Equation 22). If, for some reason, intramolecular hydrogen transfer cannot be realized, the excited carbonyl group reacts intermolecularly and abstracts hydrogen from other molecules of the reaction medium. The latter

$$\tag{22}$$

reaction found wide application in the generation of free radicals in the condensed phase. The reaction (Equation 23) of benzophenone with hydrogen donors belongs to the most frequent examples.[72,73]

$$C_6H_5COC_6H_5 + RH + h\nu \rightarrow C_6H_5\overset{\cdot}{C}(OH)C_6H_5 + R\cdot \tag{23}$$

Several of these photochemical processes (Equations 24 and 25) rely upon initial electron transfer between an excited sensitizer (S*) and an added substrate.[74]

$$S^* + D\text{--}Y \rightarrow \overset{\cdot}{S}^- + D\cdot + Y^+ \tag{24}$$

$$S^* + A\text{--}X \rightarrow \overset{\cdot}{S}^+ + A\cdot + X^- \tag{25}$$

Equation 24 describes oxidative fragmentation of an electron-rich species D-Y, which quenches a sensitizer by donation of an electron to produce a kinetically unstable cation radical (D-Y^+) which rapidly dissociates to radical D· and cationic fragment Y^+. Reductive fragmentation follows from quenching of S* by the relatively nonefficient quencher A-X.

Photoinduced electron transfer reactions can occur because of the increase of the electron donor and acceptor properties when molecules are electronically excited. These processes taking place between electron donor (D) and acceptor (A) molecules in a picosecond to nanosecond time produce high energy intermediates: exciplexes (A^-D^+), radical ion pairs ($A^- \vdots D^+$) and separated free radical ions, which under suitable energetic conditions undergo subsequent reactions leading by electron back-transfer to a singlet or triplet product.[75]

Ketone/amine quenching pairs are examples of oxidative fragmentation.[76,77] Quenching of aromatic compounds by disulfides, quenching of ketones by thiols, and quenching of photoreductive dyes by organometallics (Si, Sn, Ge, Pb compounds) also proceed by the mechanism of oxidative fragmentation.[78] A key feature of these systems is that they require the use of a quencher possessing a low ionization potential.

The photochemistry of benzil in the presence of triethylamine (Equation 26) involves hydrogen abstraction.

$$C_6H_5COCOC_6H_5^* + (C_2H_5)_3N \rightarrow$$

$$C_6H_5CO\overset{.}{C}OHC_6H_5 + CH_3\overset{.}{C}HN(CH_2CH_3)_2 \qquad (26)$$

The reaction of an α-aminoalkyl radical with benzil giving a benzil radical anion or ketyl radical takes place with rate constants of the 10^9 $dm^3 \cdot mol^{-1} \cdot sec^{-1}$ range and belongs among the fastest known radical-molecule reactions.[79] From absorption spectra, it has been concluded that ketone as an electron acceptor forms the molecular complex with amines as donors.[80] Photolysis of this complex results in the intermolecular hydrogen atom transfer from amine to ketone.

An example of photoreductive generation of radicals is the photolysis of H_2SO_4 glasses containing Fe^{2+} ions (Equation 27).

$$Fe^{2+} + h\nu \rightarrow Fe^{3+} + e^- \qquad (\lambda > 250 \text{ nm}) \qquad (27)$$

The electron is transferred from an excited state of Fe^{2+} to the surrounding water ligands from which H atoms are subsequently formed[81] (Equation 28).

$$H_2O + e^- \rightarrow {}^-OH + H· \qquad (28)$$

The hydrogen atoms in H_2SO_4 glasses have a mean half-time of 15 min at 95 K. The same half-time for more rigid H_3PO_4 glasses is obtained at a higher temperature (115 K).

The similar reductive fragmentation is a reaction of a photogenerated electron in solid semiconductors, such as TiO_2 or CdS, with organic halides as 4-nitrophenyl chloride or bromide which decompose upon the capture of electron.[74,82] The photooxidation of the semiconductor pigment particles results in electron-hole pair formation. The hole can be trapped by electron donor D-Y as, e.g., by sodium 4-toluene-sulfonate or triethanol amine.

The photochemistry of analogical systems has important implications in imaging technology, in producing printed circuit boards, integrated circuits, and printing plates.

Production of radicals accompanying the quenching of excited states without the observation of ions may be demonstrated on the reaction of excited Hg with hydrogen or hydrocarbons[83] (Equations 29 and 30).

$$Hg^* + H_2 \rightarrow Hg + 2 H· \qquad (29)$$

$$Hg^* + RH \rightarrow Hg + R· + H· \qquad (30)$$

A similar mechanism is probably valid also for photosensitized decomposition of N_2O (Equation 31)

$$Hg^* + N_2O \rightarrow Hg + N_2 + \cdot O \cdot \tag{31}$$

which is a very convenient method of generation of oxygen atoms.

The main advantage of photolytic methods consists in the generation of free radicals in the low temperature region where primarily formed radicals may often be trapped and detected. Also the possibility of regulating the rate of free radical formation by the change of light intensity is worth noticing. The efficient application of photolytical methods requires the overlapping of absorption bands of photosensitizing compounds with the emission bands of the light source.

It should be pointed out that besides the purposeful photolytical generation in nature, free radicals come into existence spontaneously by the effect of solar radiation in cosmic space, in water vapors in the atmosphere, in the sea,[84] as well as in plants and in animals exposed to light.

V. RADIOLYSIS

The exposure of investigated compounds to the effect of ionization irradiation of energy from 10^3 to 10^6 eV belongs to the very efficient and frequently used methods of free radical generation.[85]

During the interaction of radiation with the molecular system, there occurs ionization due to the knocking out of electrons from the system. Energy transfer of energetic electrons to matter results mainly from inelastic collision with the atomic or molecular electrons of the target. Inelastic scattering produces, initially, further electrons which in turn cause the variety of processes leading to excitation and ionization.[86] The released electron e^- leaves the rest of the molecule (cation radical $M^{.+}$) and reacts either with some other cation radical (Equation 32)

$$e^- + M^+_. \rightarrow M^* \tag{32}$$

or with another molecule (Equations 33 and 34)

$$e^- + M \rightarrow R \cdot + X^- \tag{33}$$

$$e^- + M \rightarrow \dot{M}^- \tag{34}$$

or it is trapped in the system of molecules (Equation 35).

$$e^- + nM \rightarrow e^-(M)_n \tag{35}$$

The molecules (M^*) formed by Reaction 32 are so energetically rich that they easily may decompose into radicals (Equation 36).

$$M^* \rightarrow \cdot R^1 + \cdot R^2 \tag{36}$$

Reactions 33, 34, and 35 illustrate the possible ways of stabilization of released electron-giving anion and radical, anion radical or solvated electron.[87,88] Besides the decomposition of excited molecules another reaction of free radical formation is fragmentation of ion radicals. Such fragmentation results in neutral radicals and corresponding ions. At the decomposition of cation radicals a proton is usually eliminated, whereas at the fragmentation of anion radicals it is halogenide or hydroxy anion.

Reactions of released electrons are very fast and take place practically immediately in the interval 10^{-13} to 10^{-6} sec so that we may identify there only radicals and solvated electrons.[91] This is, however, possible only at that time when the temperature of the reactor medium is sufficiently low or the reaction medium is so rigid that it hinders the mutual deactivation of more bulky free radicals and the fast reactions of solvated electrons. The extent of electron solvation depends on the medium polarity. It is, e.g., known that in hydrocarbon glasses the electron is almost free, nonsolvated, whereas in polar media solvation occurs very efficiently.

Most of the studies dealing with the effect of ionizing radiation on free radical formation were performed with polymers, hydrocarbons, acids, and water solutions frozen into the solid glasses or ice. The physical structure of such solid systems usually depends on the way and the rate of cooling. The difference in experimental conditions of freezing and consequently in physical structure significantly affect the yield of free radicals, even in the case of chemically identical systems. (Radiation yield [G] is defined as the number of respective particles formed after absorbtion of 100 eV of radiation energy by a given system.) Estimating radiation yield by a number of free radicals formed, some deactivation reactions should be taken into account above a certain concentration of free radicals. The reliable value of the yield may be obtained from the slope of the initial part of increasing concentration of free radicals plotted against time of radiation. The chemical structure of the irradiated compound has the decisive influence on the yield of free radicals. Aromatic compounds give, e.g., from 10 to 100 times lower yield of free radicals than saturated hydrocarbons.

The more numerous are the conjugated double bonds in aromatic hydrocarbon, the lower the yield of free radicals obtained. Comparing the effect of radiation on benzene, cyclo-hexane, and phenylcyclohexane, the often-observed fact of protection of irradiated unstable molecules by some molecular moieties which are more resistant to the effect of γ-irradiation can be demonstrated. Radiolysis of alkylaromatic hydrocarbon derivatives proceeds with the yield which lies between the yield of free radicals for aliphatic (G = 5) and aromatic hydrocarbons (G = 0.04 to 0.2).

In general, the efficiency of absorbed radiation in radical homolysis of chemical bonds is relatively low, particularly for aromatic compounds.[64] Taking into consideration the energy of a broken bond from 4 to 5 eV, then the efficiency is 25% for saturated hydrocarbons, while that for aromatic compounds is only 0.2%.

This may be due to the low stability of corresponding alkyl cation radicals which, in the case of saturated hydrocarbons, decompose at very early times (10^{-11} sec) to radicals and ions before a substantial fraction of the geminate ions undergoes neutralization.[92] In irradiated aromatic hydrocarbons, the deficit of electrons is partially compensated by the shift of π-electrons to the radical center of corresponding cation radicals which thus have a relatively long lifetime and can be deactivated by free electrons forming primarily in the irradiation of the organic compound.

The relative stability of molecules exposed to radiolysis follows from the difference of energy levels for the lowest excited singlet state and ground state of the bond in the molecule. When compared to photolysis, radiolysis populates higher excited states with higher probability, but taking into account the very short time of their existence (from 10^{-12} to 10^{-14} sec) they can only participate a little in chemical reactions.

When irradiating hydrocarbons at the temperature of liquid nitrogen (77 K), we detect there predominantly free radicals having unpaired electrons on carbon. Hydrogen atoms which are also formed in radiolysis remain trapped in the sample scarcely. This is brought about by the different mass of the organic radical and the hydrogen atom. In the case of such simple molecules as, e.g., benzene, the difference is almost two orders, which should reflect in much higher mobility and consequently in higher frequency of mutual collisions of hydrogen atoms in contrast to benzene. If they meet, hydrogen atoms dimerize to molecular hydrogen. Hydrogen atoms in irradiated hydrocarbons may be identified provided that radiolysis is performed at the temperature of liquid helium (4 K) or when hydrogen atoms arising in the systems are stabilized in some other way.

The high mobility and reactivity of hydrogen atoms has its impact on the surprising selectivity of the formation of certain types of free radicals in the mixture of compounds with similar structure. If, e.g., in neopentane $[C(CH_3)_4]$ there is dissolved a small amount (about 0.5%) of cyclohexane and the mixture is cooled down to 77 K and γ-irradiated, the hydrocarbon free radicals may be observed by the ESR spectrometer. The concentration of formed cyclohexyl radicals is, however, 100 times higher than should correspond to the ratio of cyclohexane in an irradiated mixture.[93] Cyclohexyl radicals are formed in transfer reaction of both the hydrogen atoms and neopentyl radicals with cyclohexane. The terminal alkyl radicals are selectively formed with a yield higher than statistically expected, regardless of the chain length.[94] This suggests that the primary event of alkyl radical formation takes place at the chain ends. The interchain radical transfer reaction takes place at 45 to 65 K only in the even homologues higher than $C_{10}H_{22}$, resulting in the conversion of the terminal alkyl radical mainly to the penultimate radical, whereas it is prohibited in the odd homologues below 77 K. The marked difference in the transfer reaction rate in the even and odd homologues can be accounted for in terms of the difference in the alignment of the two neighboring chains in the triclinic for the even and the orthohombic unit cell for the odd homologues.

Radiolysis of halogen hydrocarbons (RX) provides higher yields of radicals than corresponding hydrocarbons. Except for other reasons, this may be due to primarily released electrons which cleave the carbon-halogen (X) bond (Equation 37).

$$R–X + e \rightarrow R· + X^- \tag{37}$$

In the case of saturated halogen hydrocarbons the electron is captured into a σ-orbital of the C-X bond and causes its dissociation. In the case of unsaturated halogen hydrocarbons, chlorobenzene (Scheme 4), benzyl chloride (5), vinyl chloride (6), and allyl chloride (7)

SCHEME 4	SCHEME 5	SCHEME 6	SCHEME 7

the electron is accommodated in the lowest unoccupied π*-orbital.[95] The Cl⁻ yield is highest when the C-Cl bond does not lie in the nodal plane of the system of π-electrons (halogen hydrocarbon 5 and 7), thus permitting the interaction between this bond and the system of π-electrons. The temporary negative ion is then formed in a repulsive state and Cl⁻ appears with considerable kinetic energy. If the C-Cl bond is directly linked to the π-system, i.e., if it lies in the nodal plane of the π-system, the symmetry of the ground states of the fragments no longer permits their formation adiabatically by direct dissociation. Longer dissociation time channels the excess energy into various internal degrees of freedom of the organic moiety, resulting in less kinetic energy of the fragments.

Radiolysis is an important method of generation of free radicals in aqueous medium.[96] In radiolysis of water, hydroxyl radicals and hydrogen atoms are formed. Part of them undergo recombination to molecular hydrogen and hydrogen peroxide already in the course of irradiation. In addition to free radicals, hydrated electrons and protons also appear in the system. The overall stoichiometry in terms of radiation yields may be designated by Equation 38

$$H_2O \rightarrow 2.7 \, HO\cdot + 2.7 \, e^-(H_2O) + 2.7 \, H_3O^+ + 0.5 \, H\cdot + 0.4 \, H_2 + 0.7 \, H_2O_2 \quad (38)$$

where individual numbers denote radiation yield. At the excess of free protons in the system, hydrated electron reacts with hydrated proton and the yield of free radicals is increased (Equation 39).

$$e^-(H_2O) + H_3O^+ \rightarrow H\cdot + 2 \, H_2O \quad (39)$$

Provided that N_2O is dissolved in water, it reacts (Equation 40) with hydrated electrons

$$e^-(H_2O) + N_2O \rightarrow HO\cdot + HO^- + N_2 \quad (40)$$

and gives a hydroxyl radical, a hydroxy anion and a molecule of nitrogen. The yield of hydroxyl radicals is then twofold higher. The almost exclusive formation of hydroxyl radicals may be achieved in an alkaline medium where hydrogen atoms give with hydroxyl anions hydrated electrons (Equation 41)

$$H\cdot + HO^- \rightarrow e^-(H_2O) \quad (41)$$

which in the presence of nitrous oxide convert to hydroxyl radicals. Since the solvated electron reacts very rapidly in the time region from psec to nsec, the experimental yield depends on the time resolution of the apparatus.[97] The yield of solvated electrons obtained at 30 psec after irradiation, was determined to be 4.8 ± 0.3 as compared to the $G(e^-_{aq})$ obtained after 100 nsec, which was 2.7. The high reactivity of the solvated electron may be seen on its reaction with H_2O_2 (Equation 42)

$$H_2O_2 + e^- \rightarrow HO\cdot + {}^-OH$$
$$k = 10^{10} \, dm^3 \cdot mol^{-1} \cdot sec^{-1}, \quad 20°C \quad (42)$$

or with water (Equation 43).

$$H_2O + e^- \rightarrow H\cdot + {}^-OH$$
$$k = 5 \cdot 10^2 \, dm^3 \cdot mol^{-1} \cdot sec^{-1}, \quad 20°C \quad (43)$$

The decay of solvated electrons in H_2O ice is very rapid, with an initial half-time of about 75 nsec even at 6 K.[98] Upon radiolysis of an aqueous oxygenated solution containing formate, the following reactions (Equations 44 to 47) take place.[99]

$$HO\cdot + HCO_2^- \rightarrow \dot{C}O_2^- + H_2O \quad k = 3.5 \cdot 10^9 \, dm^3 \cdot mol^{-1} \cdot sec^{-1} \quad (44)$$

$$CO_2^- + O_2 \rightarrow CO_2 + O\dot{}_2^- \quad k = 2.4 \cdot 10^9 \, dm^3 \cdot mol^{-1} \cdot sec^{-1} \quad (45)$$

$$e^-(H_2O) + O_2 \rightarrow O\dot{}_2^- \quad k = 2 \cdot 10^{10} \, dm^3 \cdot mol^{-1} \cdot sec^{-1} \quad (46)$$

$$H\cdot + O_2 \rightarrow HO\dot{}_2 \quad k = 2 \cdot 10^{10} \, dm^3 \cdot mol^{-1} \cdot sec^{-1} \quad (47)$$

This occurs in a solution that contains predominantly $HO_2^{\cdot}/O_2^{\cdot-}$ radicals, where the proportion of $HO_2^{\cdot}$ to $O_2^{\cdot-}$ depends upon the acid-base equilibrium[100,101] (Equation 48).

$$HO_2^{\cdot} \leftrightarrow H^+ + O_2^{\cdot-} \qquad pK = 4.8 \qquad (48)$$

As sources of γ-irradiation, radioisotopes of Co and Cs and X-rays are used. Cobalt-60 emits γ-rays of energy 1.25 MeV and has a half-time decay of 5.2 years. Cesium-137 is a softer source with a longer half-time (0.66 MeV, 30 years).

The large choice exists in γ-cyclotrones (γ-radiolysis) with energy of radiation from 1 to 100 MeV. Very fast reactions may be suitably investigated by means of cyclotrons having tunable frequency of electron pulses of several MeV. The time of duration of individual pulses is of the order of microseconds, often only nanoseconds or even picoseconds. The fate of formed free radicals is monitored again by the very fast spectrometric method. The pulse method of free radical generation allowed researchers to reinvestigate thousands of elementary reactions and to measure their rate constants. Its importance consists in the possibility of separating fast from slow reaction steps in the overall complex system of subsequent processes.

Radiolytic production of radicals makes it possible to treat samples which are not transparent (biological objects, samples under high pressure, in steel reactors, etc.) Another advantage is the wide extent of radiation intensities which allow one to produce different concentrations of free radicals in a reaction medium.

On the other hand, the method has low selectivity in the production of radicals and its usage requires one to take necessary precautions to protect living organisms in the neighborhood of the energy source against the harmful effects of radiation.

VI. ELECTROLYTIC GENERATION OF FREE RADICALS

At the heterolytic dissociation of the molecule two ions are formed with opposite charges. If a direct electric current passes through the system, anions move towards the positive electrode-anode, whereas cations move to the negative electrode-cathode. At the anode, anions are oxidized so that they transmit the electron to the electrode and are transformed to radicals. At the cathode, cations accept the electrons and are reduced to radicals.[102-104]

Electrolysis of potassium acetate in water is one of the best-known electrode reactions (Equation 49). Acetate anions are oxidized there to acetoxy radicals.

$$CH_3CO_2^- - e^- \rightarrow CH_3CO_2^{\cdot} \qquad (49)$$

In the next stage, the radical decomposes to methyl radicals and CO_2; methyl radicals recombine to ethane.

In the electric field, even nondissociated molecules, such as some aromatic halogenides RX_1 may decompose onto radicals.[105] The electrolytic decomposition of the latter compounds begins by transfer of electrons from the electrode to the molecule. Whenever the reduction pathway involves a discrete RX^- anion radical rapidly cleaving to R· and X^- or a concerted electron transfer bond breaking process leading directly to R· and X^-, radicals R· are formed very close to the electrode surface. Provided that the radicals accept electrons more easily than starting RX, they will be reduced at the electrode surface (Equation 50).

$$R\cdot + e^- \leftrightarrow R^- \qquad (50)$$

before having time to undergo any other reaction. On the electrode surface, even ion radicals may be formed which, similarly to neutral radicals, dimerize. This type of procedure of the electrolytic reductive preparation of ion radicals is industrially used to produce adiponitrile from acrylonitrile (Equation 51)

$$2\ CH_2{=}CHCN \xrightarrow{+2e} CN{-}\overset{\ominus}{C}H(CH_2)_2\overset{\ominus}{C}H{-}CN \xrightarrow{+2H^+} CN(CH_2)_4CN \qquad (51)$$

Since at the electrochemical generation of free radicals their nonuniform distribution is created in the system, the method is used for preparative purposes in organic synthesis of dimers particularly, and not for initiation of chain reactions.[106] The method is, however, unique when it comes to the generation of a required type of free radical from more complex molecules. Another important aspect stems from the fact that it is a convenient means for the investigation of the activation driving force of free energy relationships characterizing electron transfer reactions.

Water, methyl alcohol, pyridine, acetonitrile, and other solvents (sometimes with dissolved supporting electrolyte) are used as the medium for electrochemical production of free radicals.

VII. GENERATION OF RADICALS BY ELECTRICAL DISCHARGE

The effect of an electric field, consisting in strong polarization and deformation of electron shells, may lead to the decomposition of gaseous molecules to ions, electrons, and radicals. The experimental conditions may be chosen so that the radical decomposition will prevail.[107] In the glowing discharge of hydrogen at a pressure from 10 to 200 Pa, a high concentration of hydrogen atoms is formed. By the effect of an electrical discharge on more than two atomic molecules different radicals are already formed. It appears, therefore, to be more convenient to generate hydrogen atoms primarily and let them react with, e.g., alkenes to alkyl radicals than to perform nonspecific discharge in corresponding alkene.

In addition to the glowing discharge, the arc spark discharge may be used at the pressure of gases 0.1 MPa. Here, the discharge zone has, however, a temperature of about 6000 K and the fragmentation of molecules due to the impact of fast electrons is already accompanied by their thermolysis.

VIII. RADICAL FORMING ELECTRON-TRANSFER REACTIONS

The reaction of transition metal ions with peroxides belongs to advantageous methods of free radical generation.[108-110] Radicals may arise either in the oxidation of metal ions M (Equation 52)

$$M^{n+} + H_2O_2 \rightarrow M^{(n+1)+} + HO^- + \cdot OH \qquad (52)$$

or in the reduction of a higher oxidation state of the corresponding metal (Equation 53).

$$M^{(n+1)+} + H_2O_2 \rightarrow M^{n+} + HOO\cdot + H^+ \qquad (53)$$

As a primary step there occurs the transfer of electrons between reacting partners. The effect of transition ions follows from the presence of nonpaired electrons in their d-orbitals. There exists a large variety of similar reactions of metal ions and different peroxides, which differ from each other in the rate of free radical elementary steps, in the ratio of assertion of side reactions of peroxides forming free radicals, and in the deactivation of catalytic ions by reaction products, respectively.

Catalytic decomposition of hydrogen peroxide by Fe ions (Fenton's reaction — Equation 52) was intensively studied for many years. In spite of this and probably just because of it opinions on the mechanisms of decomposition are not uniform but are quite antagonistic.[111] One group of research workers accepts the Haber-Weiss scheme, according to which the radical decomposition of H_2O_2 may be illustrated by reactions (Equations 54 to 57).

Initiation

$$Fe^{2+} + H_2O_2 \rightarrow Fe^{3+} + {}^-OH + HO \cdot \qquad (54)$$

Propagation

$$HO \cdot + H_2O_2 \rightarrow H_2O + HO_2^{\cdot} \qquad (55)$$

$$HO_2^{\cdot} + H_2O_2 \rightarrow O_2 + H_2O + HO \cdot \qquad (56)$$

Termination

$$HO \cdot + Fe^{2+} \rightarrow Fe^{3+} + {}^-OH \qquad (57)$$

The mutual interaction of hydroperoxy radicals is apparently not important, which was substantiated by experiments with hydrogen peroxide having one isotopically labeled oxygen. No isotopically asymmetrical O-O bond was formed. Worthy of detailed discussion is the key reaction of $HO_2^{\cdot}$ radicals with H_2O_2. The rate constants found for this interaction lie in the range of six orders of magnitude which indicates that the reaction is influenced by some unknown and, therefore, noninvestigated factors. This may be the accidental amounts of transition metal ions which are always present in trace quantities as the reaction background of each real system.[111]

Another alternative is that radicals in Fenton's reaction are formed even in reduction of Fe^{3+} ions by H_2O_2 (Equation 58), and Fe ions enter propagation reactions (Equation 59).

$$Fe^{3+} + H_2O_2 \rightarrow Fe^{2+} + HOO \cdot + H^+ \qquad (58)$$

$$HO_2^{\cdot} + Fe^{3+} \rightarrow Fe^{2+} + H^+ + O_2 \qquad (59)$$

The decay of hydroxyl and $HO_2^{\cdot}$ radicals may proceed by oxidation of Fe^{2+} ions (Equations 57 and 60) and by a disproportionation reaction (Equation 61 and 62).

$$HO_2^{\cdot} + Fe^{2+} \rightarrow HO_2^- + Fe^{3+} \qquad (60)$$

$$2\, HO_2^{\cdot} \rightarrow H_2O_2 + O_2 \qquad (61)$$

$$HO \cdot + HOO \cdot \rightarrow H_2O + O_2 \qquad (62)$$

Both ideas are mutually compatible at some qualitative assumptions. The proposals of mechanisms concerning nonradical decomposition of hydrogen peroxide in the coordination sphere of ligands surrounding the catalyst ions are somewhat different and will not be involved here.

The solving of these problems is complicated by the high sensitivity of hydrogen peroxide to trace concentrations of transition metals in the system and the catalyst itself. Regardless of uncertainties in the exact description of mechanisms of H_2O_2 decomposition, Fenton's reaction has found wide application for production of hydroxyl radicals in aqueous media.

The widely used redox system[112] for radical generation consists of Ti^{3+}-H_2O_2 acting in aqueous solution at pH = 2. Reactions occurring in this system are, namely (Equations 63 to 66).

$$Ti^{3+} + H_2O_2 \xrightarrow{\sim 10^3} Ti^{4+} + HO\cdot + HO^- \tag{63}$$

$$Ti^{3+} + \cdot OH \xrightarrow{3 \cdot 10^9} Ti^{4+} + HO^- \tag{64}$$

$$H_2O_2 + \cdot OH \xrightarrow{3 \cdot 10^7} HOO\cdot + H_2O \tag{65}$$

$$Ti^{4+} + H_2O_2 \xrightarrow{6 \cdot 10^2} Ti^{4+} - H_2O_2 \tag{66}$$

$$k/dm^3 \cdot mol^{-1} \cdot sec^{-1}, \qquad 20°C$$

The radicals ($\cdot OH$ and $\cdot OOH$) have quite different dependences of concentration on time and thus are not even approximately in equilibrium during reaction.[113]

Hydroxyl and hydroperoxyl radicals in the gas phase may be prepared at room temperature also by the rapid redox reaction (Equations 67 to 70) of hydrazine with ozone.[114]

$$H_2NNH_2 + O_3 \rightarrow H_2N\dot{N}H + \cdot OH + O_2 \tag{67}$$

$$H_2N\dot{N}H + O_2 \rightarrow HN{=}NH + HO_2^{\cdot} \tag{68}$$

$$HN{=}NH + O_3 \rightarrow HN{=}\dot{N} + \cdot OH + O_2 \tag{69}$$

$$HN{=}\dot{N} + O_2 \rightarrow N_2 + HO_2^{\cdot} \tag{70}$$

Transition metal ions such as Co and Mn decompose hydroperoxides (Equation 71 and 72) at ambient temperatures to alkoxyls,

$$ROOH + M^{n+} \leftrightarrow [ROOH\ M^{n+}] \rightarrow RO\cdot + M^{(n+1)} + {}^-OH \tag{71}$$

and alkyl peroxyls, too.

$$ROOH + M^{n+} \leftrightarrow [ROOH\ M^{n+1}] \rightarrow RO_2^{\cdot} + H^+ + M^{n+} \tag{72}$$

The radicals produced in these two reactions may then undergo a variety of reactions such as reaction with uncomplexed hydroperoxide or metal ion, organic substrate or with themselves.[115]

Rates of reaction depend on the nature of the transition metal ion and the ligand. Reaction kinetics is complex and involves reversible association of the metal ion and the hydroperoxide taking place before electron transfer. The reaction is furthermore retarded by reaction products such as alcohol and water. The reactive metal complex is destroyed by reaction with $RO_2^{\cdot}$ or $RO\cdot$ which usually occurs at the metal center.[116]

In addition to metal ions, some organic compounds having atoms with a free electron pair may act as accelerators of the decomposition of peroxides. As an example, the activated decomposition of dibenzoyl peroxide by dimethyl aniline may be given.[117] The main pathway for this reaction involves the initial interaction of both compounds (Equation 73)

$$C_6H_5N(CH_3)_2 + (C_6H_5COO^-)_2 \rightarrow C_6H_5N^+(CH_3)_2OOC_6H_5 + C_6H_5COO^- \tag{73}$$

which is responsible for the accelerated production of free radicals arising at homolysis of the intermediate (Equations 74 and 75).

$$C_6H_5N^+(CH_3)_2OOCC_6H_5 \rightarrow C_6H_5COO\cdot + C_6H_5\overset{\cdot}{N}{}^+(CH_3)_2 \tag{74}$$

$$C_6H_5\overset{\cdot}{N}{}^+(CH_3)_2 \rightarrow C_6H_5N(CH_3)\overset{\cdot}{C}H_2 + H^+ \tag{75}$$

At an equimolar ratio of both reactants and an overall concentration of 0.002 mol dm^{-3} the mean halftime of peroxide decomposition is about 10 min at 20°C. Compared with thermal dissociation of peroxide at the same temperature, the catalyzed decomposition reaction is faster by 5 to 6 orders. The efficiency of the escape of free radicals to the reaction medium is often, however, very low. In the above reaction it reaches about 20%, but in other similar reactions it is only 1% or less.

Redox initiation systems are important from the practical viewpoint, since they allow generation of free radicals at ambient temperatures, and the rate of free radical production may be controlled by the initial concentration of reactants.

Sometimes unwanted generation of free radicals in organic materials is implemented by the relative slow reaction of organic substrate with oxygen in its ground triplet state (Equation 76).

$$RH + O_2 \rightarrow R\cdot + HOO\cdot \tag{76}$$

In biological systems and in many technological and industrial operations there, e.g., occurs relatively rapid oxidation of monosaccharides.[118] The production of free radicals begins here probably by electron transfer from the anionic form of the ene-diol on oxygen (Equation 77).

$$\begin{matrix} OHO^- & & & OHO\cdot \\ |\ \ | & & & |\ \ | \\ R\text{-}C\text{=}CH & + & O_2 \rightarrow & R\text{-}C\text{=}CH & + & O_2^{\cdot -} \end{matrix} \tag{77}$$

Since the formation of the ene-diol (Equation 78)

$$\begin{matrix} CHO & & CHOH & & CH_2OH \\ | & & \| & & | \\ CH\text{-}OH & \leftrightarrow & CHOH & \leftrightarrow & C\text{=}O \\ | & & | & & | \\ R & & R & & R \end{matrix} \tag{78}$$

is necessary for monosaccharide oxidation, the production of free radicals may be dependent on the ratio of the aldehyde form in the parent monosaccharide.

At the same time, furanose (Scheme 8) and pyranose (Scheme 9) forms

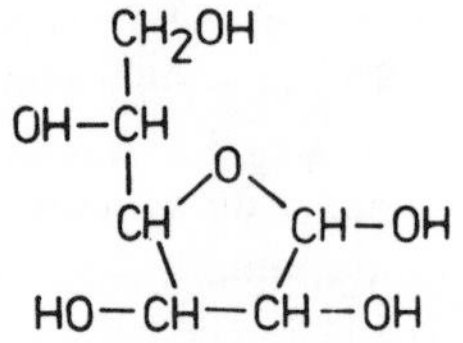

SCHEME 8

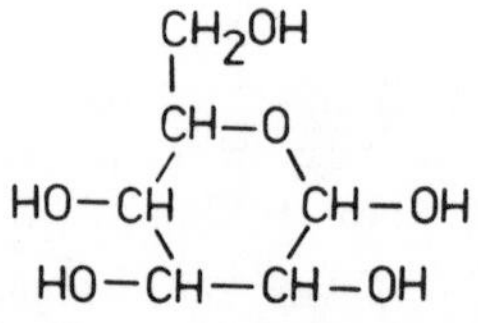

SCHEME 9

are not susceptible to oxidation under mild conditions. In such a way we may also interpret the nonenzymatic oxidation of vitamine C (ascorbic acid) in living organisms (Equation 79).

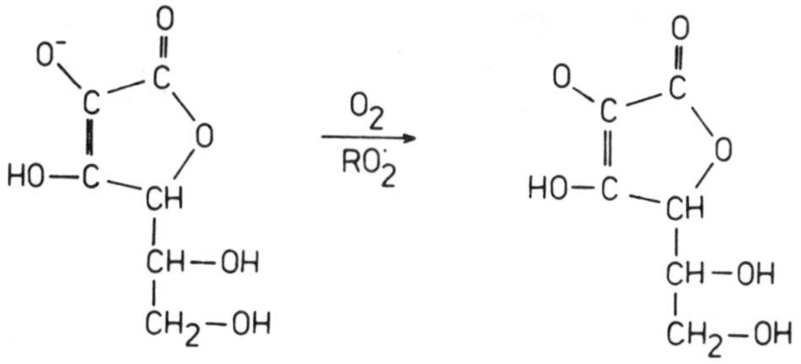

$$(79)$$

Ascorbic acid is metabolically active, while the ascorbyl radical is a relatively inert nontoxic product which has the role of an antioxidant in biological systems.[119] Alkaline degradation of polysaccharides (in solution of potassium hydroxide) at 100°C and in the presence of air gives aromatic semidone radicals (Scheme 10)

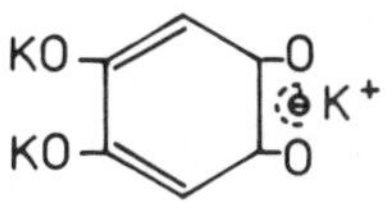

SCHEME 10

resulting from the degradation of saccharides to two-, three-, and four-carbon fragments which cyclize in base catalyzed aldol condensation into semiquinones or semidones.[120]

During the iron catalyzed oxidation of cystein (cys) by molecular oxygen hydroxyl radicals are produced.[121] For their formation the following reactions (Equations 80 and 81) may be proposed (cys− is $HOOCCH(NH_2)CH_2S^-$ forming strong complexes with iron ions).

$$Fe^{2+} (cys^-)_2 + O_2 \rightarrow Fe^{3+} (cys^-)_2\, O_2^{\cdot -} \tag{80}$$

$$Fe^{3+} (cys^-)_2\, O_2^{\cdot -} + 2\ Fe^{2+} (cys^-)_2 + 2\ H_2O$$

$$\rightarrow 3\ Fe^{3+} (cys^-)_2\, (^-OH) + {\cdot}OH + O_2 \tag{81}$$

The occurrence of radicals in vivo in the vicinity of vitally important molecules may lead to serious damages in functions of tissue. Oxygen free radicals have been suggested to be involved in several pathological processes including mutagenesis, carcinogenesis, arthritis, and aging.

IX. PREPARATIVE METHODS OF GENERATION OF STABLE FREE RADICALS

The beginning of radical chemistry dates from the first unaware synthesis of triphenyl-methyl radical. M. Gomberg (1900) originally intended to prepare hexaphenyl ethane by condensation reaction of triphenylchlormethane and Ag dust. Analysis of the product obtained, however, surprisingly showed that the synthesized hydrocarbon contained oxygen. Only after the repeated condensation reaction in the atmosphere of CO_2, the compound was obtained, which had the ratio of hydrogen and carbon corresponding to hexaphenylethane.

Benzene solution of the synthesized product rapidly absorbed oxygen and a peroxide compound was formed. Presumed hexaphenylethane also entered the reaction with halogens yielding triphenyl halogenides. Dissolving under conditions of an inert atmosphere, the solutions of hexaphenylethane gradually turned from yellow to brown. Evaporation of solvent brought about the discoloration of the reaction product. The results, which were quite unusual at that time, were interpreted by the existence of triphenylmethyl radicals.[122]

By a similar reaction (Equation 82)

$$(Aryl)_3CCl + Ag \rightarrow (Aryl)_3C\cdot + Ag^+Cl^- \tag{82}$$

other radicals of the triaryl methyl type may also be prepared.[123] The triphenylmethyl radical was also prepared by heat treatment of triphenylmethane.[124] The triarylmethyl radicals are in equilibrium with their dimer (Equation 83).

$$[(Aryl)_3C\cdot]_2 \rightarrow 2\,(Aryl)_3C\cdot \tag{83}$$

The degree of dissociation of the dimer depends on the stability of the radical.

Quoting just a few possibilities of preparation of stable radicals, worth mentioning is especially the treatment of α, γ-bisdiphenylene-β-phenylallyl chloride with mercury which gives the red brown radical, α, γ-bisdiphenylene-β-phenylallyl[125] (Scheme 11).

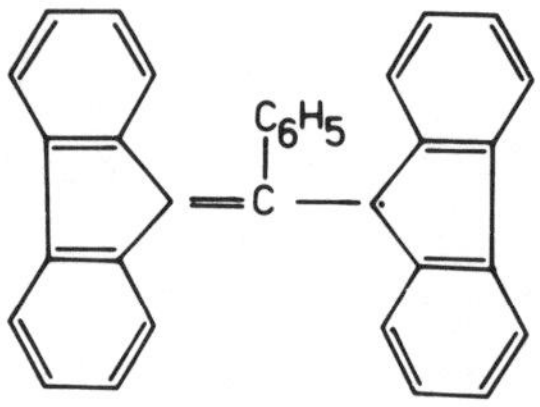

SCHEME 11 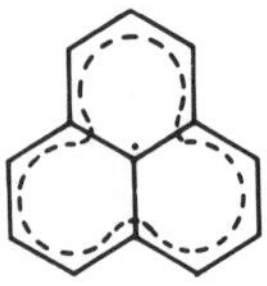   SCHEME 12

The perinaphthyl radical (Scheme 12) may readily be prepared by treating perinaphthalene with methoxide ions and shaking the resulting perinaphthenide anion in the presence of oxygen.[126]

A stable radical, 1-ethyl-4-carbomethoxypyridinyl was prepared[127] by reduction of 1-ethyl-4-carbomethoxy pyridinium iodide with sodium dispersion or sodium amalgam[128] (Equation 84)

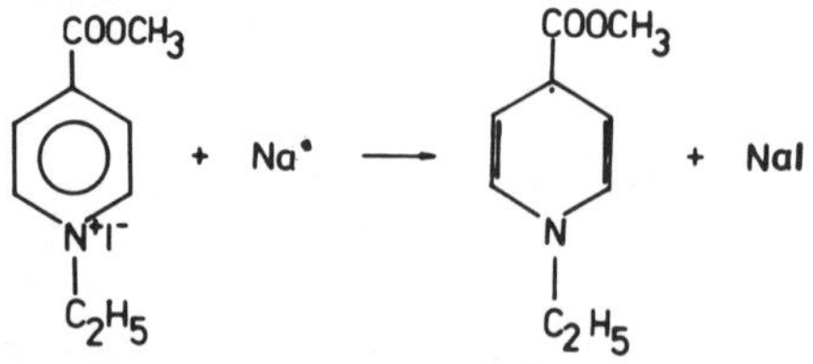

$$\tag{84}$$

while the polymeric analog of the pyridinyl radical was synthesized by the reduction of the corresponding pyridinium salt with 1-benzyl-1,4-dihydronicotinamide.[129]

Persistent alkyl radicals may be synthesized by the addition of transient radicals to an appropriate spin trap.[130] One of many examples is the chain addition of trimethylsilyl radicals to *tert*-butyl acetylene.[131]

$$(CH_3)_3Si \cdot (a) + (CH_3)_3CC\equiv CH \rightarrow (CH_3)_3C\dot{C}=CHSi(CH_3)_3(b) \qquad (85)$$

$$b + (CH_3)_3SiH(c) \rightarrow (CH_3)_3CCH=CHSi(CH_3)_3(d) + a \qquad (86)$$

$$a + d \rightarrow (CH_3)_3C(CH_3)_3SiCH\dot{C}HSi(CH_3)_3(e) \qquad (87)$$

$$a + e \rightarrow (CH_3)_3C(CH_3)_3SiC=CSi(CH_3)_3(f) + c \qquad (88)$$

$$a + f \rightarrow (CH_3)_3C(CH_3)_3SiCCH[Si(CH_3)_3]_2 \qquad (89)$$

The half-time of the decay of the persistent radical formed in the last step (Equation 89) and at the concentration $3 \cdot 10^{-6}$ mol·dm^{-3} is 23 hr at 50°C.

The second group of stable radicals has an unpaired electron located on oxygen. There belong especially aryloxy and nitroxy radicals. Stable aryloxy radicals are produced by oxidation of the ortho, para-trisubstituted phenols.[132-134] Galvinoxyl (Scheme 13) is a particularly stable representative of this group.

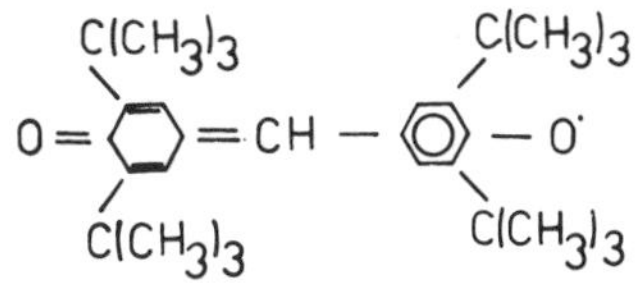

SCHEME 13

Nitroxide radicals can often be prepared by oxidation of N,N-disubstituted hydroxylamines[135] (Equation 90)

$$R_2NOH \rightarrow R_2NO\cdot \qquad (90)$$

or oxidation of 4 H-imidazol di-N-oxides with O_2 or PbO_2 in the presence of nucleophilic reagents.[136] A further method is based on the addition of reactive radicals to a C-nitroso-compound (Equation 91)

$$R\cdot + R^1NO \rightarrow R\overset{\overset{\textstyle O\cdot}{|}}{N}R^1 \qquad (91)$$

or to a nitrone (Equation 92)

$$R\cdot + R^1\overset{\oplus}{\underset{\underset{\textstyle O^-}{|}}{N}}=CHR^2 \rightarrow R^1\overset{\overset{\textstyle O\cdot}{|}}{N}CHRR^2 \qquad (92)$$

2-methyl-2-nitroso butane-2-on, tert-nitrosobutane, 2,3,4,6-tetramethylnitrosobenzene and β-phenyl-N-tert-butyl nitrone are used as compounds trapping reactive free radicals.[137]

Nitrogen-centered free radicals are also very numerous. Semistable diaryl aminyl radicals (Aryl)$_2$N· can be synthesized from the parent amine by oxidation (Equation 93) with silver oxide or electrochemically.[138]

$$(Aryl)_2NH - e \rightarrow (Aryl)_2N\cdot + H^+ \qquad (93)$$

Hydrazyl radicals $-\overset{\bullet}{N}-N=$ are produced by oxidation of triarylhydrazines with PbO_2. A remarkable long-lived hydrazyl radical is 2,2-diphenylpicrylhydrazyl (DPPH) (Scheme 14) often used as a concentration standard for radicals in reaction systems.[139]

By oxidation (sometimes only O_2 is sufficient) of hydrazidines, very stable tetraazapentenyl (Scheme 15), verdazyl (Scheme 16), and tetrazolinyl (Scheme 17) radicals are prepared.[140]

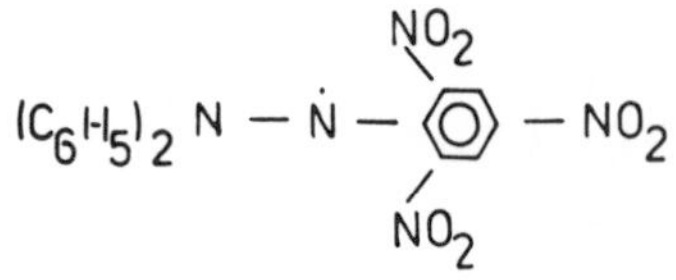

SCHEME 14

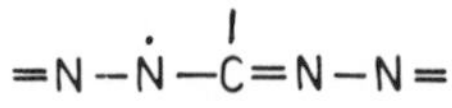

SCHEME 15

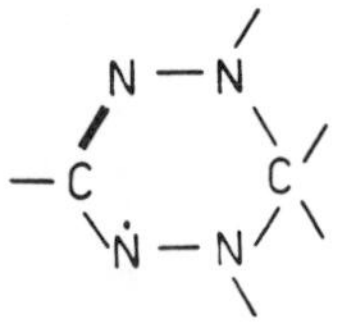

SCHEME 16

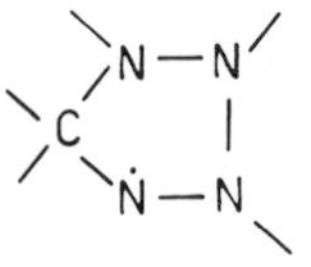

SCHEME 17

From the viewpoint of preparation, purity, and large variability in the choice of radical structure, the combined method of reduction of monohalogen compounds or nitroalkanes yields relatively good results.[141,142] The procedure involves the thermal or light-stimulated decomposition of initiators (Equation 94) where primary radicals are generated at first

$$Y_2 \rightarrow 2\, Y\cdot \tag{94}$$

which react (Equation 95) with extremely reactive trialkylsubstituted stannous hydride.

$$Y\cdot + R_3SnH \rightarrow R_3Sn\cdot + YH \tag{95}$$

This reaction is immediately followed by specific reduction of the halogen organic compound or nitroalkane ($X = NO_2$) XZ and the required radical $Z\cdot$ is thus formed (Equation 96)

$$R_3Sn\cdot + XZ \rightarrow R_3SnX + Z\cdot \tag{96}$$

The method is very frequently used because of its experimental simplicity and extraordinary selectivity.

It may be concluded that although the same type of radicals may be generated in different procedures, reactivity cannot be necessarily the same. The difference consists in different excitation of formed radicals. The reactivity of the same kind of radical generated from different precursors is unified by thermalization, which means that the free radical usually undergoes several collisions with other particles of the system before its further reaction. As we have seen, free radicals arise either from molecules which have absorbed some form of energy, or they come into existence in the process of reorganization of reacting molecules and ions. There are numerous methods of the production of free radicals. The choice of the method depends on the type of generated radical from its precursor and on the purpose of generation. Regarding the possibilities of chemical transformations of free radicals and of reactions of radicals with molecules, there practically does not exist any limitation to the preparation of any type of free radical. Complications begin, however, at that time when we put more strict requirements on the selectivity of preparation of certain type of radicals. It should be taken into account that each method of free radical generation gives in parallel with the required radicals also other radicals formed by competition and subsequent reactions of the free radical precursor with molecules in the medium, which may complicate the matter considerably.

REFERENCES

1. **Zhurkov, S. N. and Korshunov, V. E.**, Atomic mechanism of fracture of solid polymers, *J. Polym. Sci., Polym. Phys. Ed.*, 12, 351, 1974.
2. **Pazonyi, T., Tudos, F., and Dimitrov, M.**, Untersuchung der beim Schnitzeln von Polymeren sich bildenke Radikale, *Angev. Makromol. Chem.*, 10, 75, 1970.
3. **Campbell, D. and Peterlin, A.**, Free radical formation in unaxially stressed nylon, *Polym. Lett.*, 6, 481, 1968.
4. **Radtsig, V. A. and Butyagin, P. Yu.**, Free radical forming in mechanical rupture of polyethylene and polypropylene, *Vysokomol. Soed.* (Russian), 9, 2549, 1967.
5. **Kurokava, N., Sakaguchi, M., and Sohma, J.**, Trapping sites of polypropylene mechano radicals, *Polym. J.*, 10, 93, 1978.
6. **Zhurkov, S. N., Zakhrevskii, V. A., and Tomashevskii, Z. E.**, The formation of free radicals at rupture and deformation of polymers containing sulfidic bonds, *Phys. Solid Materials* (Russian), 6, 1912, 1964.
7. **Takeuchi, Y., Yamamoto, F., Konaka, T., and Nakagawa, K.**, Radical and void formation in ultradrawn polyoxymethylene, *J. Polym. Sci., B, Polym. Phys.*, 24, 1067, 1986.
8. **Sakaguchi, M., Kinpara, H., Shimada, S., and Kashiwabara, H.**, Ionic products from the mechanical fracture of solid polymers; polyethylene and polypropylene, *Polym. Commun.*, 26, 142, 1985.
9. **Dickinson, J. T. and Jensen, L. C.**, Fracto-emission from filled and unfilled polybutadien, *J. Polym. Sci., Polym. Phys. Ed.*, 23, 873, 1985.
10. **Dickinson, J. T., Jahan-Latibari, A., and Jensen, L. C.**, Electron emission and acoustic emission from the fracture of graphite epoxy composites, *J. Mater. Sci.*, 20, 229, 1985.
11. **Dodson, B. W. and Graham, R. A.**, *Shock Waves on Condensed Matter*, Nellis, W. J., Seaman, L., and Graham, R. A., Eds., American Institute of Physics, New York, 1982, 42.
12. **Urbanski, T.**, Formation of free radicals by mechanical action, *Nature (London)*, 216, 577, 1967.
13. **Napier, D. H. and Subrahmanyan, N.**, Pyrolysis of methane in a single pulse shock tube, *J. Appl. Chem., Biotechnol.*, 22, 303, 1972.
14. **Benson, S. W.**, *Thermochemical Kinetics*, 2nd ed., John Wiley & Sons, New York, 1976, 85.
15. **Robinson, P. J. and Holbrook, K. A.**, *Unimolecular Reactions*, John Wiley & Sons, London, 1972.
16. **Sanderson, R. T.**, *Chemical Bonds and Bond Energy*, Academic Press, New York, 1976.
17. **Sanderson, R. T.**, Electronegativity and bond energy, *J. Am. Chem. Soc.*, 105, 2259, 1983.
18. **Sanderson, R. T.**, The interrelationship of bond dissociation energies and contributing energies, *J. Am. Chem. Soc.*, 97, 1367, 1975.
19. **Fujimoto, G. T., Selwyn, G. S., Keiser, J. T., and Lin, M. C.**, Temperature effect on the removal of hydroxyl radicals by a polycrystalline platinum surface, *J. Phys. Chem.*, 87, 1906, 1983.
20. **Canning, N. D. S. and Madix, R. J.**, Towards an organometallic chemistry of surface, *J. Phys. Chem.*, 88, 2437, 1984.
21. **Schonhammer, K. and Gunarson, O.**, Energy dissipation at metal surfaces: electron versus vibrational excitation, *J. Electron Spectrosc. Relat. Phenom.*, 29, 91, 1983.
22. **Duke, C. B.**, Hot atoms and cold facts: mysteries and opportunities in vibration-assisted surface chemistry, *J. Electron Spectrosc. Relat. Phenom.*, 29, 1, 1983.
23. **Beckhaus, H. D. and Ruchardt, C.**, Die sterische Beschleunigung der Termolyse hochverzweigter Alkane, *Chem. Ber.*, 110, 878, 1977.
24. **Beckhaus, H. D., Hellmann, G., and Ruchardt, C.**, Thermolabile Kohlerwassestoffe. Tetra-tert-buty-lethan, *Chem. Ber.*, 111, 72, 1978.
25. **Hellmann, S., Beckhaus, H. D., and Ruchardt, C.**, Zusamenhang zwischen termischer Stabilitat und Spannung nichtsymmetrisch hochverzweigter Kohlenwasserstoffe, *Chem. Ber.*, 116, 2238, 1983.

26. **Cremer, D.,** General and theoretical aspects of peroxide group, in *The Chemistry of Peroxides,* Patai, S., Ed., John Wiley & Sons, 1983, 1.
27. **Batt, L. and Robinson, G. N.,** Reaction of Methoxyradicals with Oxygen, *Int. J. Chem. Kinetics,* 11, 1045, 1979.
28. **Howard, J. A.,** Free-radical reaction mechanism involving peroxides in solution, in *The Chemistry of Peroxides,* Patai, S., Ed., John Wiley & Sons, London, 1983, 235.
29. **Zuravlev, M. V., Burmakov, A. J., Bloscica, F. A., Sass, V. P., and Sokolov, S. V.,** Thermal decomposition of 2,2-difluoropropyl peroxide in hydrocarbon solvents (Russian), *Z. Org. Chim.,* 18, 1825, 1982.
30. **Plesnicar, B.,** Organic polyoxides, in *The Chemistry of Peroxides,* Patai, S., Ed., John Wiley & Sons, London, 1983, 483.
31. **Hiatt, R.,** *Organic Peroxides II,* Swern, D., Ed., Wiley Interscience, New York, 1971, chap. 8.
32. **Reetz, M. T.,** Anchimerically accelerated bond homolysis, *Angew. Chem. Ed. Engl.,* 18, 173, 1979.
33. **Swern, D., Ed.,** *Organic Peroxides* I (1970), II (1971), III (1972), Wiley Interscience, New York.
34. **Barrtlett, P. D. and Lahav, M.,** Crystalline Di-tert-butyltrioxide and dicumyl trioxide, *Isr. J. Chem.,* 10, 101, 1972.
35. **Dulong, L. and Stahlberg, H.,** Reaktion von Hydroperoxiden mit Radikalen, *Angew. Makromol. Chem.,* 74, 285, 1978.
36. **Barter, J. B. and Kellar, D. E.,** Suspension polymerization of vinyl chloride initiated by in situ generated diisobutyryl peroxide, *J. Polym. Sci., Polym. Chem. Ed.,* 15, 2545, 1977.
37. **Macho, V., Rusina, M., Orsagova, M., and Porubsky, J.,** Suspension copolymerization of vinylchloride with vinyl-acetate initiated by diisopropyl peroxydicarbonate formed in situ, *Chem. Papers,* 35, 817, 1981.
38. **Joshino, K., Ohkatson, J., and Tsuruta, T.,** Study of the modes of thermal decomposition of several azo type initiators, *Polym. J.,* 9, 275, 1977.
39. **Hinz, J., Oberliner, A., and Ruchardt, C.,** Zur frage des Einstufigen oder Mehrstufigen Verlaufs der Azoalkanthermolyse, *Tetrahedron Lett.,* 22, 1975, 1973.
40. **Neuman, R. C., Jr. and Lockyer, G. D., Jr.,** One-bond azo initiators. Thermal decomposition of substituted phenyl azotriphenylmethanes, *J. Am. Chem. Soc.,* 105, 3982, 1983.
41. **Daanenberg, J. J. and Rocklin, D.,** A theoretical study of the thermal decomposition of azoalkanes and 1,1-diazenes, *J. Org. Chem.,* 47, 4529, 1982.
42. **Baudlish, B. K., Garner, A. W., Hodges, M. L., and Timberlake, J. W.,** Substituent effects in radical reactions. Thermolysis of substituted phenylazomethanes, 3,5-diphenyl-1-pyrazolines, and azopropanes, *J. Am. Chem. Soc.,* 97, 5856, 1975.
43. **Steel, C. and Trotman-Dickenson, A. F.,** Kinetics of the thermal decomposition of azomethane, *J. Chem. Soc.,* 975, 1959.
44. **Cohen, S. G. and Wang, Ch. H.,** Azo-bis-diphenylmethane and the decomposition of aliphatic azo compounds. The diphenylmethyl radical, *J. Am. Chem. Soc.,* 77, 2457, 1955.
45. **Prochazka, M.,** Synthesis of branched azoalkanes and kinetics of their thermal decomposition, *Collect. Czechoslov. Chem. Commun.,* 41, 1557, 1976.
46. **Barbe, W., Beckhaus, H. D., Lindner, H. J., and Ruchardt, C.,** Struktur und Spanungsenthalpie tetrasubstituierter Bernstein saure dinitrile, *Chem. Ber.,* 116, 1017, 1983.
47. **Huang, R. L., Goh, S. H., and Ong, S. H.,** *The Chemistry of Free Radicals,* Edward Arnold, London, 1974.
48. **Alt, H. G.,** Photochemistry of alkyltransition metal complexes, *Angew. Chem. Int. Ed. Engl.,* 23, 766, 1984.
49. **Dowd, P., Shapiro, M., and Kang, J.,** The mechanism of action of vitamine B_{12}, *Tetrahedron,* 40, 3069, 1984.
50. **Halpern, J.,** Mechanistic aspects of coenzyme B_{12} dependent rearrangement. Organometalics as free radical precursors, *Pure Appl. Chem.,* 55, 1059, 1983.
51. **Dowd, P. and Shapiro, M.,** A nonenzymic model for the coenzyme B_{12}-dependent isomerization of methyl malonyl-SCoA to succinyl-SCoA, *Tetrahedron,* 40, 3063, 1984.
52. **Margulis, M. A.,** Contemporary ideas about the nature of sonochemical reactions, *Zh. Phys. Chim.* (Russian), 50, 1, 1976.
53. **Hart, E. J. and Henglein, A.,** Free radical and free atom reactions in the sonolysis of aqueous iodide and formate solutions, *J. Phys. Chem.,* 89, 4342, 1985.
54. **Henglein, A. and Fischer, Ch. H.,** Sonolysis of chloroform, *Ber. Bundes Ges. Phys. Chem.,* 88, 1196, 1984.
55. **Ronn, A. M.,** Laser chemistry, *Sci. Am.,* 5, 114, 1979.
56. **Moore, C. B., Ed.,** *Chemical and Biochemical Applications of Lasers,* Academic Press, New York, 1977.
57. **Danen, W. C. and Jang, J. C.,** *Laser Induced Chemical Processes,* Steinfield, J. I., Ed., Plenum Press, New York, 1981, chap. 2.
58. **Pola, J.,** Laser infrawave photochemistry, *Chem. Listv.* (Czech), 75, 907, 1981.

59. **Evans, J. K., Mc Alpine, R. D., and Adams, H. M.,** The multiphoton absorption and decomposition of fluoroform-d, laser isotope separation of deuterium, *J. Chem. Phys.,* 77, 3551, 1982.
60. **Danen, W. C., Rio, V. C., and Setser, D. W.,** Multiphoton laser induced absorption and reactions of organic esters, *J. Am. Chem. Soc.,* 104, 5431, 1982.
61. **Manenkov, A. A. and Prochorov, A. M.,** Laser degradation of transparent materials, *Usp. Phys. Nauk.* (Russian), 148, 179, 1986.
62. **Pappas, S. P.,** UV Curing — Science and Technology, Technology Marketing Corporation, Stanford, 1978.
63. **Shimo, N., Nakashima, N., Ikeda, N., and Yoshihara, K.,** Laser flash photolysis of 1-alkenes at 193 nm in the gas phase: effect of molecular size on the formation yield of allyl radical, *J. Photochem.,* 33, 279, 1986.
64. **Plotnikov, V. G. and Ovchinikov, A. A.,** Photo and radiation stability of molecules; reactions of molecular abstraction of hydrogen atoms, *Usp. Khim.* (Russian), 47, 444, 1978.
65. **Ledwith, A., Russel, P. J., and Sutcliffe, L. M.,** Radical intermediates in the photochemical decomposition of benzoin and related compounds, *J. Chem. Soc., Perkin Trans. II,* 1925, 1972.
66. **Holmes, B. E., Paisley, S. D., Rakestraw, D. J., and King, E. E.,** Generation of ground electronic state haloalkyl radicals in the gas phase, *Int. J. Chem. Kinet.,* 18, 639, 1986.
67. **Kropp, P. J.,** Photobehaviour of alkyl halides in solutions, radical carbocation and carbene intermediates, *Acc. Chem. Res.,* 17, 131, 1984.
68. **Sonawane, H. R., Nanjundiah, B. S. B. S., and Rajput, S. I.,** Photochemistry of vinyl halides, *Indian J. Chem.,* 23B, 331, 1984.
69. **Neckers, D. C. and Wagenaar, F. L.,** Unimolecular heteroatomic photoinitiators, *Tetrahedron Lett.,* 25, 2931, 1984.
70. **Ingold, K. V., Johnston, L. J., Lusztyk, J., and Scaiano, J. C.,** Triplet. Quenching by diacyl peroxides, *Chem. Phys. Lett.,* 110, 433, 1984.
71. **Tatahara, S., Urano, T., Kitamura, A., Sakuragi, H., Kikuchi, O., Yoshida, M., and Totumaru, K.,** The role of aroyloxy radicals in the formation of solvent derived products in photodecomposition of diaroylperoxides, *Bull. Chem. Soc. Jpn.,* 58, 688, 1985.
72. **Murai, H., Jinguji, H., and Obi, K.,** Activation energy of hydrogen atom abstraction by triplet benzophenone at low temperature, *J. Phys. Chem.,* 82, 38, 1978.
73. **Wagner, P. J., Truman, R. J., and Scaino, J. C.,** Substituent effects on hydrogen abstraction by phenyl ketone triplets, *J. Am. Chem. Soc.,* 107, 7093, 1985.
74. **Eaton, D. F.,** Electron transfer induced photofragmentation as a route to free radicals, *Pure Appl. Chem.,* 56, 1191, 1984.
75. **Weller, A.,** Mechanism and spindynamics of photoinduced electron transfer reactions, *Z. Phys. Chem. Neue Folge,* 130, 129, 1982.
76. **Wamser, C. C., Hammond, G. S., Chang, C. T., and Baylor, C., Jr.,** Photo reaction of Michler's ketone with benzophenone. A triplet exciplex, *J. Am. Chem. Soc.,* 92, 6362, 1970.
77. **Inbar, S., Linchitz, H., and Cohen, S. G.,** Nanosecond flash studies of reduction of benzophenone by aliphatic amines. Quantum yields and kinetic isotope effects, *J. Am. Chem. Soc.,* 103, 1048, 1981.
78. **Guttenplan, J. B. and Cohen, S. G.,** Triplet energies, reduction potentials, and ionization potentials in carbonyl-donor partial charge-transfer, *J. Am. Chem. Soc.,* 94, 4040, 1972.
79. **Scaiano, J. C.,** Photochemical and free-radical processes in benzil-amine systems. Electron-donor properties of α-aminoalkyl radicals, *J. Phys. Chem.,* 85, 2851, 1981.
80. **Arimitsu, S. and Tsubomura, H.,** Intermolecular hydrogen transfer reactions in the excited states of EDA-complexes of benzophenone with aromatic amines at low temperature, *Bull. Chem. Soc. Jpn.,* 45, 1357, 1972.
81. **Moan, J., Kaalhus, O., and Hovik, B.,** H atoms in acid low temperature glasses containing Fe^{2+} ions, *Chem. Phys. Lett.,* 61, 481, 1979.
82. **Kornblum, N.,** Substitutionreaktionen uber Radikalanionen, *Angew. Chem.,* 87, 797, 1975.
83. **Bremer, N., Brown, B. J., Morine, G. H., and Willard, J. E.,** Mercury photosensitized production of free radicals in organic glasses, *J. Phys. Chem.,* 79, 2187, 1975.
84. **Zafirion, O. C. and True, M. B.,** Nitrite photolysis as a source of free radicals in productive surface water, *Geophys. Res. Lett.,* 6, 81, 1979.
85. **Swalow, A. J.,** *Radiation Chemistry,* Longmans Green, New York, 1973.
86. **Huttermann, J.,** Physical mechanisms of electron interaction with organic solids, *Ultramicroscopy,* 10, 7, 1982.
87. **Symons, H. C. R.,** Solutions of metals:solvated electrons, *Chem. Soc. Rev.,* 5, 337, 1976.
88. **Calvo-Perez, V., Beddard, G. S., and Fendler, J. H.,** Hydrated electrons in surfactant solubilized water pools in heptane, *J. Phys. Chem.,* 85, 2316, 1981.
89. **Willard, J. E.,** Some properties of trapped electrons, radicals and H atoms in glassy hydrocarbons, *Int. J. Radiat. Phys. Chem.,* 6, 325, 1974.

90. **Kevan, L.,** Solvated electron structure in glassy matrices, *Acc. Chem. Res.,* 14, 138, 1981.
91. **Baba, H. and Fueki, K.,** Electron scavenging by bromobenzene in the radiolysis of heptane, 1-hexane and their mixtures, *Bull. Chem. Soc. Jpn.,* 48, 3039, 1975.
92. **Isildar, M. and Schuler, R. H.,** On the Precursor of fragment radicals in the radiolysis of normal alkanes, *Radiat. Phys. Chem.,* 11, 11, 1978.
93. **Miyazaki, T., Kinigawa, K., and Kasagai, J.,** Selective formation of solute radical by hydrogen atom in radiolysis and photolysis of alkene mixtures, *Radical Phys. Chem.,* 10, 152, 1977.
94. **Iwasaki, M., Toriyama, K., Fukaya, M., Muto, H., and Nunome, K.,** 4 K radiolysis of linear alkanes as studied by electron spin resonance spectroscopy: selective formation of terminal alkyl radicals in the primary process, *J. Phys. Chem.,* 89, 5278, 1985.
95. **Dressler, R., Allan, M., and Haselbach, E.,** Symmetry control in bond cleavage processes. Dissociative electron attachment to unsaturated halocarbons. *Chimia,* 39, 385, 1985.
96. **Baxendale, J. H. and Rodgers, M. A. J.,** Contributions of pulse radiolysis to chemistry, *Chem. Soc. Rev.,* 7, 235, 1978.
97. **Sumiyoshi, T., Tsugaru, K., Yamada, T., and Katayama, M.,** Yields of solvated electrons at 30 proseconds in water and alcohols, *Bull. Chem. Soc. Jpn.,* 58, 3073, 1985.
98. **Trudel, G. J., Gillis, H. A., Klassen, N. V., and Theather, G. G.,** Localized excess electrons in H_2O glasses and ice, *Can. J. Chem.,* 59, 1235, 1981.
99. **Bielski, B. H. J., Cabelli, D. E., and Arudi, R. L.,** Reactivity of HO_2/O_2 radicals in aqueous solution, *J. Phys. Chem. Ref. Data,* 14, 1041, 1985.
100. **Bielski, B. H. J.,** Reevaluation of the spectral and kinetic properties of HO_2 and O_2 free radicals, *Photochem. Photobiol.,* 28, 645, 1978.
101. **Getoff, N. and Prucha, M.,** Spectroscopic and kinetic characteristics of HO_2 and O_2 species studied by pulse radiolysis, *Z. Naturforsch.,* 38A, 589, 1983.
102. **Gilde, H. G.,** Electrolytically generated radicals, in *Methods in Free-Radical Chemistry,* Vol. 3, Huyser, E. S., Ed., Marcel Dekker, New York, 1972, 1.
103. **Evans, D. H.,** Voltametry. Doing chemistry with electrodes, *Acc. Chem. Res.,* 10, 313, 1977.
104. **Wendt, H.,** The reactivity of primary free radicals and radical ions, mass transfer, and electrosorption — the fundamental factors for selectivity in electrochemical syntheses of organic compounds, *Angew. Chem. Int. Ed. Engl.,* 21, 256, 1982.
105. **Andrieux, C. P., Merz, A., and Saveant, J. M.,** Dissociative electron transfer, *J. Am. Chem. Soc.,* 107, 6097, 1985.
106. **Feldhues, M. and Schafer, H. J.,** Selective mixed coupling of carboxylic acids, *Tetrahedron,* 41, 4195, 1985.
107. **McTaggart, T. K.,** *Plasma Chemistry in Electrical Discharges,* Elsevier, New York, 1976.
108. **Vinogradov, M. G.,** Redox methods of generation of free radicals, *Iz. Vsesojuz. Obschestva* (Russian), 24, 175-180, 1979.
109. **Skibida, I. P.,** Kinetics and mechanisms of decomposition of organic hydroperoxides in the presence of transition metals, *Usp. Khimii* (Russian), 44, 1729, 1975.
110. **Hajdu, P., Nemes, I., Sumegy, L., Vidoczy, T., and Gal, D.,** On the kinetics of the transition metal catalyzed decomposition of secondary hydroperoxides, *Int. J. Chem. Kinetics,* 13, 1191, 1981.
111. **Lunak, S. and Veprek-Siska, J.,** Mechanisms of Thermal and Photoinitiated decomposition of hydrogen peroxide, *Chem. Listy* (Czech), 77, 1121, 1983.
112. **Ranby, B. and Rabek, J. F.,** *ESR Spectroscopy in Polymer Research,* Springer-Verlag, Berlin, 1977, 122.
113. **Irvine, M. J. and Wilson, I. R.,** Rates of formation and decay of radical species during reaction of titanium (III) and hydrogen peroxide. Concentrations of radicals during reaction of titanium (III) with hydrogen peroxide, *Aust. J. Chem.,* 32, 2131, 2283, 1979.
114. **Tuazon, E. C., Carter, W. P. L., Atkinson, R., and Pitts, J. N., Jr.,** The gas phase reaction of hydrazin and ozone: nonphotolytic source of ˙OH radicals for measurement of relative ˙OH radical rate constants, *Int. J. Chem. Kinetics,* 15, 619, 1983.
115. **Black, J. F.,** Metal-catalyzed autoxidation. The unrecognized consequences of metal-hydroperoxide complex formation, *J. Am. Chem. Soc.,* 100, 527, 1978.
116. **Howard, J. A. and Tong, S. B.,** Metal complexes as antioxidants. 7. Kinetics and product study of the reaction of tertiary alkylperoxy radicals with cobalt (II) acetylacetonate, *Can. J. Chem.,* 58, 1962, 1980.
117. **Pryor, W. A. and Hendricson, W. H., Jr.,** The mechanism of radical production from the reaction of *N,N*-dimethylaniline with benzoyl peroxide, *Tetrahedron Lett.,* 24, 1459, 1983.
118. **Thornalley, P. J. and Stern, A.,** The production of free radicals during the autoxidation of monosaccharides by buffer ions, *Carbohydr. Res.,* 134, 191, 1984.
119. **Swartz, H. M. and Dodd, N. J. F.,** *Oxygen and Oxy Radicals in Chemistry and Biology,* Rodgers, M. A. J. and Powers, E. L., Eds., Academic Press, New York, 1981, 161.

120. **Simkovic, I., Tino, J., Placek, J., and Manasek, Z.**, ESR study of alkaline, oxidative degradation of saccharides: identification of 2,5-dihydroxy-p-benzosemiquinone, *Carbohydr. Res.*, 116, 263, 1983.

121. **Searle, A. J. F. and Tomasi, A.**, Hydroxyl free radical production in iron-cysteine solutions and protection by zinc, *J. Inorg. Biochem.*, 17, 161, 1982.

122. **Ihde, A. J.**, The history of free radicals and M. Gomberg's contributions, *Pure Appl. Chem.*, 15, 1, 1967.

123. **Forrester, A. R., Hay, J. M., and Thompson, R. H.**, *Organic Chemistry of Stable Free Radicals*, Academic Press, London, 1968.

124. **Lewis, I. C. and Singer, L. S.**, EPR and ENDOR study of thermally produced triphenylmethyl radicals, *Org. Magn. Reson.*, 22, 761, 1984.

125. **Kolsch, C. F.**, Synthesis with triarylvinyl magnesium bromides, alfa, gamma-bisdiphenylene-beta-phenylallyl, a stable free radical, *J. Am. Chem. Soc.*, 79, 4439, 1957.

126. **Reich, D. H.**, Stable π-electron systems and new aromatic structures, *Tetrahedron*, 3, 339, 1958.

127. **Kosower, E. M. and Poziomek, E. J.**, Isolation and distillation of 1-ethyl-4-carbomethoxypyridinyl, *J. Am. Chem. Soc.*, 86, 5515, 1964.

128. **Ikegami, Y.**, Kinetic ESR and CIDEP studies on the monomer-dimer equilibrium systems of pyridinyl radicals, *Rev. Chem. Intermediates*, 7, 91, 1986.

129. **Endo, T. and Okawara, M.**, Production of a poly(pyridinyl) radical by reduction with 1-benzyl-1-,4-dihydronicotinamide and *N,N'*-Dicyclohexylalloxan, *J. Polym. Sci.*, 19, 1591, 1981.

130. **Malatesta, V., Forrest, D., and Ingold, K. V.**, Persistent radicals from di-tert-butyldiazomethane and di-tert-butylketene, *J. Phys. Chem.*, 82, 2370, 1978.

131. **Griller, D. Cooper, J. W., and Ingold, K. U.**, Kinetic applications of electron paramagnetic resonance spectroscopy. 18. Persistent vinyl, alkyl, and allyl radicals, *J. Am. Chem. Soc.*, 97, 4269, 1975.

132. **Muller, E. and Kiedaisch Ley K. W.**, Über ein stabiles sauestoffradikal, das 2,4,6-Tri-tert.butyl-phenoxyl-(1). Weitere Herstellungs methoden und Lebensdauer des Aroxyls, *Chem. Ber.*, 87, 1605, 1954.

133. **Bartlett, P. D. and Funahashi, T.**, Galvinoxyl 2,6-Di-tert-α-(3,5-di-tert-butyl-4-oxo-2,5-cyclohexadiene-1-ylidene)-p-tolyloxy) as a scavenger of shorter-lived free radicals, *J. Am. Chem. Soc.*, 84, 2596, 1962.

134. **Neunhoeffer, O. and Heitmann, P.**, Uber freie Radikale mit Betain-Grenz Strukturen., *Chem. Ber.*, 96, 1027-1034, 1963.

135. **Rozantsev, E. G.**, *Free Nitroxyl Radicals*, Plenum Press, New York, 1970.

136. **Khramtsov, V. V., Weiner, L. M., Gogolev, A. Z., Grigoriev, I. A., Starichenko, V. F., and Volodarsky, L. B.**, ESR and ^{1}H NMR studies of a new class of nitroxyl, nitronylnitroxyl and iminonitroxyl radicals, *Magn. Reson. Chem.*, 24, 199, 1986.

137. **Torssel, K.**, Investigation of radical intermediates in organic reaction by use of nitroso compounds as scavengers, *Tetrahedron*, 26, 2759, 1970.

138. **Balaban, A. T., Frangopol, P. T., Frangopol, M., and Nigoita, N.**, Preparation of stable sterically shielded, diaryl nitrogen radicals with donor and acceptor aryl groups in the same molecule, *Tetrahedron*, 23, 4661, 1967.

139. **Hutchison, C. A., Pastor, R. C., and Kowalsky, A. G.**, Paramagnetic resonance absorption in organic free radicals fine structure, *J. Chem. Phys.*, 20, 534, 1952.

140. **Neugebauer, F. A.**, Hydrazidinyl radicals: 1,2,4,5-tetra-azapentenyls, verdazyl and tetrazolinyls, *Angew. Chem. Int. Ed. Engl.*, 12, 455, 1973.

141. **Chatyilialoglu, C., Ingold, K. U., and Scaiano, J. C.**, Rate constants and arrhenius parameters for the reactions of primary, secondary, and tertiary alkyl radicals with tri-n-butyl hydride, *J. Am. Chem. Soc.*, 103, 7739, 1981.

142. **Ono, N., Miyake, H., Kamimura, A., Hamamoto, I., Tamura, R., and Kaji, A.**, Dinitrohydrogenation of alifatic nitro compounds as radical precursors, *Tetrahedron*, 41, 4013, 1985.

Chapter 3

ELEMENTARY REACTIONS OF FREE RADICALS

I. INTRODUCTION

Chemical reactions which proceed via radical intermediates are of various character. Their common features may, however, be expressed in terms of similar elementary steps.

Primarily formed free radicals are often capable of changing their structure by rearrangement of atoms (isomerization), by splitting off some groups of atoms (fragmentation), or by linking of a radical with some reactive molecules (addition). Also transfer (substitution) of an atom or group of atoms from a neighboring molecule to a radical may be considered as an elementary reaction; the original radical is changed to a molecule and some other molecule to a new radical. Substitution reaction takes place either as a simple elementary transfer or it is composed of two elementary reactions, namely of addition and of subsequent fragmentation of a formed unstable radical. Even though the resulting products of both types of substitution reactions may be identical, the different mechanism is well distinguishable experimentally. During the transfer reaction, only the electron from the reaction partner is sometimes displaced to a free radical. In such a case, the molecule is transformed to a cation radical and the radical, functioning as an electron acceptor, to an anion. Radicals may, however, be also donors of electrons.

Each radical transformation of compounds should necessarily include the elementary reactions of free radical decay. This is the matter of either linkage of two radicals (recombination) or of such deactivation where each pair of decaying radicals gives at least two stable molecules (disproportionation).

In the course of radical reactions, several elementary reactions usually take place simultaneously. They mutually compete or are consecutive. The full understanding and quantitative modeling of complex radical reactions requires, therefore, the knowledge of rate constants of individual elementary steps. Most of these rate constants cannot be measured directly but only relative to other reactions such as, e.g., to self termination or monomolecular radical rearrangement which serve then as an internal "clock". The importance of individual elementary reactions in the reaction mechanism is determined by reactivity and mobility of reactants. The reaction medium where elementary reactions are taking place is also of significance from the viewpoint of reaction conditions.

In the gaseous phase, free radicals and surrounding molecules are further away from each other than in the condensed phase. At atmospheric pressure and at ambient temperature the distance between two reacting molecules in the gaseous phase is on average ten times higher than the largest size particle. Filling free space in the gas with other molecules requires a thousand fold increase in the number of particles. The difference in the density of molecules of gas and in the condensed phase has its consequence in the frequency of collisions among individual particles of the system. The higher density of the condensed phase leads to the considerably faster deactivation of excited states of reactants. Intermolecular interactions may, moreover, change the reactivity of molecules. This is especially valid for reactions in polar systems. As an example we may give the reaction of methyl radicals with methyl alcohol. In the gaseous phase, there is abstracted hydrogen bound to oxygen of methyl alcohol, whereas in the condensed phase hydrogen is bound to the carbon atom. The difference is brought about by the partially oriented associated molecules of methyl alcohol in the liquid phase due to hydrogen bond formation. In the liquid phase, there are also more probable weak interactions among radicals and molecules of solvent. When compared to the gaseous phase the difference increases with the increasing polarity of radicals and of the reaction medium.

The change in the reactivity is not, however, the only manifestation of the transition of the reaction system from the gaseous to the liquid state. In the liquid phase, radicals and molecules displace more slowly to the new positions. Two particles may collide from 100 to 1000 times before they separate definitely and lose direct contact. The slower displacement of reactants slows down such reactions which are controlled by diffusion; the change in viscosity of the reaction medium will also change the rate of individual elementary reactions. At the reactions in the solid state, free radicals mostly disproportionate or fragmentate directly in the place of their generation.

II. THE ISOMERIZATION OF POLYATOMIC RADICALS

The structure of the radical may be changed and regenerated back and nothing will apparently happen with the original radical.

The most simple structural isomerizations of free radicals may be realized by partial rotation of a group of atoms around chemical bonds from one equilibrium position to another. For cyclohexyl and cycloheptyl the most stable conformer is a twisted chair structure which interconverts to their mirror image via a low energy process ($\sim$ 15 kJ/mol).[1] The barrier for ring inversion of cycloalkyl radicals is lower than for corresponding cycloalkanes.[2] The reason for the decrease of the rotational barrier of C-C bonds may consist in weakening and lengthening of corresponding β-bonds. The structural changes taking place during the inversion of these cycloalkyl radicals are very fast.[3] The average time of existence of the cyclohexyl isomer is on the order of nanoseconds at 300 K.

The preferred conformation of the cyclohexene ring is the half-chair which interconverts more rapidly than the chair conformation of cyclohexane. Such conformation isomerism was observed for cyclohex-2-enyl methyl radicals.[4] At high temperatures there is a preponderance of the quasi-equatorial conformer (Scheme 1, a), however, at T>170 K the quasi-axial conformer (Scheme 1, b) predominates.

SCHEME 1

The free-energy difference between these two conformers is small (>1 kJ/mol).

Conformation isomers of free radicals would have different reactivity. In most cases this difference will not be very high.

Larger difference in the stability of conformers may be expected in radicals where interactions among nonvicinal atoms of a radical occur. In the case of the 1-naphthoyl radical two rotamers (Scheme 2) are present in the conformer

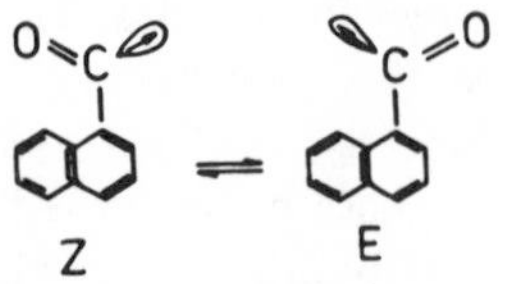

Z   E

SCHEME 2

ratio 55:45 at 148 K (Reference 5). In the 1-naphthaldehyde the Z-conformer (80 to 90%) is more stable and both conformations are separated by the rotation barrier of 27 kJ/mol. In the radical, the barrier for rotational isomers is only about 10 kJ/mol. A stabilizing effect on the ground state of the Z-conformer follows from a hydrogen-bond type interaction of the carbonyl group with the peri-hydrogen, while that of the E-conformer from interaction of the sp^2 orbital containing the unpaired electron with the peri C-H bond.

The sterical effect of the methyl group may be seen in the 2,3-dimethyl 1-naphthoyl radical (Scheme 3) which exists as a single Z-conformer.

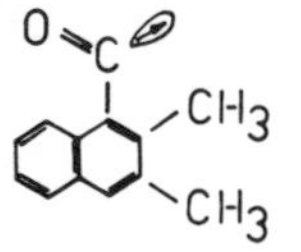

SCHEME 3

On the other hand, because of a strong internal OH...O hydrogen bond, the 2-hydroxy-1-naphthoyl radicals have essentially the same E-structure as found for the parent 1-naphthaldehyde.

The temporary rupture and subsequent regeneration of the same π-bond is assumed to occur at *cis-trans* isomerization of the vinyl radicals (Scheme 4)

(R is alkyl group or hydrogen and X is halogen or other substituent.)

SCHEME 4

Activation energy of the cis-trans transition which depends on the type of substituents may achieve 85 kJ/mol.[6] The half-life of the configuration isomers is then considerably prolonged and in reactions, radical isomers may, thus, exist as independent particles. The more forward or even irreversible isomerization of radicals takes place only at the disappearance of old and at the formation of new chemical bonds among atoms of a radical.

The idea of radical isomerization leading to the displacement of hydrogen atoms was first put forward for explanation of different products forming at the pyrolysis of hydrocarbons in the 1940s (Rice, Kossiakoff). In spite of the stepwise accumulation of equivalent results, the hypothesis of intramolecular isomerization could not be accepted for a long time because of the lack of unambiguous evidence. There were doubts whether isomerization really proceeds as an intramolecular process or whether experimentally found analytical data are simply the result of subsequent reactions. Now, the situation is somewhat different.

Taking into account the rate of the formation of tautomery isomers and the appropriate difference in the energetical state of parent and isomerized radicals, the self evidence of intramolecular isomerization of many radicals is quite apparent. The idea of the atom transfer inside of one radical mediated by five and especially by six or seven centered cyclic transition states (Scheme 5), was commonly accepted.

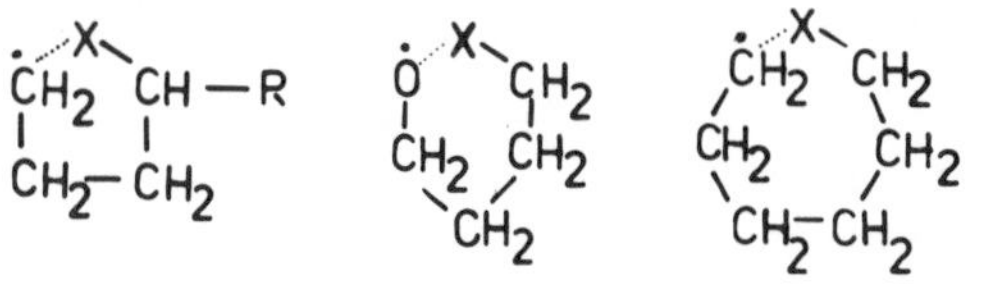

(X denotes H or halogen. R is alkyl or aryl.)

SCHEME 5

As may be seen later, the reaction resembles a current elementary transfer reaction of an atom from reacting molecule to a radical. Some complications may arise from intramolecular transfer of atoms in the radical to shorter distances such as it is, e.g., in 1,2 or 1,3-migrations.* The probability of the course of such isomerizations was the matter of many discussions and controversies. The problem was of importance mainly in the case of alkyl radicals. It seems, now, that the intramolecular 1,2-isomerization has to be considered seriously in reaction mechanisms, especially in diluted or gas systems.

Isomerization of radicals as in any other process may be realized due to the fact that the new arrangement is thermodynamically more favorable for the systems of atoms. The isomerization may, however, be also induced by the excess of energy from the surrounding molecules. In such a case, the resulting structure is energetically equivalent or much richer than the initial state.

A. Tautomery of Radicals

The chemical rearrangement of atoms inside one radical is a very frequent reaction in each case when the formation of the cyclic activated complex composed of five, six, or seven atoms may easily be realized, and structurally the same radical is formed.[7] Isomerization with fast hydrogen jumps may be demonstrated on stable 2-hydroxy-3,6-*ditert*-butyl phenoxy radicals (Scheme 6). In tautomeric isomers of a radical, the spin of an unpaired

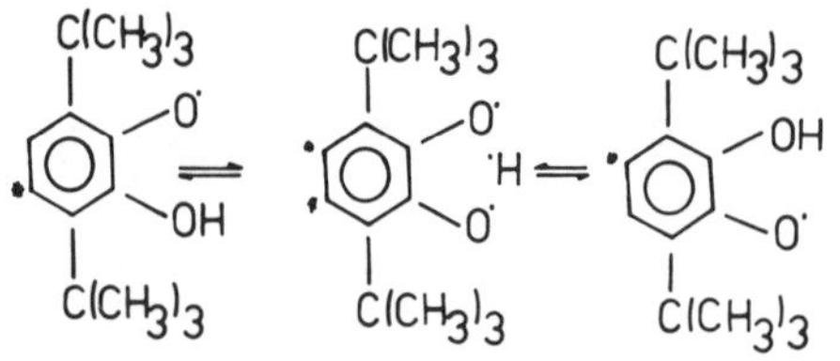

SCHEME 6

electron is in the interaction with one hydrogen (denoted by asterisk) in position 4 on the phenyl ring and with hydrogen of the hydroxyl group. These radicals give at lower temperatures a two-line ESR spectrum split to two doublets. The frequency of hydrogen atom migration in the radical may be increased so significantly by the increased temperature that the ESR spectrum changes to three lines with doublet splitting. This spectrum is brought about by the interaction of two equivalent protons of the aromatic ring and of hydrogen of the hydroxyl group. The radical has, thus, symmetrically distributed density of spin of the unpaired electron and that of hydrogen on both oxygens. This type of radical cannot be satisfactorily represented by a classical structural formula; its structure may rather be understood as time averaged above structures including that representing the activated complex. The frequency of hydrogen atom displacement in this radical is $3.10^9 \cdot sec^{-1}$ at the ambient temperature; the activation energy of the migration is 12 kJ/mol.

Tautomery is not the specific property of phenoxy radicals only, but may be observed also in other radicals. In the case of the properly substituted nitroxy radicals (Scheme 7) the initially observed three-component ESR spectrum is changed to quintet by the increased temperature.

---

* Numbers denote mutual positions of both centers of isomerization reactions, namely, of unpaired electron and of transferred atom.

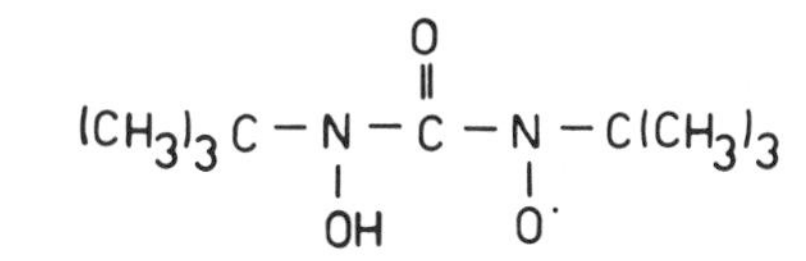

$$(CH_3)_3 C - N - \overset{\overset{\displaystyle O}{\|}}{C} - N - C(CH_3)_3$$
$$\underset{OH}{|} \qquad \underset{O^{\cdot}}{|}$$

SCHEME 7

This indicates that there occurs the interaction of an unpaired spin with two equivalent nitrogen atoms. Due to interaction with hydrogen of the hydroxyl group, the quintet ESR spectrum, as in the initial triplet, also has each line split to doublet.

In addition to the migration of hydrogen between two oxygen atoms we also know "polyhole" tautomeric radicals (Scheme 8) where the optimum position of the unpaired electron may be ascribed equivalently to more oxygen atoms.

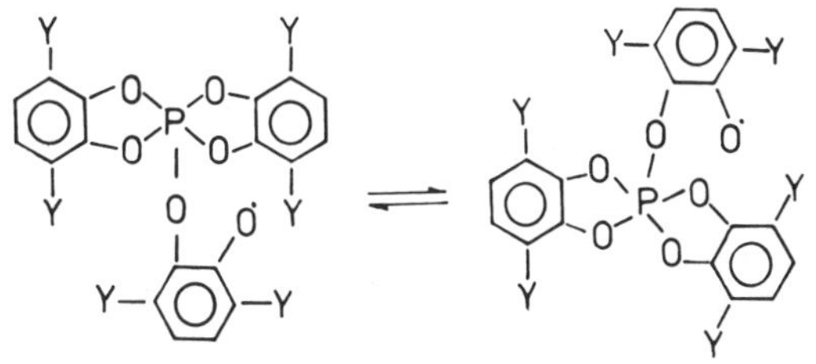

$(Y = C(CH_3)_3)$

SCHEME 8

Kinetic features of the tautomery of such radicals perform great dynamics of the hydrogen atoms transfer or of the scission and of the back formation of bonds linking oxygen and central phosphorus atom. The rate constant of the tautomery transition on P-O bonds (k/s = 3.4 10^{12} exp($-18 000/RT$)), indicates that the lifetime of the respective radical is only 35 nsec at the ambient temperature. The above radicals permanently isomerized even when they apparently keep the same structure.

Similar radicals having wandering radical centers were synthesized also for other central atoms as, e.g., for Si, B, and Al.

B. Transfer of Hydrogen to a Nonequivalent Position

The rate of hydrogen atom transfer in 3-chloro-2-hydroxy-4,6-*ditert*-butyl phenoxy radicals (Equation 1)

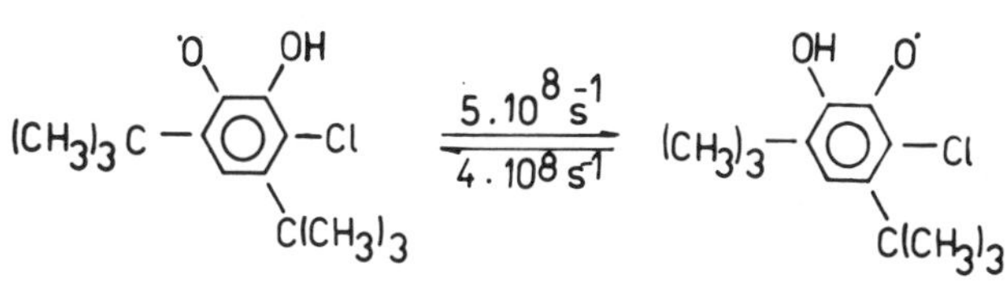

$$(1)$$

is comparable with the rate of tautomery transitions.[7] The essential fact is, however, that the rate is different in the backward direction. The phenoxy radical with the chlorine atom in an adjacent position will be more stable and, consequently, will occur in higher equilibrium concentrations. The difference in activation energies of reverse isomerization reactions is about 10 kJ/mol; the reaction leading to a more stable isomer has lower activation energy, namely 18 kJ/mol. The higher the difference in the energetical content of the isomeric radical, the more different will be the rates of forward and backward isomerizations and, in the limiting case, the process becomes irreversible.

The advantage of certain mutual orientation of attacking orbital of the unpaired electron and of displacing atom becomes more distinct when we compare the ratio of isomerization of phenylalkoxy radicals to the positions denoted by arrows (Scheme 9). Despite the fact that in the case of the seven-membered ring complex a more stable α-phenylalkyl radical is formed, the isomerization reaction obviously prefers the six-numbered activated complex.[8]

$$C_6H_5CH_2-CH_2-CH_2-CH_2(CH_3)_2C-O\cdot$$

$$\uparrow \qquad \uparrow$$

$$10\% \qquad 90\%$$

SCHEME 9

The highest probability of the formation of a six-centered activated complex in the isomerization reaction is coherent with the minimum strain forces in the six-membered cyclic structure and with the optimal angle of the radical center to the decaying bond. It should be recalled that at reactions of atoms with biatomic molecules, the probability of the mutual reaction is usually highest when three atoms in forming the activated complex are colinear. Rough quantum mechanical calculations show that the activation energy for a perpendicular approach is approximately twofold higher than that for a colinear approach. In isomerization reactions the colinear approach cannot be attained in many cases and, generally, the activation energy of isomerization is greater than for the corresponding intermolecular transfer reaction.

Another important feature of the intraradical six-membered activated complex may be demonstrated on frequent isomerization of alkoxy or peroxy radicals well known in combustion and atmospheric chemistry.[9,10] Isomerization of the 2-pentoxy radical by 1,5-hydrogen shift via a low-strain six-membered ring transition state (Equation 2) has the activation energy 40 kJ/mol and a relatively high frequency factor of the order $10^{11} \cdot sec^{-1}$.

$$\overset{\textstyle O}{\underset{\textstyle |}{}}$$
$$CH_3CHCH_2CH_2CH_3 \rightarrow CH_3CH(OH)CH_2CH_2\dot{C}H_2 \qquad (2)$$

The kinetic data show that under atmospheric conditions at 298 K and 102 kPa, both the isomerization and the decomposition of the radical to acetaldehyde and propyl radical are quite comparable.[11]

By isomerization of, e.g., peroxy radicals a hydroperoxy group is formed and the radical center is shifted to carbon (Equation 3). This type of process is faster than intermolecular transfer between the neighboring hydrocarbon molecule and peroxy radical. Provided that the oxidized hydrocarbon molecule is long enough and has reactive hydrogens, comb-like hydroperoxy groups are formed and arranged along one chain, whereas some surrounding molecules are not oxidized at all.[12]

$$\overset{\textstyle O-O\cdot}{\underset{\textstyle |}{}} \qquad \qquad \overset{\textstyle OOH}{\underset{\textstyle |}{}}$$
$$(CH_3)_2C-CH_2-CH(CH_3)_2 \rightarrow (CH_3)_2-C-CH_2-\dot{C}(CH_3)_2 \qquad (3)$$

The intramolecular course of the reaction is supported mainly by more frequent collisions of reacting centers, when compared with intermolecular transfer reactions.

Table 1
KINETIC PARAMETERS OF
ISOMERIZATION REACTIONS OF
ALKYL RADICALS OF GENERAL
FORMULA $R^1R^2\overset{*}{C}H(CH_2)_3CH_2\cdot$

R^1	R^2	k(20 °C) (sec^{-1})	log A.s	E (kJ/mol)
$CH_3\cdot$	H	51	7.7	34
C_2H_5	H	17	8.2	39
C_3H_7	H	15	7.9	38
C_5H_{11}	H	5	8.2	42
C_7H_{13}	H	7	8.3	42
C_9H_{19}	H	5	8.2	42
CH_3	CH_3	804	5.9	17
C_4H_9	CH_3	441	6.5	22
C_4H_9	C_5H_{1-}	153	6.9	26

Note: The transferred hydrogen atom is denoted by an asterisk. 1,5 Migration of radicals takes place in the solution
of bromalkanes at a temperature range from 30 to 150°C.

The displacement of hydrogen in alkyl radicals (1,5-migration) depends on the type of substituents in close proximity to transferred hydrogen and especially of those on the carbon of the radical center.[13] Increasing bulkiness of corresponding alkyl substituents decreases the rate of isomerization since it requires higher activation energy for the realization of the activated complex (Table 1). The rate constant is much higher for the migration of tertiary hydrogen, which is due to the lower strength of the bond, tertiary carbon-hydrogen. The rate of migration is determined also by the strength of the newly formed chemical bond. This is the main reason why for 1,5-migration of hydrogen in alkoxy radicals the rate constant is four orders higher than that in similar alkyl radicals.

1,4-Migration of hydrogen in alkyl and alkenyl radicals via five-membered cyclic transition state has activation energy in the range of 60 to 90 kJ/mol.[14]

1,3-Migration requires even higher activation energy and occurs only rarely at reactions of thermalized radicals. The convincing proof of its existence may be demonstrated on the formation of side ethyl groups in polyethylene during the radical polymerization of ethylene. For each 2000 methylene units of the main polymer chain there may be found 3 ethyl side groups. There were, moreover, identified 11 butyl (1,5 migration), 4 pentyl groups and some 1,3-paired diethyls (Scheme 10), 2-ethylhexyls, 1,3 paired ethylbutyls, and other anomalous groupings.[15]

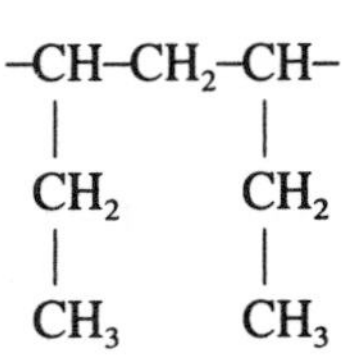

$$-CH-CH_2-CH-$$
$$\quad|\qquad\qquad|$$
$$CH_2\qquad CH_2$$
$$\quad|\qquad\qquad|$$
$$CH_3\qquad CH_3$$

SCHEME 10

The reactions of intraradical transfer in the growing polyethylene radicals will be more frequent when the radical center has a lower opportunity of reaction with the present mon-omer. It seems, therefore, quite obvious that with increasing concentration of ethylene in

Table 2

EXPERIMENTAL AND CALCULATED VALUES[17] OF ACTIVATION ENERGY E FOR 1,2 HYDROGEN MIGRATION IN ALKYL RADICALS AND RELATIVE CONTRIBUTIONS C OF INDIVIDUAL TERMS GIVING $E_{cal} = E_T + + E_B + E_D$

E C	Reaction Activation energy E, kJ/mol			
	$CH_3CHCH_2C_2H_5$ ↓ $CH_3CH_2CHCH_2H_5$	CH_3CD_2 ↓ CH_2CD_2H	$CH_2CH_2CH=CH_2$ ↓ $CH_3CHCH=CH_2$	CH_3CHCH_3 ↓ $CH_3CH_2CH_2$
E_{exp}	138 ± 5	170 ± 17	84; 138[a]	152 ± 9
E_{calc}	146	149	118	154
$E_T/\%$	75.0	74.8	73.8	70.5
$E_B/\%$	17.8	17.9	19.1	24.3
$E_D/\%$	7.2	7.3	7.1	5.2

Note: E_T, repulsive energy in the transition complex brought about by parallel electron spin on C_α and C_β during the transfer reaction; E_B, bond energy term for the formation of the activated complex due to breaking (or formation) of the C_α .. H and C_β .. H fractional bonds; E_D, deformation energy of the valence angles for the formation of cyclic transition state.

[a] Range of experimental values.

the polymerization system the extent of branching and also the number of side ethyl groups are reduced. A relatively high ratio of ethyl groups may be interpreted by the 1,3-isomerization of growing radicals activated by the reaction heat. The approximate excess of energy of such radicals which are formed just in the instant of ethylene addition to alkyl macroradical is 150 kJ/mol.[16] At isomerization of vibrationally excited alkyl radicals, the transfer of hydrogen from the first carbon atom of the radical center is preferred. Also, the finding that among short side branches of polyethylene chain there are lacking propyl groups as potential products of 1,4-migration supports such an idea.

As it was already pointed out the shift of hydrogen to the neighboring carbon atom proceeds with higher activation energy (about 150 kJ/mol in the case of 1,2-hydrogen migration). The higher value of activation energy may be interpreted in terms of both the energy of the triplet state of carbon atoms in the transition complex (Scheme 11) and the energy necessary for the deformation of the valence angles in the three centered cyclic complex.[17]

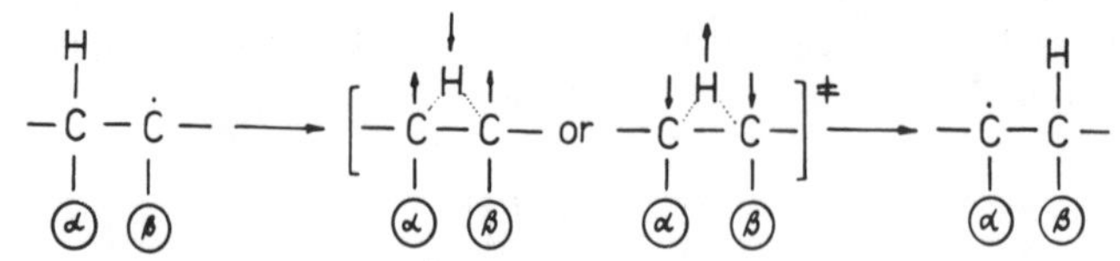

SCHEME 11

The main part of the calculated activation energy belongs to the triplet repulsive term (Table 2).

The process of the 1,2-hydrogen transfer reaction can be envisaged in the following way. The minimum energy reaction pathway involves the carbon-hydrogen bond stretching and bending in a direction towards the three valent C_β carbon. The maximum energy on the optimal reaction pathway corresponds to the configuration of the hydrogen and of the two carbon atoms into an isosceles triangle. The hydrogen atom in such a configuration has the following possible ways of rearrangement: addition to the carbon C_β leading to 1,2-isomerization, addition to the carbon C_α giving the parent radical and finally, the elimination from the hydrocarbon molecule (fragmentation).

C. Tunnel Shift of Hydrogen

The more reactive radical is transformed during the spontaneous isomerization to the less reactive one. The higher the difference in the Gibbs energy between the resulting and parent radical, the more easily isomerization takes place. The isomerization of alkylphenyl radical (Equation 4) has therefore a relatively low activation energy, even though a strong bond between hydrogen and primary carbon should be broken. Energetical requirements are compensated by the formation of a stronger bond between hydrogen and carbon of the aromatic ring. The above isomerization achieves an unusually high isotope effect since the deuterated radical rearranges with a rate 13,000 times lower than the nondeuterated one. The high influence of the mass of the migrating atom on the rate of isomerization may be interpreted by the tunnel effect.[18]

$$ \text{(CH}_3)_3\text{C} - \underset{\text{C(CH}_3)_3}{\overset{\text{C(CH}_3)_3}{\bigcirc}} \cdot \quad \xrightarrow{k/s = 2.10^5 \exp(-18\,800/RT)} \quad \underset{\text{C(CH}_3)_3}{\overset{\text{C(CH}_3)_3}{\bigcirc}} - \underset{\text{CH}_3}{\overset{\text{CH}_3}{\text{C}}} - \text{CH}_2^{\cdot} \tag{4} $$

According to the most simple conception, the activation energy represents a barrier which has to be surmounted by the reacting particle. Only a particle which has enough energy may overcome it, otherwise the reaction does not take place. In the tunnel reaction, particles with lower energy than the critical value may penetrate through the barrier. The probability of tunneling depends on the shape and on the height of the energetical barrier and is highest for particles with low mass, such as an electron, proton, or hydrogen atom. Provided that the tunnel effect occurs in some chemical reaction, then the resulting rate of the process will be higher than without the contribution of the tunnel effect. The importance of the tunnel effect increases with decreasing temperature. This may also explain why activation energies determined for some reactions are lower in the region of lower temperatures (Figure 1). For the same reasons also preexponential factors of the Arrhenius equation determined by extrapolation of the rate constants to $1/T \to 0$ will be dependent on the temperature interval of investigated rate constants. The tunnel effect was observed predominantly for reactions taking place in the condensed phase at low temperatures.

D. Isomerization of Halogenalkyl Radicals

Most of the data in the literature concerning the migration of halogens involve reactions of 1,2-shift.[19] The different behavior of halogens compared to hydrogen is a consequence of their more complex electronic structure. The electron orbitals of halogens may easily participate in the formation of a three-centered cyclic complex.

Migration of chlorine in a 2,2,2,-trichloroethyl radical (Equation 5) was observed by several methods.

$$ \cdot\text{CH}_2\text{CCl}_3 \to \text{ClCH}_2\overset{\cdot}{\text{CCl}}_2 \tag{5} $$

It is of interest that under comparable conditions and for corresponding radicals no migration of chlorine to more distant carbons was observed.

1,2-Migration of chlorine in alkyl radicals (Equation 6) may well be documented by the appearance of the anomalous chlormethyl side groups in polyvinyl chloride macromolecules.[20]

$$ -\text{CH}_2-\text{CHClCHCl}-\overset{\cdot}{\text{CH}}_2 \to -\text{CH}_2-\text{CHCl}\overset{\cdot}{\text{CH}}\text{CH}_2\text{Cl} \tag{6} $$

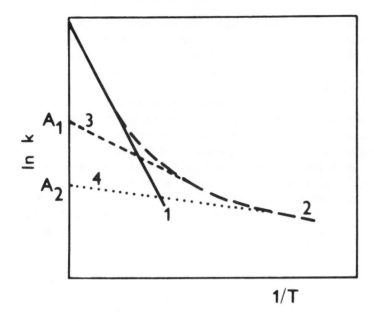

FIGURE 1. Schematic plot of the rate constants of chemical reactions in dependence on temperature in Arrhenius coordinates (1). The same dependence modified by tunnel effect (2). By the extrapolated lines 3 and 4, the influence of the temperature region of observed rate constants on the value of preexponential factor A (A_1, A_2) and activation energy is illustrated.

Worth noticing is also the low activation energy ($\sim$ 5 kJ/mol) of the process which indicated that the isomerization reaction probably proceeds via vibrationally excited radicals forming in the anomalous (tail to tail) addition step of polymerization. 1,3-Migration of hydrogen atom in the polyvinyl chloride radical is about 20 times slower at 55°C and has an activation energy of 30 kJ/mol; 1,5-migration of hydrogen is only 6 times slower (E = 22 kJ/mol). 1,2-Migration of bromine and iodine takes place more easily than it does in the case of chlorine, while fluorine does not migrate at all. The reaction is explained by the bridged radical (Scheme 12).

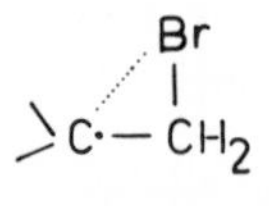

$$\underset{}{\overset{\displaystyle \overset{Br}{\underset{|}{}}}{\rangle C\cdot - CH_2}}$$

SCHEME 12

The interaction of halogen with the unpaired electron protects it against reactions with the medium and mediates simultaneously the shift of the halogen.[21,22]

E. Shift of Polyatomic Groups

Regarding the intramolecular tautomery shift (Equation 7) of different polyatomic groups

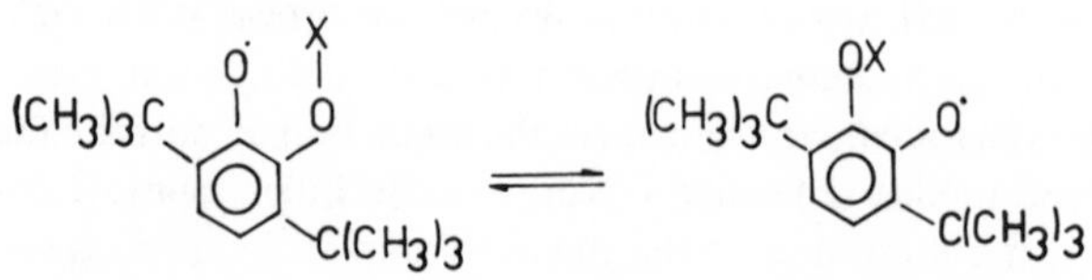

$$(CH_3)_3C \underset{C(CH_3)_3}{\overset{\overset{\textstyle O}{}\quad \overset{\textstyle X}{\underset{|}{}}}{\bigcirc}} O \;\;\rightleftharpoons\;\; (CH_3)_3C \underset{C(CH_3)_3}{\overset{OX}{\bigcirc}} O \qquad (7)$$

X, we see that it is strongly influenced by the character of the respective group.[7] Migration of the methyl group ($X = CH_3$) was not observed even at 200°C. On the other hand, if X is acyl group R-CO, the rate constant of the tautomery shift is already on the order 10^4 sec^{-1} at room temperature. Provided that the central atom of the migrating group has free coordination sites (X is, e.g., $Sn(CH_3)_3$), the rate constant of tautomery transitions increases up to 10^9 sec^{-1} under the same experimental conditions.

The migration of only the alkyl group is questionable. The unambigous experimental proof of its course is lacking even in such cases where the migrating alkyl should displace to an energetically more favorable position, i.e., at exothermic isomerizations. It suffices, however, to replace the central carbon atom with silicon, and isomerization of substituted phenoxy radicals (Equation 8) takes place quite easily and has a relatively high ratio within other possible reactions.[23]

$$
(CH_3)_3Si\text{—}\underset{\underset{Si(CH_3)_3}{|}}{\overset{\overset{\dot{O}}{|}}{\bigcirc}}\text{—}Si(CH_3)_3 \quad \longrightarrow \quad (CH_3)_3Si\text{—}\underset{\underset{Si(CH_3)_3}{|}}{\overset{\overset{O\text{—}Si(CH_3)_3}{|}}{\bigcirc}}\cdot
$$

$$(8)$$

Rearrangement of groups in this radical seems to be mediated by unoccupied d-orbitals of silicon. At the same time, the driving force of the isomerization is the replacement of bond Si-C (326 kJ/mol) by the stronger bond Si-O (451 kJ/mol). As a consequence, the trimethylgermanyl group migrates even more readily than the trimethylsilyl group. The higher reactivity of germanyl-substituted phenoxy radicals may be explained also by the higher sizes of germanium atoms which decrease the distance of potential centers of the reaction. Because of the coordination saturation of carbon atoms the carbon analog of corresponding phenoxy radical thus belongs to one of the most stable radicals. It should be remembered, however, that in hydrocarbon radicals also the migration of polyatomic groups cannot be entirely ruled out.

There exist, e.g., studies dealing with isomerization of 2-methyl-2-phenyl-1-propyl radical to 1-methyl-1-benzyl-ethyl radical (Equation 9).

$$(CH_3)_2CC_6H_5\dot{C}H_2 \xrightarrow{\ k\ =\ 60\ sec^{-1},\ 25°C\ } (CH_3)_2\dot{C}CH_2C_6H_5 \qquad (9)$$

The reaction is slightly exothermic (33 kJ/mol).[18] We know, of course, that exothermicity itself cannot be a decisive factor in the course of the migration reaction. This may well be documented by the example of tautomeric migration of phenyl groups of isotopically labeled phenylethyl radicals (Equations 10 and 11) which takes place regardless of the fact that the reactions are thermoneutral.

$$C_6H_5CH_2{}^{14}CH_2\cdot \rightarrow \dot{C}H_2{}^{14}CH_2C_6H_5 \qquad (10)$$

$$(C_6H_5)_2CH^{14}\dot{C}HC_6H_5 \rightarrow C_6H_5CH^{14}(C_6H_5)_2 \qquad (11)$$

Although the shift of hydrogen gives the more stable radical, and the reaction is exothermic, the migration of the phenyl group prevails. As will be shown later, the slower course and lower probability of 1,2-hydrogen migration is due to its lower electron affinity when compared with phenyl groups, which hinders the formation of polarized activated complex. Provided that in the 2-methyl-2-phenyl 1-propyl radical, the isopropyl group is replaced by the carbonyl group, the migration of the phenyl group is 5 times slower. This reaction is likely to be due to the partial delocalization of the unpaired electron in the radical.

The naphthyl group in the 2-methyl-2-naphthyl-1-propyl radical migrates 50 times faster when compared with its phenyl analog. This is probably brought about by the larger system of π-electrons in the naphthyl group. An incomparably higher rate of phenyl group displacement occurs in the triphenylethyl radicals (Equation 12) which takes place even in the solid state.[24]

$$(C_6H_5)_3C\dot{C}H_2 \xrightarrow{\quad k\,=\,1.7\,\cdot\,10^6\ s^{-1},\ 25°C\quad} (C_6H_5)_2\dot{C}H_2C_6H_5 \qquad (12)$$

The reaction in the solid state has an activation energy of about 40 kJ/mol and is 10^2 to 10^4 times slower than the corresponding process in the solution. The higher rate of the isomerization compared to 2-methyl-2-phenyl-1-propyl radicals is, moreover, due to a three times higher number of potentially replaceable phenyl groups and also the increased repulsion of relatively bulky groups on one carbon atom.

The significance of the steric pushing away of the migrating group by neighboring substituents may be seen in the ratio of 1,2-migration in the course of the generation of different β-phenylethyl radicals. The influence of methyl groups and hydrogen atoms is of special interest. As a matter of fact, the electron-withdrawing substituents on phenyl groups favor rearrangement. This is due to the ability of these substituents to impart the partial carbonium ion character of the transition state.[25] The similar effect of substituents may be observed in the 1,2-migration of the aryl group in 2-phenyl-*p*-substituted-2-ethoxy-2-phenylethyl radicals.[26] When the *para*-substituent X was an electron-withdrawing group (X = CN), only the substituted phenyl group migrated. On the other hand, when the substituent was an electron-releasing group (X = CH$_3$), the unsubstituted phenyl group migrated preferentially. The substituent effect observed in rearrangement was interpreted in terms of the slightly polar transition state (Scheme 13).

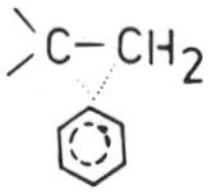

$$\begin{array}{c} \diagup \\ \diagup\!\!\!C-CH_2 \\ \ominus \end{array}$$

SCHEME 13

The polarization of the transition state can be ascribed to the electron-releasing character of the cyclopropane ring. Electron-withdrawing groups stabilize the transition state and support, thus, the shift of the aryl group. The tendency of carbonium ions to isomerization is well known. During ionic isomerization only two electrons enter the transition state and occupy the lowest bonding molecular orbital. In the slightly polar transition state, the unpaired electron of the radical occupies the antibonding MO orbital, which, however, leads to destabilization of the transition state and to the increase of its energy. As a consequence, the reaction has a higher activation energy and proceeds with difficulties. The comparison of 1,2-migration of different polyatomic groups R in reaction (Equation 13) shows that the phenyl group does not shift with the highest rates.[27]

$$RC(CH_3)_2\dot{C}H_2 \rightarrow (CH_3)_2\dot{C}H_2R \qquad (13)$$

As may be seen in the activation energies (in kJ/mol) for the rearrangement of the radicals, they increase in the sequence of R as follows: CH$_2$ = CH – (24), (CH$_3$)$_3$C – C = O (32), C$_6$H$_5$ – (49), (CH$_3$)$_3$C – C ≡ C – (53), and – C ≡ N (69) while the preexponential factors are in the range $10^{10.9}$ to 10^{12} sec^{-1}. It should be borne in mind, however, that the migration of respective polyatomic groups probably does not occur via the identical transition states. The vinyl group, e.g., displaces by the addition-fragmentation mechanism (Scheme 14).

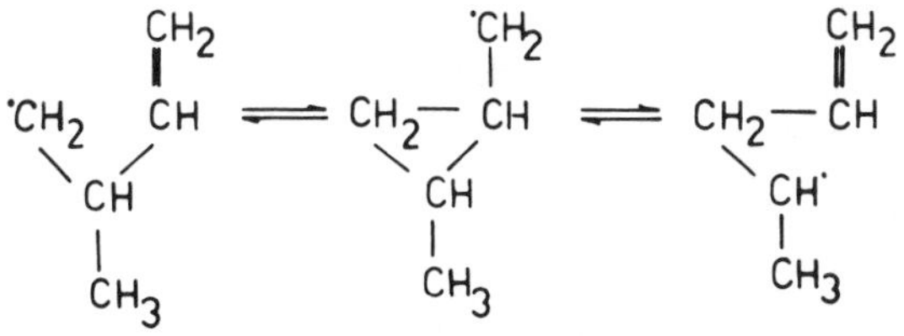

SCHEME 14

The transitory existence of methylcyclopropyl radicals was evidenced by the appearance of 1,2-dimethylcyclopropane in reaction products as well as by the loss of specific stereo geometry in vinyl group of deuterated derivatives. A similar sequence of reactions may also represent the course of migration of other groups.

Displacement of the acetoxy group takes place after the interaction of carbonyl oxygen with a radical center (Scheme 15). In this specific case, the acetoxy shift is not reversible because of the high strength of the forming Si-O bond.[28]

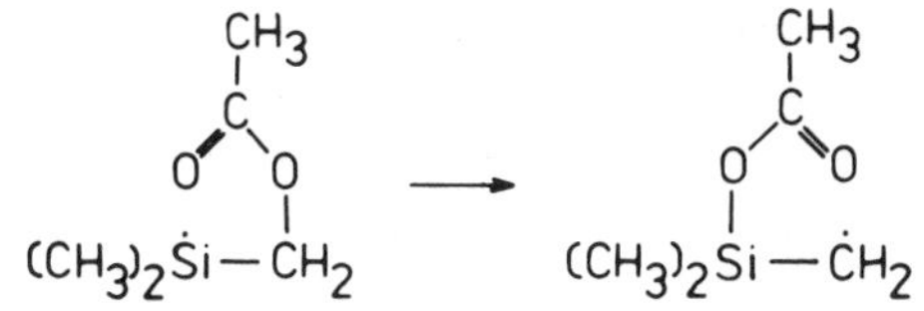

SCHEME 15

The migration of the alkylthio or phenylthio group (Equation 14) is more difficult to explain.

$$
\begin{array}{c}
CH_3 \\
| \\
C_6H_5-C-\dot{C}H_2 \\
| \\
S-C_6H_5
\end{array}
\rightarrow
\begin{array}{c}
CH_3 \\
| \\
C_6H_5-\dot{C}-CH_2SC_6H_5
\end{array}
\qquad (14)
$$

From the higher rate of phenylthio group migration when compared to the phenyl group, it can be deduced that d-orbitals of sulfur mediate 1,2-migration more efficiently than the electron system of the phenyl group. The conclusions should, however, be reserved since β-phenyl-thio radicals are known to split off the thiophenoxy radicals[29] (Equation 15).

$$
\begin{array}{c}
S-C_6C_5 \\
| \\
-\dot{C}-C- \\
| \; |
\end{array}
\leftrightarrow
\;\; \diagdown C=C \diagup \; + \; \cdot SC_6H_5
\qquad (15)
$$

The mechanism of the apparent migration may thus be composed of radical fragmentation and of the back addition of formed radical to the second carbon of the double bond.

At this connection, the enzyme catalyzed, coenzyme B_{12}-dependent carbon skeleton rearrangements may be worth interest. The mechanism of the action of the B_{12} vitamin has been a source of puzzlement for a long time.[30,31] It may, however, be assumed that some peculiarities in its metabolism may be explained by the 1,2-migration (Equation 16) of groups X = OH, NH_2, $C(CH_3)COOH$, $CH(NH_2)COOH$ in participating molecules proceeding either by fragmentation-addition or addition-elimination mechanisms in the enzyme cage.

$$
\begin{array}{cccc}
X & H & H & X \\
| & | & | & | \\
-C^{1'}\!-\!C^2 & \leftrightarrow & -C^1\!-\!C^2- \\
| & | & | & |
\end{array}
\tag{16}
$$

F. Cyclization and Decyclization Reactions

Provided that a free radical has a double bond and is of such structure that makes possible the course of an intraradical addition reaction, the cyclization + isomerization may then occur. As expected, this is the case of the cyclization of e.g., the 1-pentene-5-yl radical (Equation 17) which has an activation

$$CH_2{=}CHCH_2CH_2\dot{C}H_2 \longrightarrow \qquad \tag{17}$$

energy of 268 kJ/mol.[32] This relatively high value compared with intermolecular additions, however, indicates that the formation of the activated complex requires overcoming a ring strain. By the addition of a carbon of a radical center to a carbon of the nitrile group the cyclization of 4-cyanobutyl radicals starts.[33] The reaction (Equation 18) is fast and has a

$$\xrightarrow{\;4 . 10^3 \ s^{-1}, \ 25\,°C\;} \tag{18}$$

relatively low activation energy (36 kJ/mol). A five-membered cycle is also formed (Equation 19) at the cyclization of a 1-hexen-6-yl radical (exocyclization) where one could expect the

$$CH_2(CH_2)_3CH{=}CH_2 \xrightarrow{\;10^5 \ s^{-1}, \ 25\,°C\;} \tag{19}$$

formation of a more stable cyclohexyl ring (endocyclization).[34] The reaction is even faster than the above intramolecular addition to the nitrile group. This is caused by the lower activation energy (28 kJ/mol). The formation of cyclohexyl radicals in an endoaddition proceeds by a rate constant two orders lower than that of the methylcyclopentyl radical in the exoaddition. In the addition of methyl radical to propene, due to the polarization effect of the alkyl groups, the attack occurs at the less substituted carbon. However, in hexenyl radical cyclization, the orientation of the interacting orbitals is important. This stereoelectronic effect favors following the transition state (Scheme 16) over that leading to the more thermodynamically stable cyclohexyl radical. The intramolecular radical exocyclization is kinetically favored if the newly formed bond and the semioccupied orbital at the new radical center become completely coplanar.[18] At the formation of the six-membered cycle from the

1-hexen-6-yl radical, the optimum orientation between the π-orbital at the end carbon of the double bond p-orbital of the unpaired electron in the transition state can be realized only with considerable difficulties. However, when the radical center is on silicon (Scheme 17), the pyramidal configuration of the radical center and

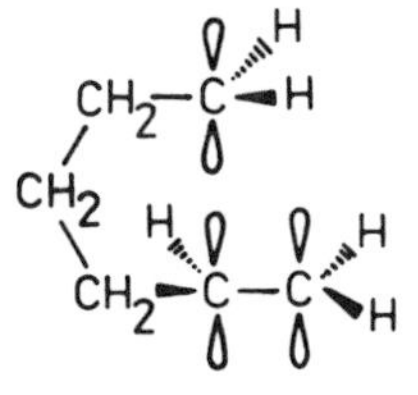

SCHEME 16 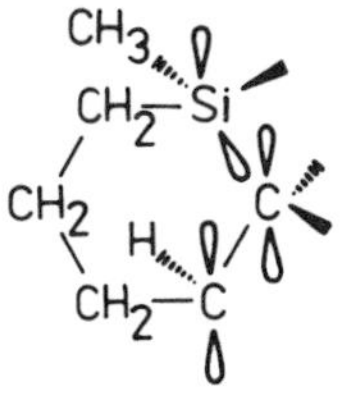   SCHEME 17

the length of the C-Si bond come into play,[35] and as a result, endocyclization giving the more stable cyclo radical is favored. Regioselectivity in the cyclization reaction of the alkenyl silyl radical can also be studied through energy-partitioning analysis.[36] The global difference is by 5 kJ/mol more favorable to the end transition state. By contrast, the strain engendered in accomodating the most favorable disposition of reactive centers in alkenyl radicals is much greater for the 1,6 transition structure than it is for the 1,5. The difference in strain energy is sufficient to outweigh those steric and thermochemical factors favoring formation of the larger ring. There is satisfactory qualitative agreement between the calculated strain energies of the transition structures and experimental activation energies for a large number of alkene radical cyclization reactions.[37]

The important role of a conformation adaptability of the radical follows also from almost one half the value of activation energy for isomerization of the 4-oxy-6-heptene-2-yl radicals,[38] (Equation 20) when compared to isomerization of the 5-hexene-1-yl radicals.

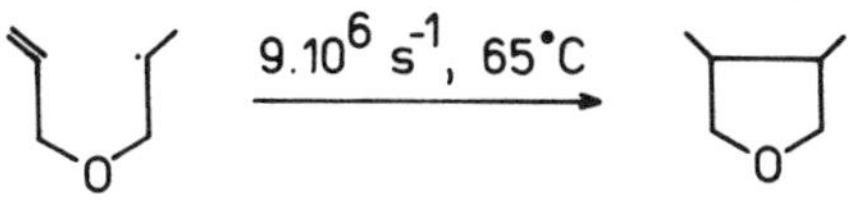

$$\xrightarrow{9.10^6 \ s^{-1}, \ 65°C}$$

(20)

The presence of substituents on the end carbon of the double bond of 5-hexene-1-yl radical supports exocyclization by lowering the activation energy.[39] The rationalization of the kinetic influence should be sought in the perturbation of the frontier orbitals of the reacting moieties. The semioccupied orbitals (SOMO) of the radical will interact with the highest occupied (HOMO) and lowest unoccupied molecular orbital (LUMO) of the alkene. An electron withdrawing group in the alkene will lower the HOMO and LUMO energies of the alkene, which leads to the increased SOMO-LUMO interaction. An electron-donating group will have the opposite effect on the HOMO and LUMO of alkenes and brings about the increased SOMO-HOMO interaction.

The regioselectivity of the cyclization reaction may be explained also in other ways. Radicals having a lower spin density on the unpaired electron on carbon give cyclohexyl radicals by cyclization, preferentially.[40] The lowered reactivity of the radical implies the higher selectivity of the endoaddition course which is thermodynamically more favorable. Provided that there are efficient donors of hydrogen atoms in the system, the less stable but more frequently forming five-membered cyclic radicals may be traced. As a matter of fact, less reactive radicals isomerize by cyclization reversibly. The kinetic product in vinyl alkene radical cyclization (Scheme 18) involving the possible formation of

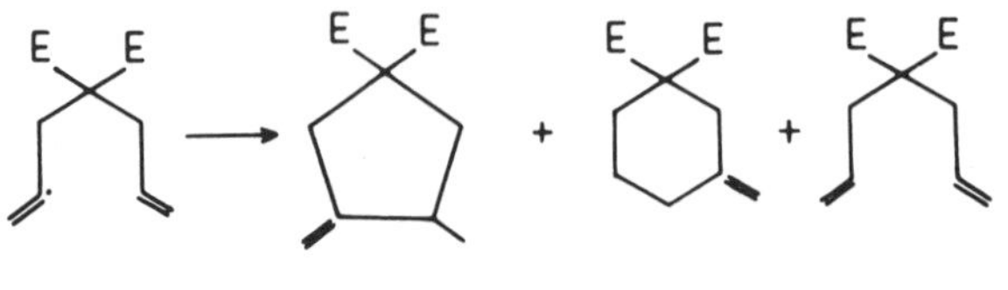

$$(E = CO_2CH_3)$$

SCHEME 18

either a five- or a six-membered ring is normally a methylenecyclopentyl radical.[41] This radical may be trapped by a hydrogen atom donor, or it may rearrange and end up as a methylenecyclohexane. The ratio of these two products depends on the concentration and on the reactivity of hydrogen donors. At low concentration of hydrogen donors in the system, the six-membered rings predominate, while at sufficiently high concentrations of hydrogen donors, 5-membered cyclic products are formed. It indicates the importance of backward reactions of the respective radicals. Cyclohexyl radicals are thermodynamically more stable than cyclopentyl radicals; the former open rings more slowly than the latter and succeed, thus, in reacting by a transfer reaction to a six-membered product. If the shorter lifetime of more numerous cyclopentyl radicals is compensated by 100-fold higher concentration of a reaction partner, then the transfer reaction gives a cyclopentyl type reaction product preferably. It is a striking illustration of the great effectiveness of vinyl radical cyclizations, that, even in neat tributyl tin hydride, the reaction still largely (4 : 1) gives the cyclized product rather than diene.

Also, from these results it may be deduced that thermodynamic factors support the formation of six-membered cycles, whereas the kinetic factors prefer rather the formation of five-membered carbon cycles.

Isomerization proceeding with the breaking of the cyclic structure occurs as the scission of chemical bond in β-position to a radical center. From cyclic hydrocarbon radicals, cyclopropylmethyl radicals decompose to 1-butene-3-yl radicals (Equation 21) very fast.[18] The reaction has the frequency

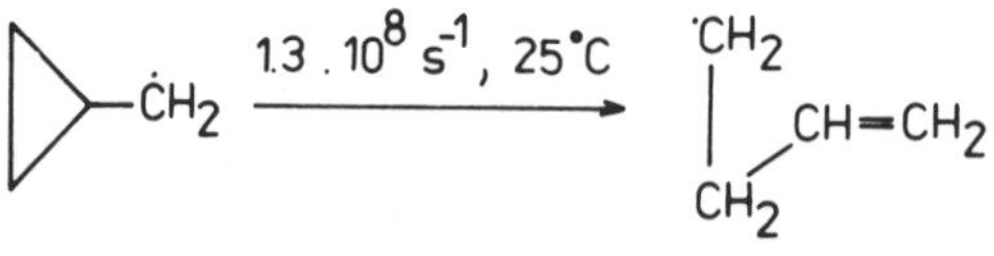

$$(21)$$

factor of $3 \cdot 10^{12}$ sec^{-1} and very low activation energy (25 kJ/mol), which indicates the significant weakening of the β-bond between carbons of the cyclopropyl ring.

The isomerization of nonsubstituted cyclopropyl radicals is not so easy (E = 84 kJ/mol) regardless of the fact that the more stable allyl radicals are formed. Also, the inner strain (125 kJ/mol) of the cyclopropane ring is by this fragmentation annulled. The explanation of this difference may consist in the stereoelectronic factor of the cyclopropyl ring scission. The more suitable position the semioccupied MO orbital with regard to the antibonding σ* orbital of the breaking bond has, the more easily the respective bond is broken. In the cyclopropylmethyl radical exists the higher degree of inner motion. The stereoelectronic approach rationalizes the selectivity of the β-scission in a number of bicyclo alkyl radicals, where the C-C bond being broken is the one which overlaps most efficiently with the semioccupied orbital, even when this affords the thermodynamically less stable product.[42]

There are, however, several examples of ring scission reactions, which occur in contra-stereoelectronic fashion.[43] For the bicyclo radical (Scheme 19) the SOMO overlaps

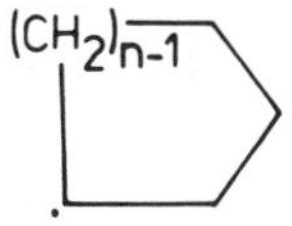

$$(CH_2)_{\overline{n-1}}$$

SCHEME 19

the outer cyclopropane C-C bond more efficiently than the inner-ring bond; this is especially valid for radicals with n = 1 and 2. In spite of it, isomerization of such radicals leads to the scission of inner-ring bond, and cycloalkenyl radicals (Scheme 20) are formed.

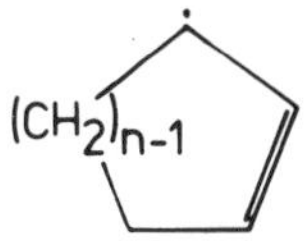

$$(CH_2)_{n-1}$$

SCHEME 20

Such a course of the reaction corresponds to about 100 kJ/mol greater relief of ring strain on fission of the intra-ring bond than outer-ring bond. It seems that the much greater release of ring strain is able to outweigh the unfavorable stereoelectronic effect. For the bicycloradical with, e.g., n = 3, the ring strain is only 20 kJ/mol which is not sufficient to outweigh the stereoelectronic effect, and the corresponding radical already rearranges by the fission of the outer cyclopropane bond. For radicals with n = 4, 5, and 6, both the stereoelectronic effect and the release of ring strain favor fission of the outer cyclopropane bond, and the preferential formation of the thermodynamically less stable cycloalkenalkyl radicals (Scheme 21) may thus be explained.

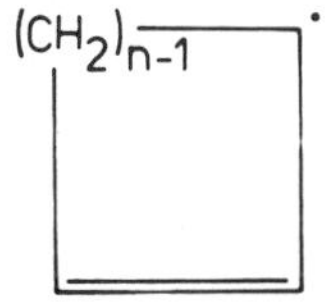

$$(CH_2)_{\overline{n-1}}$$

SCHEME 21

The scission of cycles occurs (Equation 22) also in the case of such radicals which have the minimum ring strain.[44]

$$(22)$$

Decyclization reactions will be rather preferred for cyclic radicals (Equation 23) which have an unpaired electron on oxygen and nitrogen atoms or in their close proximity.

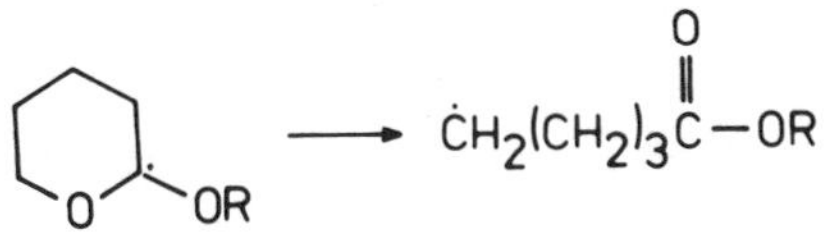

$$\dot{C}H_2(CH_2)_3\overset{O}{\overset{\|}{C}}-OR$$

$$(23)$$

The high energy of bonds in arising $C = O$ and $C \equiv N$ groups and the high exothermicity of the reaction facilitates the cycle cleavage.

Isomerization of free radicals represents a large group of elementary reactions. Some isomerizations may, of course, be analyzed in terms of other more simple reactions. This may be especially demonstrated on reactions which are connected with the closure and the cleavage of cyclic structures and may alternatively be classified as addition or fragmentation reactions. Their alignment to the group of isomerizations is due to the fact that they occur inside of one radical, whereas the addition is a bimolecular reaction of radicals and neutral molecules. Similarly, at elementary fragmentation, the parent radical is decomposed to a neutral molecule and a smaller radical. The same approach can be used for classification of migration of atoms or polyatomic groups. At their shift to the larger distance than corresponds to the distance of four carbon atoms, the isomerization may be considered as a transfer reaction of a radical to a molecule, particularly in such cases where the structures of the activated complex are similar.

The relevant moment of isomerization reactions consists in the direction of the approach of an unpaired electron orbital to the axes of the breaking σ bond. As was already pointed out, if the atom attacks the biatomic molecule, the probability of the reaction is usually highest when the activated complex is linear. The increasing angle of the bond attack in the biatomic molecule leads to the increase of activation energy of the transfer reaction. In isomerization reactions only the nonlinear activated complex, with respect to other chemical bonds in the radical, may sometimes be formed, and the reaction of the atom shift to shorter distances particularly requires a higher activation energy when compared with intermolecular reactions.

The shift of an atom or a group to a short distance inside of one radical belongs to the particular type of elementary isomerization reactions. Such intrinsic rearrangement of the radical structure is not so easy nor so frequent as it is, e.g., in the case of organic carbocations, but it is not so exceptional as for corresponding anions. Since free radicals may be considered as some intermediates on the pathway from cations to anions, they may be well put in midway of the reactivity in isomerization reactions.

Reorganization of atoms in the sense of radical isomerization to shorter distances is much more probable for excited radicals. Thermalized radicals have to accept some energy in collisions with other molecules before isomerization, and their rearrangement is rather rare. Keeping this fact in mind as well as that the termination and transfer reaction will always compete with isomerization, we see that only at that time when the mutual fast reactions of radicals are suppressed, isomerization may successfully come into play. This is especially the case of low concentrations of reactive molecules and radicals which may be realized in the gas phase systems. In the gas phase, the rate of deactivation of excited free radicals decreases and the isomerization reaction becomes a natural part of the whole complex of elementary reactions.

The importance of the knowledge about isomerization of free radicals consists in the fact that it may be the starting point of the synthesis of new compounds as well as the basis of anticipation of possible side reactions of reactants. This aspect is significant in the evaluation and regulation of the purity of organic compounds prepared by radical reactions.[45,46]

III. FRAGMENTATION OF FREE RADICALS

Provided that they absorb energy exceeding the dissociation energy of the weakest bond, polyatomic radicals can split to smaller particles. When compared to neutral molecules, however, radicals decompose much more easily. This may be seen in the example (Equations 23 and 24) of dimethylether and the corresponding alkyl radical.

$$CH_3OCH_3 \rightarrow CH_3O\cdot + \cdot CH_3 \quad A = 10^{15} \text{ sec}^{-1}, E = 347 \text{ kJ/mol} \tag{24}$$

$$-CH_2OCH_3 \rightarrow CH_2O + \cdot CH_3 \quad A = 10^{13} \text{ sec}^{-1}, E = 106 \text{ kJ/mol} \tag{25}$$

During the decomposition of both particles the C-O bond is primarily broken and methyl radicals are formed in both cases, while the second products are different: methoxy radicals and formaldehyde. Arrhenius parameters indicate that the decomposition of a neutral molecule requires considerably higher activation energy than a radical.[47] The higher value of the frequency factor shows on higher activation an entropy change accompanying the formation of the activated complex at the decomposition of a molecule. At the fragmentation of the radical, the activated complex has a lower degree of freedom but is more easily formed. The rate of decomposition of dimethylether is about 25 orders of magnitude lower at 500 K than it is in the case of the methoxy radical. The above comparison also implies that the removal of one hydrogen, which represents only about 2% w. from a molecule, leads to a decisive change of stability.

The first step of decomposition of the molecule and corresponding radical does not differ only in the rate but there also may be a quite different mechanism. At the decomposition of ethane, e.g., the C-C bond is broken and two methyl radicals are formed (Equation 26), while the ethyl radical splits out the

$$CH_3-CH_3 \rightarrow 2 \ CH_3\cdot \quad A = 10^{16} \text{ sec}^{-1}, E = 360 \text{ kJ/mol} \tag{26}$$

hydrogen atom and yields ethylene (Equation 27).

$$\cdot CH_2CH_3 \rightarrow H\cdot + CH_2=CH_2 \quad A = 3.10^{14} \text{ sec}^{-1}, E = 170 \text{ kJ/mol} \tag{27}$$

The difference in the rate constants due to the lower activation energy of decomposition of the radical is again enormous.

A. β-Fragmentation

The dissimilarity of the decomposition of ethane and the ethyl radical consists in the fact that the former does not have the type of C-H bond which could be easily broken. Ethyl radicals have three β-C-H bonds (Scheme 22) which, as other bonds in

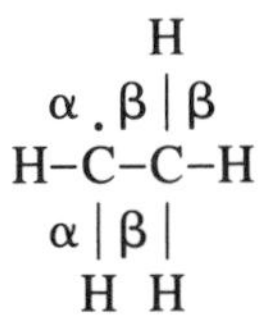

H
α . β | β
H–C–C–H
α | β |
H H

SCHEME 22

β-position to an unpaired electron site, can split out more easily than α-bonds. The relative instability of β-bonds is brought about by negative spin density on β-carbon appearing there as a consequence of spin polarization, i.e., of the effect of an unpaired electron on neighboring occupied orbitals. The electrons in originally equilibrated MO orbitals are thus partially dispaired. Outwardly, it manifests itself as the decrease of the strength of β-bond (Table 3). Another reason for β-fragmentation of free radicals is the possibility of the formation of a multiple bond in one of the reaction products, which is an exothermic process and contributes to the faster course of a reaction. One should bear in mind, however, that not all bonds in radicals are weakened when compared to neutral molecules but some may even

Table 3

DISSOCIATION ENERGY (D) OF INDICATED BONDS IN RADICALS AND PERCENTAGE OF THE STRENGTH OF THE SAME BOND FORMED AFTER THE LINKAGE OF RADICAL WITH HYDROGEN OR X GROUP[48-50]

Bond	Bond in radical	X	D kJ/mol	%
C–C	$\dot{C}H_2CH_2-CH_3$	H	131	38
	$O=\dot{C}-C_6H_5$	C_6H_5	125	31
	$C_6H_5\dot{C}HCH_2-CH_2C_6H_5$	H	117	39
	$O\dot{C}-CH(CH_3)_2$	H	67	20
	$\cdot O\ CH_2-CH_3$	H	54	16
	$O=\dot{C}-CH_3$	H	42	14
	$\cdot OC-CH(CH_3)(C_6H_5)$	H	36	10
	$\cdot O(OC)-C_6H_5$	C_6H_5	40	10
C–H	$H_2\dot{C}-H^a$	H	355	83
	$H_2\dot{C}H_2C-H$	H	170	42
	$O=\dot{C}-H$	H	128	39
	$HOH\dot{C}-H$	H	117	29
	$\dot{O}CH_2-H$	H	100	25
C–O	$C_6H_5\dot{C}HCH_2-OC_6H_5$	H	117	42
	$\cdot OO-C(CH_3)_3$	H	117	35
	$\cdot OO-CH_3$	H	109	33
	$H_2\dot{C}O-CH_3$	H	106	30
	$C_6H_5\dot{C}HO-CH_2C_6H_5$	H	63	22
	$\dot{O}O-CH_2C_6H_5$	H	50	15
	$\dot{O}O-CH_2CH=CH_2$	H	50	15
	$O=\dot{C}-OC_6H_5$	C_6H_5	29	11
	$O=\dot{C}O-C_6H_5$	C_6H_5	25	6
	$\dot{O}-C_6H_5$	H	420	140
O–H	$\dot{O}-H$	H	426	86
	$\dot{O}OO-H$	H	284	77
	$\dot{O}O-H$	H	205	56
O–O	$\dot{O}-OH$	H	276	130
	$\cdot O-OCH_3$	H	245	115

ᵃ In polycarbon radicals α-bond C–C is stronger than in the respective neutral hydrocarbon.

be strengthened. This fact is the reason for a certain selectivity of fragmentation reactions observed for polyatomic radicals. Provided that at β-fragmentation either the C-C or C-H bond may be broken alternatively, the usually weaker C-C bond is more likely to be impaired.[51] The 1-propyl radical thus fragmentates (Equation 28) to a methyl radical and ethylene

$$\cdot CH_2CH_2CH_3 \rightarrow \cdot CH_3 + CH_2=CH_2 \tag{28}$$

predominantly while the minor part includes the fragmentation to hydrogen atoms and propylene (Equation 29).

$$\cdot CH_2CH_2CH_3 \rightarrow H\cdot + CH_2=CHCH_3 \tag{29}$$

Although the rule of β-fragmentation of radicals is relatively general and very instructive, it cannot be accepted absolutely. The 2-propyl radical, e.g., should dehydrogenate only to

Table 4
ARRHENIUS PARAMETERS AND RATE CONSTANT OF
RADICAL FRAGMENTATION AT 227°C[54-56]

Reaction	log(A·s)	E kJ/mol	K/sec^{-1}
$(CH_3)_3C\cdot \rightarrow H\cdot + (CH_3)_2{=}CH_2$	15.5	182	$2\cdot10^{-4}$
$(CH_3)_2CH\cdot \rightarrow H\cdot + (CH_3)CH{=}CH_2$	13.3	154	$1\cdot10^{-3}$
$(CH_3)_2CHCH_2 \rightarrow CH_3 + CH_3CH{=}CH_2$	12.4	130	$6\cdot10^{-2}$
$CH_3\dot{C}HCH_2CH_3 \rightarrow CH_3 + CH_3CH{=}CH_2$	14.6	136	2
$CH_3CH_2CH_2\dot{C}H_2 \rightarrow CH_3CH_2 + CH_2{=}CH_2$	13.6	120	10
$ClCH_2CH_2 \rightarrow Cl\cdot + CH_2{=}CH_2$	13.0	100	$3\cdot10^2$
$((CH_3)_3C)_2C{=}N\cdot \rightarrow (CH_3)_3\;\dot{C} + (CH_3)_3CCN$	14.4	72	$7\cdot10^5$
$(CH_3)_2CHO\cdot \rightarrow CH_3 + CH_3CHO$	10.6	67	$4\cdot10^3$
$(CH_3)_3CO\cdot \rightarrow CH_3 + CH_3COCH_3$	12.5	58	$3\cdot10^6$
$F_2NCH_2CH_2 \rightarrow \cdot NF_2 + CH_2{=}CH_2$	12.9	57	$8\cdot10^6$
$C_6H_5CO_2{}^a \rightarrow C_6H_5 + CO_2$	13.1	58	$9\cdot10^6$
$(CH_3)_3CCO_2 \rightarrow (CH_3)_3C\cdot + CO_2$	10.8	32	$3\cdot10^7$
$(CH_3)_3C\dot{C}O \rightarrow (CH_3)_3C\cdot + CO$	11.9	39	$7\cdot10^7$

[a] Radical which has alkyl group instead of phenyl group, immediately after its formation splits out CO_2.

propylene according to such an approach, but in reality there may be observed also its rather large fragmentation to ethylene and the methyl radical occurring, of course, with a lower rate than in the case of the 1-propyl radical. The fragmentation reaction is slowed down here by an intramolecular transfer followed by β-fragmentation of the C-C bond (Equation 30).

$$CH_3{-}\overset{\displaystyle H}{\underset{\displaystyle \cdot}{CH}}{-}CH_2 \rightarrow CH_3\cdot + CH_2{=}CH_2 \qquad (30)$$

The ejection of the hydrogen atom from various hydrocarbon radicals has a similar A factor ($\sim 10^{13}$ sec^{-1}). Activation energies of fragmentation may, however, differ for individual radicals.[52] The elimination of hydrogen atoms from alkyl radicals requires, e.g., activation energy of about 145 kJ/mol, while the same reaction of cyclohexadienyl radical only 105 kJ/mol.

If there are several possibilities in the course of the fragmentation reaction, the selectivity of the respective reaction pathway will be higher, the higher difference in the strength of bonds in β-position to the radical center. This is the reason why β-phenylthioalkyl radicals eliminate only the thiophenyl group and not the hydrogen or methyl radical[53] (Equation 31).

$$(CH_3)(C_6H_5S)CH\dot{C}HCH_3 \rightarrow C_6H_5S\cdot + CH_3CH{=}CHCH_3 \qquad (31)$$

The effect of the structure of alkyl or alkoxy radicals on the β-fragmentation may be seen in corresponding kinetic parameters (Table 4).

The complementary information follows from the composition of products formed during pyrolysis of polymer hydrocarbons. Primarily formed macroradicals may fragmentate or undergo other reactions. Since the depolymerization is a controlled β-scission of macroradicals, the yield of a monomer shows on the extent of β-fragmentation compared with other transformation reactions. Nonsubstituted alkyl radicals decompose thus relatively slowly; at

400°C monomeric ethylene, e.g., appears in negligible amounts in the products of polyethylene pyrolysis. Phenylalkyl radicals fragmentate more easily; styrene is about one half of the low molecular products of the thermal decomposition of polystyrene. The branched phenylalkyl radicals of poly (α-methylstyrene) undergo β-scission almost exclusively, and monomeric α-methyl styrene is the predominant component of the decomposition products. From this, it ensues that the more stable the radical formed in β-scission, the faster β-fragmentation of alkyl radical takes place. This is valid also for other types of radicals. Methylethylisopropylmethoxy radicals give mainly isopropyl radicals (Scheme 23). Other possible reactions occur

$$
\begin{array}{l}
\underset{\displaystyle \underset{\displaystyle CH(CH_3)_2}{\overset{\displaystyle |}{\underset{|}{CH_3CH_2CO\cdot}}}}{\overset{\displaystyle CH_3}{\overset{|}{}}}
\end{array}
\qquad
\begin{array}{ll}
\longrightarrow \cdot CH_3 + CH_3CH_2COCH(CH_3)_2 & (<0.5\%)\\
\longrightarrow \cdot CH_2CH_3 + CH_3COCH(CH_3)_2 & (3\%)\\
\longrightarrow \cdot CH(CH_3)_2 + CH_3CH_2COCH_3 & (95\%)
\end{array}
$$

SCHEME 23

to a considerably lesser extent.[57] From the quantitative data on stabilization energy[58] of the respective free radicals forming in this reaction ($\cdot CH_3$ [8 kJ/mol], $\cdot CH_2 CH_3$ [-4 kJ/mol], $\cdot CH(CH_3)_2$ [-8 kJ/mol]), ethyl and isopropyl radical should fragmentate 37 times and 125 times faster than the methyl radical at 100°C.

At this estimation we assume that the difference in the stabilization energy of radicals reflects the strength of the broken bond and, consequently, the activation energy of the fragmentation reaction. Even though the picture is somewhat simplified, the rough features of the observed course of fragmentation of methyl isopropylmethoxy radicals are rather easy to understand.

A more detailed view on the relative rate of fragmentation of alkoxy radicals to methyl or ethyl radicals (Scheme 24)

$$
\begin{array}{l}
\underset{\displaystyle \underset{\displaystyle CH_3}{\overset{\displaystyle |}{\underset{|}{H-CO\cdot}}}}{\overset{\displaystyle CH_2CH_3}{\overset{|}{}}}
\end{array}
\qquad
\begin{array}{ll}
\overset{k_1}{\longrightarrow} \cdot CH_2CH_3 + CH_3CHO\\
\overset{k_2}{\longrightarrow} CH_3\cdot + CH_3CH_2CHO\\
\overset{k_3}{\longrightarrow} H\cdot + CH_3CH_2COCH_3
\end{array}
$$

where $k_3 \ll k_1$ or k_2 and $k_1/k_2 = 19$

SCHEME 24

may be obtained from analysis of the products of 2-butoxy radicals decomposition at 373 K.[59]

The lower ratio of the elimination of ethyl when compared to methyl radicals than corresponds to their stabilization energies, follows from the found frequency factors of both reactions ($A_1/A_2 = 0.6$). Taking this fact into account, the accordance of the data is surprisingly good.

Substituted phenylalkoxy radicals preferentially eliminate methyl before phenyl radicals (Equation 32), which again agrees

$$C_6H_5C(CH_3)_2CO\cdot \rightarrow \cdot CH_3 + C_6H_5COCH_3 \quad A = 2.10^{12} \; sec^{-1} \; E = 36 \; kJ/mol \quad (32)$$

well with the higher stability of the methyl radical when compared to the alternative possiblity of formation of the phenyl σ-radical.

At 373 K the rate of methyl radical formation from the α-cumyloxy radical is 100 times higher than that from *tert*-butoxy radical. This difference, which at lower temperatures is even more enhanced, is due to both the weaker binding and sterical pushing out of the methyl radical by the phenyl group.

There are, however, examples, which serve as evidence against the idea of preferable release of more stable radicals.[61] This is, e.g., the case of dimethyl trifluoromethylmethoxy radicals (Scheme 25), where instead of methyl

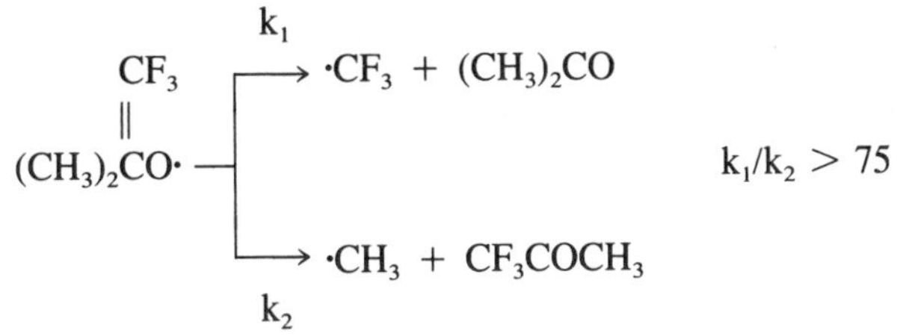

$$
(CH_3)_2CO \cdot \quad
\begin{array}{l}
\xrightarrow{\quad k_1 \quad} \cdot CF_3 + (CH_3)_2CO \\[3ex]
\xrightarrow[\quad k_2 \quad]{\quad} \cdot CH_3 + CF_3COCH_3
\end{array}
\qquad k_1/k_2 > 75
$$

with CF$_3$ double-bonded to the central carbon.

SCHEME 25

radicals, as would be expected, only the release of trifluoromethyl radicals was observed in the temperature interval from 360 to 600 K. Comparing dissociation energies of C-H bonds in trifluoromethane and methane we see that the stabilization energy of trifluoromethyl radicals is somewhat lower than that of methyl radicals, and methyl radicals could be formed more easily than trifluoromethyl radicals. The expected dependence follows from the studies of competitive formation of trichlormethyl, trifluoromethyl, and methyl radicals.[62] At the fragmentation of substituted *tert*-butoxy radicals at 433 K, the relative rates of formation of the above radicals can be approximately expressed as $CCl_3 : CH_3 : CF_3 = 600 : 10 : 1$. In such a case, $\cdot CH_3$ is formed much more readily than $CF_3\cdot$ from a common precursor.

The exact nature of the effect of α-substitution by fluorine remains as a challenging puzzle. There is a possibility of investigating the fragmentation of *tert*-butoxy radicals with a different number of fluorine atoms in the abstracted methyl radical.

At the decomposition of excited radicals, the effect of stability of formed radical on reaction selectivity is diminished. Due to the excess energy in the system, the strength of linkage of the radical in its parent precursor is less important, and a new selectivity rule based upon the predominant fragmentation of those radicals which are the most mobile becomes valid.

The stability of the released radical cannot be taken as decisive factor when comparing fragmentation reactions of alkyl and oxy radicals. The elimination of methyl from the alkyl radical (Equation 33) at 500 K, is, e.g., by 50,000 times slower

$$(CH_3)_2 CH\dot{C}H_2 \rightarrow \cdot CH_3 + CH_3CH{=}CH_2 \tag{33}$$

than that from the alkoxyradical (Equation 34).

$$(CH_3)_2CHO\cdot \rightarrow \cdot CH_3 + CH_3CHO \tag{34}$$

The facility of the fragmentation in this second case is due partially at least to the higher strength of the newly formed bond C=O when compared with the C=C bond. The stereoelectronic effect on the splitting bond may contribute to the reaction acceleration, too.

Considerable weakening of the C-C β-bond in acyloxyradicals leads to the fast decarboxylation reaction (Equation 35)

Table 5
SELECTED RATE CONSTANTS OF
DECARBOXYLATION OF ACYLOXY RADICALS
AT 25°C.[56]

Radical	k/sec^{-1}	Radical	k/sec^{-1}
$C_6H_5C\equiv CCO_2\cdot$	3.7	$c\text{-}C_3H_5CO_2\cdot$	$2.5\cdot10^4$
$C_6H_5CO_2\cdot$	$6.1\cdot10^2$		$6.5\cdot10^4$
$2,6\text{-}(CH_3)_2C_6H_3CO_2\cdot$	$1.4\cdot10^3$	$CH_3CO_2\cdot$	$6.1\cdot10^5$
		$(CH_3)_3CCO_2\cdot$	$6\cdot10^7$
	$2.2\cdot10^4$	$C_6H_5CH_2CO_2\cdot$	$\sim10^8$

$$RCOO\cdot \rightarrow R\cdot + CO_2 \tag{35}$$

which rate decreases with the decrease of the participation of the p-orbital in the hybridized AO sp-bond $C\text{-}CO_2\cdot$ (Table 5). The slower fragmentation of cycloaliphatic acyl radicals is evidence of the more rigid structure of the respective radical which hinders the formation of the activated complex of the decarboxylation reaction.

It is of interest that decarboxylation of isomeric radicals, which have the CO_2 group bound via an oxygen atom and an unpaired electron localized on carbon, is by several orders slower. The explanation obviously consists in the higher strength of the C-O bond when compared to the C-C bond and in the delocalization of the unpaired electron. In acyl radicals (Scheme 26), the unpaired electron is delocalized on oxygen atoms and both positions of unpaired spin support β-fragmentation connected with the release of CO_2. In isomeric radicals (Scheme 27), the carbon

SCHEME 26   SCHEME 27

atom has lower spin density, and fragmentation is slower. The effect of delocalized spin density is not focused here to one reaction site only. The spin of the electron on the carbon atom supports decarboxylation while the partial unpaired spin on the oxygen atom weakens the neighboring C-O bond. The elimination of CO_2 from the *tert*-butoxy carbonyl radical[63] has an activation energy of 49 kJ/mol and frequency factor of $6.3\cdot10^{13}$ sec^{-1}. The rate constant extrapolated to a temperature of 25°C is thus 300 times lower than that for the decarboxylation reaction of the trimethylacetyloxy radical.

β-Fragmentation of the peroxy radical leads to the back release of molecular oxygen. Its course depends again on the stabilization energy of the parent radical R·. In the sequence of several $RO_2\cdot$ radicals, the increase of the stabilization energy decreases the temperature characterizing the same values of equilibrium constants of the reaction (Equation 36).

$$RO_2 \leftrightarrow R\cdot + O_2 \tag{36}$$

Halogenalkyl radicals which may eliminate halogen atoms from the β-position decompose very easily. The release of the hydroxy radical from the hydroperoxide group bound to carbon with the radical center is very fast, too.

At the decomposition of unstable radicals derived from α-diols (Equation 37) there is the supposed synchronous β-scission of two bonds.[64,65] The reaction is a demonstration of the key step at the radical dehydration of biologically important low molecular and polymer carbohydrates.[66]

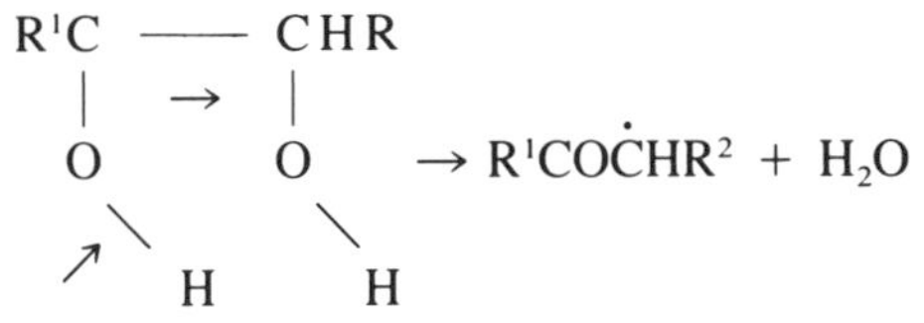

$$
\begin{array}{ccc}
R^1C & \text{——} & CHR \\
| & \rightarrow & | \\
O & & O \\
\diagdown & & \diagdown \\
\nearrow \quad H & & H
\end{array}
\rightarrow R^1CO\overset{\centerdot}{C}HR^2 + H_2O \tag{37}
$$

The exothermic decomposition of vinylperoxy radicals is an example of multistage β-fragmentation (Scheme 28) which

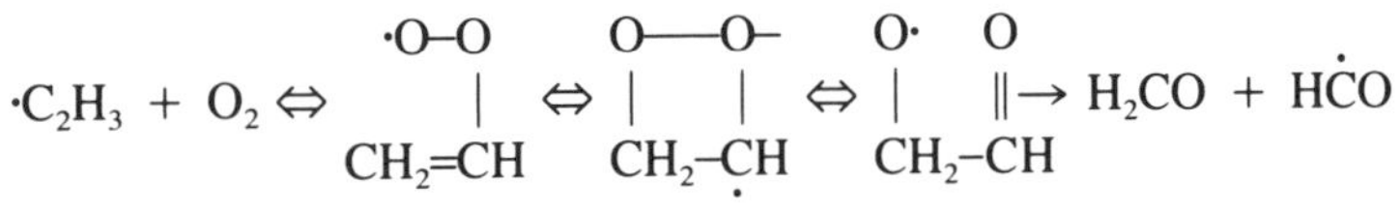

$$
\cdot C_2H_3 + O_2 \Leftrightarrow
\begin{array}{c}
\cdot O\text{–}O \\
| \\
CH_2\text{=}CH
\end{array}
\Leftrightarrow
\begin{array}{cc}
O\text{—}O \\
| \quad | \\
CH_2\text{–}CH
\end{array}
\Leftrightarrow
\begin{array}{cc}
O\cdot \quad O \\
| \quad \| \\
CH_2\text{–}CH
\end{array}
\rightarrow H_2CO + H\overset{\centerdot}{C}O
$$

SCHEME 28

represents the important elementary step of ethylene and other hydrocarbons combustion.[67] We recall that ethylene is formed by β-scission of various hydrocarbon alkyl radicals. The magnitude of the overall rate constant of the reaction of vinylperoxy radicals, $k = 4 \cdot 10^9 \exp(1045/RT)$ $dm^3 \cdot mol^{-1} \cdot sec^{-1}$ and its slight temperature dependence indicate that the reaction proceeds by an addition mechanism. The formed adduct rapidly rearranges and decomposes into the observed products.

B. Other Types of Fragmentation

The less frequent pathway of decomposition of radicals having an unpaired electron on the carbon atom is α-scission. It occurs at that time when the radical center is situated on the carbonyl group or if the radical has only one carbon atom.

Provided that the unpaired electron is localized on carbon of the carbonyl group, α-scission dominates β-scission even in cases when β-scission is potentially possible (Scheme 29).

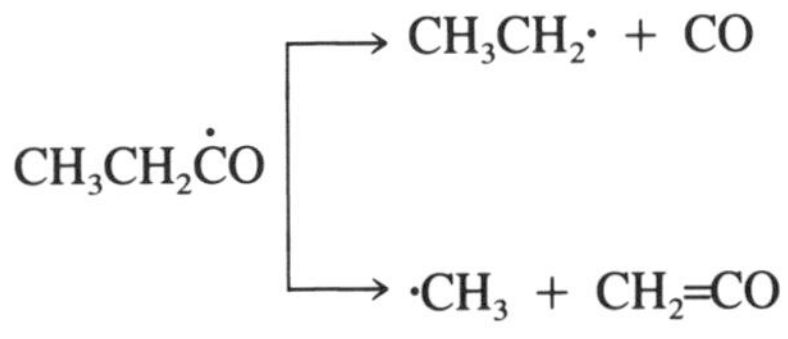

$$
CH_3CH_2\overset{\centerdot}{C}O
\begin{cases}
\rightarrow CH_3CH_2\cdot + CO \\
\\
\rightarrow \cdot CH_3 + CH_2\text{=}CO
\end{cases}
$$

SCHEME 29

The weakening of the C-C α-bond is connected with the delocalization of the spin density on oxygen. Noncompensated spin on carbon impairs, of course, also β-bonds but its effect is dissipated over three C-H and one C-C β-bonds. From the comparison of the stability of alkoxy and alkyl radicals, we already know that the destabilization effect of an unpaired electron on neighboring bonds is higher when the spin is located on oxygen. This is also

Table 6
ARRHENIUS PARAMETERS AND RATE
CONSTANTS OF DECARBONYLATION
RĊO → R· + CO OF ACYL RADICALS AT
40°C[56,68-72]

R	k_{313} K, sec^{-1}	log A,s	E, kJ/mol
C_6H_5	10^{-6}	14.6	123
$C_5H_5{}^a$	10^{-4}	12.0	100
CF_3	0.2	13.3	83
CH_3	4	13.5	72
CH_3CH_2	$9.8 \cdot 10^2$	13.3	62
$(CH_3)_2CH$	$3.9 \cdot 10^3$	14.0	54
$C_{10}H_{15}{}^b$	$1.9 \cdot 10^4$	—	—
$(CH_3)_3C$	$9.0 \cdot 10^6$	13.5	45
$HOCH_2$	$6.3 \cdot 10^5$	11.0	37
$C_6H_5CH_2$	$1.5 \cdot 10^7$	12.0	29
$C_6H_5(CH_3)CH$	$7.6 \cdot 10^7$	12.2	26
$C_6H_5(CH_3)_2C$	$2.0 \cdot 10^8$	11.2	17

a   Cyclopentadienyl — formed by isomerization of phenoxy
radicals; the actual rate of fragmentation may be higher.
b   1-Adamantyl.

the reason why, apart from the higher spin density of an unpaired electron on carbon, the resulting effect of the C-C bond weakening is due to the spin delocalization on oxygen. From this viewpoint, decarbonylation may be considered as a "hidden" β-scission. Taking this into account, it is not then surprising that fragmentation of acetyl radicals is by six orders slower at 20°C than that of acetyloxy radical.

The activation energy and the rate of decarbonylation decrease with the decrease of the dissociation energy of the broken bond (Table 6). The lower the strength of the bond, the more stable is the radical formed in the reaction. Regardless of the fact that decarbonylation of acyloxy radicals is slower than decarboxylation of alkylacyl and arylacyl radicals, it is still relatively fast and may be used as the inner reaction standard in the system of other competition reactions.

Isoelectronic alkyl and aryldiazo radicals fragmentate (Equation 38) with the same rate approximately. Their decomposition should be, however, classified as β-scission.[73]

$$R-N=N· → R· + N_2 \tag{38}$$

The decomposition of monocarbon radicals (Equation 39)

$$·CHXBr → ·CX + H· + Br·   \text{(X is halogen)} \tag{39}$$

is α-scission as such, but it occurs only with hot radicals. The vibration excitation of halogenmethyl radicals is a consequence of polyhalogenmethane excitation taking place already in the first step of the C-halogen bond scission where primary radicals are formed.[74]

Unimolecular α- and β-scission are the most important products-forming step of phosphoranyl radicals decomposition.[75] *Tert*-butoxy triethoxyphosphoranyl radicals decompose via C-O bond β-scission (Equation 40). The activation energy for β-scission of

$$(CH_3)_3COP·(OCH_2CH_3)_3 → (CH_3)_3C· + OP(OCH_2CH_3)_3 \tag{40}$$

various $(CH_3)_3COP\cdot(CR)_3$ to *tert*-butyl radicals is in the range 34 to 41 kJ/mol, whereas that for $(CH_3CH_2O)_4$ P$\cdot$ giving ethyl radicals is 54 kJ/mol. The faster fragmentation, the weaker bond is broken. The difference in the stabilization energy of the ethyl and *tert*-butyl radical is 17 kJ/mol. The release of the *tert*-butyl radical from the *tert*-butoxytriphenyl phosphoranyl radical having the activation energy of 52 kJ/mol is already more difficult than it is in the case of Equation 40. The fragmentation reaction is impeded here probably by the unpaired spin density delocalized over the phenyl rings and only to a lower extent over phosphorus.[76] The delocalization of the unpaired electron and difference in the stabilization energy explain the relatively seldom α-scission observed in the case of phosphoranyl radicals.

Provided that the stabilization energy of the released radical is sufficiently high, α-scission may occur (Equation 41).

$$(CH_3)_3COP\cdot(OCH_2CH_3)_2OC_6H_5 \rightarrow C_6H_5O\cdot + (CH_3)COP(OCH_2CH_3)_2 \qquad (41)$$

We remind that the stabilization energy of the phenoxy radical is -121 kJ/mol, while that for the ethoxy radical $+4$ kJ/mol. The competition between α- and β-scission is thus determined by relative strengths of bonds undergoing cleavage and by delocalization of the unpaired spins. The fragmentation of the quoted phosphoranyl radical seems thus to be a formal consequence of α- bond cleavage.

β-Hydroperoxyalkyl or β-peroxyalkyl radicals may decompose by different ways (Equation 42).

$$\begin{array}{ccc} OOH & & O \\ | & & \diagup\diagdown \\ R-CH-CH_2\cdot & \rightarrow & R-CH-CH_2 + \cdot OH \end{array} \qquad (42)$$

From the formal viewpoint, the reaction is γ-scission. The reaction of β-peroxyalkyl radicals is the key step in the formation of oxiranes, the decomposition of dialkyl peroxides, the oxidation of alkenes, and in the cool flame combustion of hydrocarbons. γ-Fragmentation of radicals is not a spontaneous decomposition reaction. Its mechanism is similar to that shown for substitution reactions of addition-elimination character. There is likely to be formed an unstable primary cyclic adduct which subsequently decomposes. γ-Scission may be also understood as an intraradical induced decomposition of peroxides. The more nucleophile the alkyl radical, the more quickly fragmentation occurs. For this reason, the rate constant of fragmentation of β-peroxyalkyl radicals increases (Scheme 30) if the alkyl radical has

$$\begin{array}{ccc} O-OC(CH_3)_3 & & O \\ | \quad \cdot & & \diagup\diagdown \\ R-CH-CHR & \rightarrow R-CH-CHR + \cdot OC(CH_3)_3 \end{array}$$

SCHEME 30

several more electron donor methyl groups (R $=$ CH$_3$) instead of hydrogens (R $=$ H).[77] For R $=$ H, the rate constant is, e.g., $4\cdot 10^3$ sec^{-1} while that for R $=$ CH$_3$, it reaches up to $2\cdot 10^6$ sec^{-1} at 298 K.

Elementary reactions of the decomposition of polyatomic radicals are governed by the destabilization effect of the semioccupied MO orbital. Specific weakening of some bonds is due to the oriented and distance-controlled effect of the unpaired electron which extends to the close vicinity of a radical site only. The aimed control of fragmentation reactions may be expected at chemisorption of radicals on heterogeneous catalysts, at solvated radicals, and at selective excitation irradiation. This is, however, more the goal of further research.

The fragmentation reaction of radicals is the process opposite to the addition reaction and occurs especially in systems where excited radicals are formed. It is the principal process of thermal and oxidation decomposition of organic compounds proceeding either spontaneously or in a controlled and desired manner (combustion and cracking of hydrocarbons, aging of polymers, organic synthesis of epoxides, etc.). The fragmentation of radicals is also important in kinetics measurements, where its course followed on the formation of products can serve as an inner standard for comparison with other competition reactions.

IV. ADDITION REACTIONS OF FREE RADICALS

The formation of a new linkage between a radical and a multiple bond occurs in such a way that one of the unpaired electrons of the polarized β-bond gives with a radical a new π-bond (Scheme 31). The transition state for the addition reaction is a σ-complex.[78] The perpendicular approach of radicals to the C=C ethylenic bond (Scheme 32) corresponds to a less stable intermediate structure than the unsymmetrical one.[79]

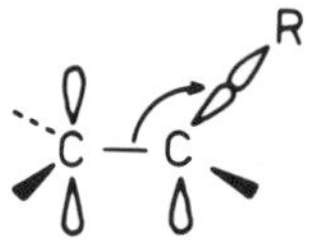

SCHEME 31 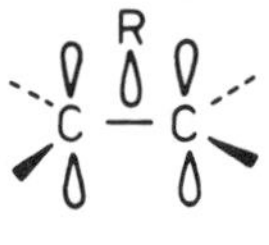   SCHEME 32

The angle φ of attack of the radical R· on the alkene (Scheme 31) is remarkably constant, varying from 102 to 109° for a variety of radicals; the nucleophilic species attack with larger angles than the electrophilic.[80] In the transition state of addition of H·, ·CH$_3$, and HO· to ethylene, the lengths of the partially formed C–H, C–C, and C–O bonds are by 85%, 46% and 33% longer than those in the final products. The extent of bond formation indicates that hydrogen atom addition occurs earlier than either R· or RO· addition, which is in correspondence with the activation energies. That is, the earlier the transition state arises, the lower the activation energy.

Unless it fragmentates back to initial reactants, the second electron of the polarized π-bond becomes the carrier of the radical character of the addition product. The reversibility of addition reactions depends on the type of attacking radical and on the possibility of stabilization of a newly formed radical. The stronger the bond between the radical and the molecule replacing the decaying π-bond, the more irreversible is the addition reaction.

Radicals having an unpaired electron on oxygen, carbon, or silicon will, therefore, add to the carbon-carbon double bond with the equilibrium constant shifted to the addition products. From the same reason, H, F, Cl, and Br atoms add to alkenes more easily than, e.g., iodine atoms, where the strength of the newly formed C-I bond is usually lower than that of the π-bond in alkene.

From this viewpoint, the addition reaction of radicals having unpaired electrons on sulfur is less obvious.[81,82] Here, the rate of back fragmentation may be so fast that the product of addition cannot be observed. The fact that addition and fast fragmentation occur indeed, is documented indirectly by cis-trans isomerization of alkene, which takes place as a stepwise process of double-bond decay due to radical addition, free rotation around the intermediately formed single bond C-C, and back regeneration of the double bond by fragmentation.

Free radicals may add not only to multiple bonds of alkenes (Equation 43)

$$R· + C=C → R-C-C·$$ (43)

but also to acetylene and its derivatives (Equation 44)

$$R \cdot + -C{\equiv}C- \rightarrow R-\overset{|}{C}{=}\overset{|}{C}\cdot \tag{44}$$

or to aromatic compounds (Equation 45).

$$R \cdot + \underset{}{\bigcirc}{-}X \longrightarrow \underset{}{\bigcirc}{-}X \tag{45}$$

Addition of alkyl radicals to ethylene and its derivatives is of importance in the synthesis of many macromolecules. To conceive the rate of this reaction, we present the Arrhenius parameters of addition of methyl radical to ethylene (A = $1.6 \cdot 10^8$ dm^3 mol^{-1} sec^{-1}, E = 32 kJ/mol)[83] from which it follows that the rate of elementary addition reaction is by 10^6 times lower at 20°C than the frequency of reactants collisions. The rate of addition reaction of identical radicals with π-bond of either ethylene or acetylene is approximately equal. More nucleophile radicals react with ethylene more quickly. The same is true for electrophile radicals and acetylene.[84]

Reactivity of multiple bonds in aromatic compounds depends on the degree of their conjugation. As may be seen from the comparison of rate constants (in dm^3 mol^{-1} sec^{-1}) for

	$R\dot{C}H_2$	$C-\dot{C}_3H_5$	$\dot{C}_6H_5$
$CH_2{=}CHC_6H_5$	10^5	2.10^7	1.10^8
C_6H_6	60	3.10^4	5.10^5

alkyl, cyclopropyl, and phenyl radicals in reaction with styrene and benzene at 300 K, the radical addition to benzene is about 3 orders slower than that to the double bond of alkene.[85]

Very reactive radicals react with benzene with such high rate constants at 20°C, (·OH : $7 \cdot 10^8$ dm^3 mol^{-1} sec^{-1}, H· : $1 \cdot 10^9$ dm^3 mol^{-1} sec^{-1}) that the difference in the rate of addition to different multiple bonds is smaller.[86,87] On the other hand, with polynuclear aromatic compounds, radicals of low and medium reactivity may be more reactive than with ethylene.

The rate and the corresponding reaction site of the radical addition to aromatic compounds depend on the spin polarizability of multiple carbon-carbon bonds. In the system of π-bonds of some compounds the nonequal delocalization of π electrons on each carbon atom is required. The lower the value of this delocalization energy, the more easily the addition reaction of the radical to the corresponding atom occurs. The energy of delocalization of electrons sinks with an increasing degree of conjugation of unsaturated bonds, and, similarly, the rate of addition to aromatic compounds increases. The rate constants of the addition of identical radicals to different aromatic compounds vary to the extent of five orders. Addition of neutral radicals to aromatic compounds is influenced also by substituents which generally activate ortho and para positions. Addition to the meta-position is preferred during the addition of silyl radicals or when the ortho-position is sterically hindered.[88] Since the rate constants of addition differ only to the extent of one to two orders, the difference in the reactivity of individual positions of benzene derivatives may be considered as relatively low.[89]

When compared to π-bond in alkenes, multiple bond carbon-oxygen is less reactive to alkyl radicals. The preferred site of attack by carbon-centered radicals is the weaker C=C

Table 7
COMPARISON OF ARRHENIUS PARAMETERS AND RATE CONSTANTS OF ADDITION OF METHYL AND TRIETHYLSILYL RADICALS TO CARBONYL GROUPS OF R^1COR^2 SUBSTRATES[83,92-94]

Radicals	R^1	R^2	log A[a]	E kJ/mol	k_{430} K dm^{3}mol^{-1}sec^{-1}
$\dot{C}H_3$	CF_3	CF_3	8.1	21	$3 \cdot 10$
	CF_3	CH_3	7.7	28	$2 \cdot 10^4$
	CH_3	$COCH_3$	7.3	26	$2 \cdot 10^4$
	H	H	8.3	32	$1 \cdot 10^4$
	H	CH_3	7.2	52	7
	CH_3	CH_3	7.5	49	30
$\dot{S}i(C_2H_5)_3$	C_2F_5	$OCOC_2F_5$	8.9	0.8	$6 \cdot 10^9$
	C_6H_5	COC_6H_5	9.3	4.3	$2 \cdot 10^9$
	H	C_3H_7	7.8	4.1	$6 \cdot 10^7$
	C_6H_5	CH_3	9.4	13	$7 \cdot 10^7$
	C_2H_5	$OCOC_2H_5$	8.0	11	$5 \cdot 10^6$
	H	OC_2H_5	8.3	21	$4 \cdot 10^6$

[a] Value A is in $dm^3 \ mol^{-1} \ sec^{-1}$.

bond (the dissociation energy of π-electrons is about 250 kJ/mol) rather than the C=O bond which is by some 60 kJ/mol stronger. Carbon radicals reacting with 1,4-benzoquinone add therefore to the C=C bond and not to the C=O bond.[90] The difference in reactivity of C=C and C=O bonds is not, however, so high and it may well turn out to change experimental conditions or reactants.[91] This is shown by the large difference in reactivity of the C=O bond of various carbonyl compounds, R^1COR^2, where alkyl radicals add preferentially to carbon of carbonyl group (Table 7).

Addition of silyl radicals to the carbonyl group may be enhanced by the possibility of the formation of a stronger Si-O bond compared with that of the Si-C bond and by the absence of substituents hindering reaction site. The competition of the addition of silyl radicals to C=C and C=O bonds is, however, determined by a more complex sequence of subsequent reactions. On one hand, the attack of silyl radicals to the carbon-carbon double bond of unsaturated carbonyl compounds is a more facile process than direct attack to C=O. Subsequently formed radicals may, however, evolve to the thermodynamically more stable oxygen adduct via internal migration of the silyl group.[95]

Provided that the oxygen of the carbonyl group is replaced by more electropositive sulfur, the reactivity to radicals increases. Alkyl radicals add to the weaker multiple C=S bond with a higher rate than to alkenes.[96,97] Addition reactions may still occur on double or triple carbon-nitrogen bonds or on double bonds nitrogen-nitrogen and nitrogen-oxygen. The addition of alkyl and alkoxyl radicals to N=O bond of nitroso compounds or nitrones with the formation of nitroxyl radicals (k [300 K] = 10^5 to 10^8 dm^3 mol^{-1} sec^{-1}) seems to be studied the most systematically.[98,99] This reaction is the fundamental elementary step in the trapping of reactive radicals from the reaction system and in identification of their structure. Alkyl radicals add either to nitrogen or oxygen of the N=O bond of nitroso compounds.[100] The usefulness of the spin trapping approach in the search for free radicals has now become particularly attractive to biological scientists. Fortunately, current nitrones are nontoxic and appear to survive within biological systems long enough to allow the detection of radicals.[101]

The reaction of alkyl radicals with molecular oxygen (Equation 46) should formally also be considered as addition reaction.

$$R \cdot + O_2 \leftrightarrow R\dot{O}_2 \tag{46}$$

Since the ground state of oxygen is triplet, the reaction may be understood also as the recombination process. This is shown by the high value of the frequency factor and by the almost zero value of the activation energy. The activation energy of the backward reaction is considerably higher, being about 130 kJ/mol. Increasing temperature, therefore, gradually reduces the concentration of peroxy radicals in the system. Reduced $RO_2\cdot$ concentration may obviously be responsible for the change in the mechanism of hydrocarbon combustion between 250 and 500 °C. Like alkyl radicals, the stability of $RO_2\cdot$ radicals may be characterized by a ceiling temperature, when the forward and backward rate of reaction (Equation 46) are equal. For isopropyl and isopropyl peroxy radicals it is bout 350°C.[102] The β-scission process of peroxy radicals that are highly resonance stabilized, such as triphenyl methyl peroxyls may be observed even at a markedly lower temperature.

The rate constants for the reactions of carbon-centered radicals with oxygen are not very sensitive to the spin delocalization.[103] For example, the rate constant (in $dm^3\ mol^{-1}\ sec^{-1}$) for tert-butyl is $4.9\cdot10^9$, benzyl $2.4\cdot10^9$ and cyclohexadienyl $1.6\cdot10^9$, respectively.

The high-pressure limiting rate constants correlate with the ionization potential of the alkyl radical; the lower the ionization potential of the alkyl radicals, the higher is their rate constant with oxygen.[104,105] For similar ionization potential values, the carbonyl radicals react more slowly with O_2 than do the alkyl radicals. The correlation does not appear to apply also to benzyl and allyl radicals. The trend of the constants does not follow from the sterical effect alone. The radical center in the tert-butyl radical is shielded by three methyl groups, and yet the rate of its reaction with oxygen is 10 times faster than the reaction of methyl radical with oxygen.

The rate constant for the reaction of the ethyl radical and oxygen increases with the increasing concentration of inert helium.[106] This shows in the possible participation of He molecules in the process of stabilization of formed peroxy radicals. The excited alkyl peroxy radicals do not fragmentate to original reactants, but rearrange to products through separate pathways.[107]

The reactions of alkyl and other hydrocarbon radicals with molecular oxygen are important elementary steps in most oxidation processes, including the combustion of fossil fuels and air pollution chemistry. The condensed phase oxidation plays the key role in the synthesis of oxygenated organic compounds and in the degradation of various materials such as plastics, oils, and living organisms.

Besides addition to π-bonds, free radicals may also react with a lone pair of electrons on the respective atom. This is the case of the addition of alkyl radicals to carbon monoxide (Equation 47)

$$\cdot CH_3 + C=O \leftrightarrow CH_3\dot{C}O \tag{47}$$

or to sulfur dioxide (Equation 48)

$$\cdot CH_3 + SO_2 \leftrightarrow CH_3\dot{S}O_2 \tag{48}$$

Addition of SO_2, as a molecule with distinct electron-acceptor properties, depends on electron-donor properties of reacting radical. While the methyl radical adds to SO_2 easily, carboxymethyl or methylperoxy radicals which have electron-acceptor properties do not react.[108] On the other hand, electron donor hydroxymethyl radicals reduce sulfur dioxide to anion radicals (Equation 49).

$$\cdot CH_2OH + SO_2 \rightarrow \dot{S}O_2^{\ominus} + CH_2O + H^+ \tag{49}$$

The outstanding influence of radical electrophility or nucleophility on the course of addition, which is not the characteristic feature of radical reactions of SO_2 only, is of much general validity.[109]

A. Regioselectivity

As was already seen, the site of attack on a multiple bond, i.e., the regioselectivity of reaction, depends on the kind of radical and on the type of reacting compound. Nucleophile radicals add to carbonyl groups via carbon atoms, whereas electrophile radicals prefer an oxygen atom. At the addition to a less polar bond the regioselectivity is determined by the interplay of several factors as, e.g., polar and steric interactions between the attacking radical and the reacting π-bond, strength of the newly formed bond, and stabilization of the radical adduct (exothermicity of the addition reaction).[110]

At the reaction of alkyl radicals with alkenes the site of addition (Equation 50) is determined by substituents (Y, Z) of

$$\text{X--}\overset{Y}{\underset{}{\dot{C}}} + \quad \overset{Y}{\underset{Z}{C=C}} \quad \rightarrow \text{X--C--C--}\overset{Y}{\underset{Z}{\dot{C}}} \tag{50}$$

alkene as well as of the radical. In order to distinguish these substituents, the group Y bound to the carbon atom of the alkene that is attacked by the radical is denoted as the α-substituent, while the neighboring group Z as the β-substituent.

The regioselectivity of the addition of free radicals to alkenes is high when the difference in the steric bulkiness of the groups at the carbon atom is large. Since all substituents are larger than hydrogen, free radicals attack preferentially the less substituted carbon atoms of alkenes. The addition to the less substituted alkenic carbon atom often leads to the more stabilized radical adduct, since the substitution of hydrogen atoms always results in a stabilization of a radical-centered carbon atom. Thus, it is difficult to separate sterical from stabilizing effects taking into account the regioselectivity data only. At the simultaneous knowledge of the rate constants of addition, it may be concluded, that β-substituents exert essentially polar effects, whereas β-substituents and radical substituents X both polar and steric effects.[110,111]

The regioselectivity decreases on the changing to the more reactive radical. The tendency of the cyclohexyl radical to attack the ester substituted vinyl carbon atom increases with the bulkiness of the Y-group. (Table 8) Whereas the unsubstituted acrylic ester (Y=H) reacts almost solely at the methylene group, the *tert*-butyl-substituted acrylic ester (Y=C(CH$_3$)$_3$) is attacked predominantly at the ester substituted vinyl carbon atom. The phenyl radicals are less sensitive to steric effects than alkyl radicals and can, therefore, add also to polysubstituted double bond such as in 4-methyl-3-penten-2-one, with the regioselectivity α/β = 0.19. The alkyl radicals do not react with this alkene for steric reasons.[112]

The effect of the structure of the attacking radical on regioselectivity may be demonstrated on substituted fluoroethylenes (Table 9). Fluorine is not of much higher sizes than hydrogen and its steric effect in free radical addition is, therefore, small. Looking at the attack of α and β positions of the double bond in fluoroalkenes by phosphine radicals we may see that the more nucleophile the radical, the more easily carbon substituted by fluorine atoms reacts. More electrophile radicals (CF$_3$)$_2$ P· react almost exclusively with carbons having hydrogen substituents. The equal rule may be deduced from the ratio of α/β addition of alkyl radicals in dependence on substituents X of carbon in the parent radical.

It follows from the overall tendency of the interaction of fluorine-substituted carbons and

Table 8

THE EFFECT OF SUBSTITUENT BULKINESS ON THE C=C BOND REGIOSELECTIVITY OF CYCLOHEXYL[110] AND 4-CHLORPHENYL[112] RADICAL ADDITION

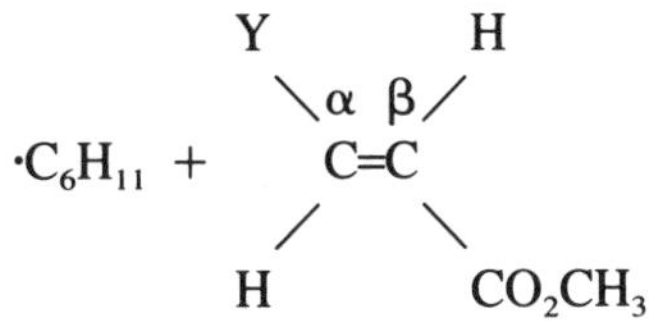

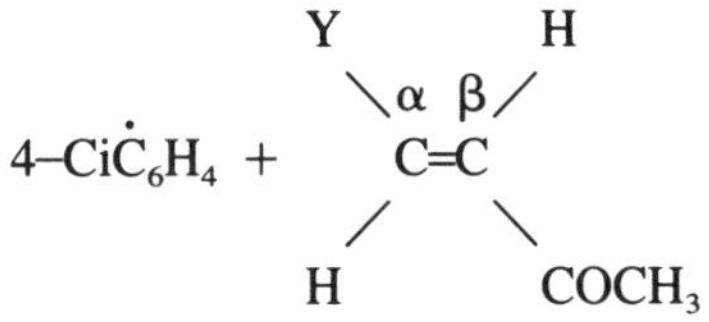

$$\cdot C_6H_{11} + \quad \underset{H}{\overset{Y}{\diagdown}} \overset{\alpha\ \beta}{C=C} \underset{CO_2CH_3}{\overset{H}{\diagup}}$$

$$4-Cl\dot{C}_6H_4 + \quad \underset{H}{\overset{Y}{\diagdown}} \overset{\alpha\ \beta}{C=C} \underset{COCH_3}{\overset{H}{\diagup}}$$

Y	α/β	Y	α/β
H[a]	~500	H[b]	24
CH_3	11.5	CH_3	3.5
C_2H_5	7.3	$CH(CH_3)_2$	1.1
$CH(CH_3)_2$	3	C_6H_5	0.32
$C(CH_3)_3$	0.25	$C(CH_3)_3$	0.25

[a] For reaction with $4-Cl\dot{C}_6H_4$ $\alpha/\beta = 3$ (Reference 112).

[b] For reaction with $C_3H_7\ \dot{C}=O$ $\alpha/\beta > 100$ (Reference 112).

Table 9

COMPARISON OF THE ATTACK OF α AND β-CARBON IN VINYLFLUORIDE, VINYLIDENFLUORIDE AND TRIFLUOROETHYLENE BY SOME RADICALS

Radical	EN_R[a]	CH_2=CHF $\alpha\ \ \beta$	CH_2=CF_2 $\alpha\ \ \beta$	CHF=CF_2 $\alpha\ \ \beta$	Ref.
		Substituted ethylene			
$(CF_3)_2P\cdot$	4.95	1:0.00	1:0.00	1:0.02	113
$F_5S\cdot$	5.44	1:0.01	1:0.01	1:0.10	113
$\cdot C_3F_7$	5.09	1:0.05	1:0.01	1:0.25	114
$\cdot CCl_3$	4.62	1:0.08	1:0.01	1:0.29	114
$\cdot CF_2Br$	4.88	1:0.09	1:0.03	1:0.46	115
$(CH_3)_2P\cdot$	3.54	1:0.09	1:0.39	1:1.08	113
$\cdot CH_3$	3.61	1:0.59	1:0.18	1:7.26[b]	114

[a] Electronegativity of radicals calculated according to Reference 58.

[b] (1:2.13) According to Reference 110.

from the electronegativity of radicals that the dependence is superimposed by the steric effect of fluorine atoms which support addition on nonsubstituted or less substituted carbons.

B. Stereoselectivity

Stereoregularity of addition is affected by the structure of both the radical and alkenes. When the two lobes of the orbitals containing p-electrons in radicals of alkenes are screened to a different extent, free radical addition occurs with the stereoselectivity corresponding to the difference in the steric polarity of substituents.[110]

Thus, the observation that the ethoxycyclopent-2-yl radical attacks the tetrasubstituted alkene exclusively trans to the ethoxy group, whereas methylacrylate reacts with lower selectivity, corresponds to the different substitution of the C=C bond in the reactive alkene (Scheme 33).

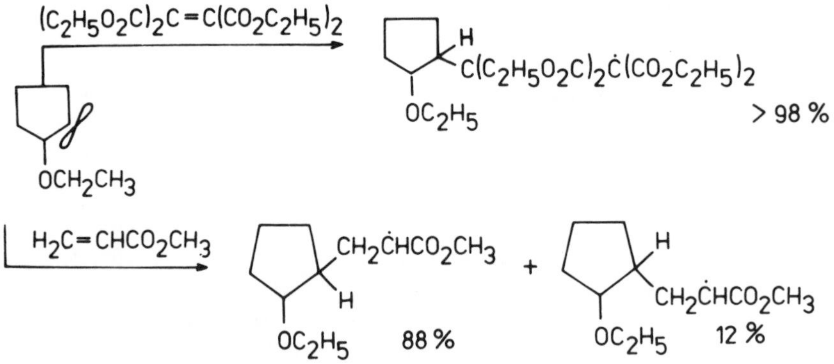

SCHEME 33

One side of the single occupied p-orbital of ethoxycyclopent-2-yl radical is shielded by a neighboring ethoxy group and by a hydrogen atom, while the second side only by two hydrogen atoms.

At the polymerization addition of the growing radical to asymmetric alkenes $CH_2 = CAB$ (Scheme 34) stereoselectivity

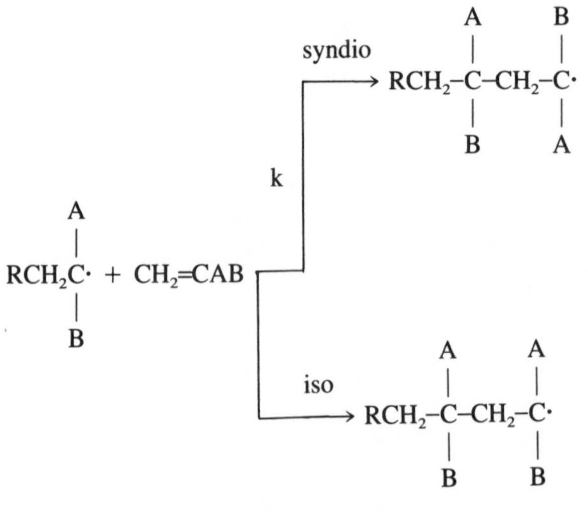

SCHEME 34

of the reaction consists in stereoisomer arrangement (tacticity) of the formed macromolecular chain.[116] Syndiotactic structural units are the more numerous in the polymer chain when the difference in the bulkiness of substituents A when compared to B is higher. The syndiotactic arrangement corresponds to such repeating addition steps of the asymmetric monomer which ensure the highest distance of bulky substituents in the formed radical adduct. The sterical effect is of the same character as in the major reaction step of ethoxycyclopent-2-yl radicals with methylacrylate. The competitive addition which leads to isotactic structural units of the polymer chain is of the minor extent.

The polar effect in addition reactions may be demonstrated on the polymerization of methacrylic acid in aqueous medium under the conditions of different pH. The radical and monomer reactants with anionic carboxylic groups, the ratio of which increases with increasing pH, prefer syndiotactic addition more than less polar reactants having nondissociated carboxyl groups. The isotactic addition reaction has a higher activation energy. The decrease of polymerization temperature, therefore, leads to the formation of the more regular structure of a polymer chain with long sequences of syndiotactic units.

C. Kinetics of Addition Reactions

Addition reaction as an aggregation process is connected with the decrease of entropy. From the thermodynamic viewpoint, only exothermic addition reactions are thus allowed. Addition reactions of polyatomic radicals have a positive change of reaction entropy in the extent of 120 to 170 J mol^{-1} K^{-1} which corresponds to the frequency factor of 10^5 to 10^9 dm^3 mol^{-1} sec^{-1}. Since single atoms do not lose the entropy of rotation after linkage with a multiple bond, their addition has a higher frequency factor (up to 10^{11} dm^3 mol^{-1} sec^{-1}). The fall of entropy in the addition reaction indicates that increasing temperature leads to the shift of the equilibrium constant towards back fragmentation.

The activation energy of addition reactions has a value between 2 to 75 kJ/mol and is inversely proportional to the reaction exothermicity. As was already shown for other reactions, the reaction heat of the addition reaction increases with the increasing difference between the overall energetical levels of the multiple bond as well as with the sum of energies of both newly formed bonds. The activated complex formed in reaction of high exothermicity resembles more the reactants than the products. The properties of reactants have then more pronounced effect on the rate of addition than the stability of a formed radical.

In the case of the reaction of alkyl radicals with monosubstituted derivatives of ethylene,[110] it may be seen that the electron acceptor substituents can increase the rate constant of addition by 3 orders of magnitude approximately, while electron donor substituents reduce it moderately. This decrease is, however, valid only for alkyl radicals. Provided that reacting radicals are more electrophile, electron donor substituents on the multiple bond increase the rate of addition, too.

For the symmetrically 1,2-disubstituted ethylene, the polar effect of the identical substituents is partially eliminated (Table 10). The effect of two electron donor methyl groups is lower than for asymmetrically substituted isobutylene or trimethylethylene. This may be seen especially at the almost thermoneutral reactions and at the high difference of donor acceptor properties of reactants. The distance between decaying and newly forming bonds in the structure of the activated complex should be large, too.

A good example of the polar effect is the addition of alkyl radicals to diethyl vinyl phosphonate.[120] The rate constant of this addition increases with the decrease of ionization potential of radicals regardless of the fact that the bulkiness of radicals increases in the given sequence. Polar interactions facilitate obviously the formation of the activated complex, and the process of transformation of reactants to products is thus accelerated.

The polarity of the substituents and the relative reactivity of substrates usually correlate in terms of Hammet's equation (Equation 51), where k_H and k denote the rate constants for

Table 10

**RELATIVE RATE OF ADDITION OF RADICALS TO ALKENES WITH
ELECTRON DONOR METHYL SUBSTITUENTS[56,117-119]**

Alkene	·CH₃	·CF₃	·C₆H₅	H·	·SiCl₃	·NF₂	:O(3p)	Br·
K	453	338	333	298	460	373	298	353
$CH_2=CH_2$	1	1	—	1	1	1	1	1
$CH_2=CHCH_3$	0.7	1.4	1	1.8	3	4	6	9
$CH_2=C(CH_3)_2$	1.1	3.8	3.7	4.4	13	20	25	151
$CH_3CH=CHCH_3$[a]	0.3	0.6	0.6	0.9	4	10	26	35
$CH_3CH=C(CH_3)_2$	0.4	—	0.2	1.5	17	34	79	—

[a]   Average values for slightly different date of *cis* and *trans* isomers.

$$\log k/k_H = \rho \cdot \sigma \tag{51}$$

nonsubstituted and substituted reactants, ρ is the relative value of electrophility of the given reaction type, and σ is the characteristic value of substituent polarity.

The positive ρ corresponds to nucleophile alkyl radicals, whereas the negative value of ρ corresponds to electrophile radicals, such as 4-nitrophenyl, peroxyl, oxygen atom, piperidine cation radical, etc.[87] The positive ρ values increase with the decreasing ionization potentials of the radicals. From the numerous experimental works approximate classification of radicals follows. The considerably larger electrophility of acylperoxy radicals $CH_3CO_3·$ and $C_6H_5CO_3·$ ($\rho = -1$ vs. σ^+) when compared to alkyl peroxyls ($\rho = -0.1$ to -0.3 vs. σ^+) may, e.g., explain by 5 orders, faster addition of acylperoxyl radicals to α-methyl styrene at 20°C.[121] Electrophility of radicals leads to the marked sink of activation energy of addition to alkenes. (Methylperoxy radical 47 kJ/mol; acetylperoxy radical 19 kJ/mol.)[122]

Correlation of the reactivity of radicals within Hammet's equation have a relatively low correlation coefficient, which indicates that the effect of reactant polarity on the rate constants is more complex than may be expressed in one parametric relation.

From the kinetic viewpoint, the addition is the second order reaction. Its rate is thus directly proportional to the product of reactant concentration. The order higher than 2 was observed namely at reactions of reactants composed of one or two atoms proceeding in the gaseous phase, such as, e.g., the addition of chlorine to NO or NO_2. The declination from the assumed dependence is due to the participation of reactants in inevitable deactivation of excited radicals which undergo fast fragmentation.[123]

V. SUBSTITUTION REACTIONS

Reactions of a radical with molecules may also proceed via the initial attachment of the attacking radical to some central or side (surface) atom of the molecule. The formed unstable adduct then subsequently decomposes into a new radical and molecule. This is, e.g., the case of the reaction of alkyl radicals with trifluormethylthiylchloride (Scheme 35).

$$R· + CF_3SCl \rightarrow RCl + CF_3S·$$

$$\rightarrow RSCF_3 + Cl·$$

SCHEME 35

Abstraction of the chlorine atom by a radical is a transfer reaction; the chlorine atom is transferred from the reacting substrate to the attacking radical. The second case represents the displacement of polyatomic radicals. The parallel course of both types of substitution reactions for one compound and the same type of radical is rather rare in chemistry.[124] More frequently, we meet either the displacement of whole groups of atoms on the central atom of a molecule or transfer reaction on side atoms occurring separately. Substitution transfer reactions are very numerous in radical reactions of hydrocarbons and their derivatives, while the displacement reactions of polyatomic radicals are current for inorganic and metaloorganic compounds.

A. Substitution Reactions on the Central Atom of Molecule

The displacement reaction occurs either on the atoms which have energetically available orbitals or on the atom with bonding π-electrons. The attacking radical is attached to the reaction site by the interaction with its unpaired electron and forms an unstable radical adduct. According to the lifetime of such an adduct, the reaction proceeds either as a synchronous or a stepwise process. In the first case, the time of existence of the activated complex is on the order 10^{-13} sec, whereas in the stepwise reaction the radical adduct has the character of a reaction intermediate.

The stability of the transiently formed radical adduct may be estimated from the difference of the strength of the forming and decaying chemical bond. Provided that the forming bond has a higher strength than the decaying, the substitution reaction is exothermic and the radical adduct has a very short lifetime. In the opposite case, the reaction is stopped in the stage of a radical adduct.

B. Substitution on Atoms with Vacant Orbitals

The essential difference in chemistry of silicon and carbon compounds consists just in the different course of their substitution reactions. Silicon has unoccupied energetically available 3d orbitals, which make possible the formation of a new bond with an attacking radical before cleaving another already existing bond. The difference may be illustrated on the radical condensation reaction according to Equation 52 and

$$CH_3SiCl_2CH_2\cdot + (CH_3)_2SiCl_2 \rightarrow CH_3SiCl_2CH_2SiCl_2CH_3 + \cdot CH_3 \tag{52}$$

on another substitution of a decomposition character (Equation 53)

$$I\cdot + (CH_3)_3SiSi(CH_3)_3 \rightarrow (CH_3)_3SiI + (CH_3)_3Si\cdot \tag{53}$$

which was not observed in hydrocarbons.

The role of the strength of formed and decaying bonds at substitution reactions may be seen on radical reactions of boranes.[125] The reaction of alkoxy radicals with alkylboranes which have the vacant p-orbital on the borine atom (Equation 54)

$$R^1O\cdot + R_3B \rightarrow R^1OBR_2 + R\cdot \tag{54}$$

is very fast since the strength of the newly formed bond is by 150 kJ/mol higher than that of the decaying bond B–C.

It may be of interest that with alkyl peroxy radicals, which in some reactions are considerably less reactive than alkoxy radicals, the homolytic substitution reaction on alkylboranes (Equation 55) is very fast, too.

$$ROO\cdot + BR_3 \rightarrow ROOBR_2 + R\cdot \tag{55}$$

If R is butyl, the rate constant of reaction is $2\cdot10^6$ dm^3 mol^{-1} sec^{-1} at the ambient temperature. The rate constant decreases by 2 orders of magnitude if alkyl substituents are branched or if one alkyl on borine is replaced by peroxyl.[124] The decrease in reactivity is brought about either by the hindered approach of more bulky radicals to the unoccupied p-orbital of borine or by partial compensation of coordination unsaturation of the central atom by the nonbonding electron of peroxidic oxygen in substituent groups.

On the other hand, during the reaction of the triphenylmethyl radical with triphenylborane (Equation 56) where the difference in the strength of the formed and the decaying bond is small, a very stable radical intermediate is formed.

$$(C_6H_5)_3C\cdot + B(C_6H_5)_3 \rightarrow ((C_6H_5)_3C\text{–}B(C_6H_5)_3)\cdot \qquad (56)$$

Substitution reactions of peroxy radicals on the central atom of a molecule takes place with many alkyl metallic compounds. Peroxidic alkyl metallic compounds decompose vigorously and branch the oxidation reactions.[126] This is also why these compounds are dangerously flammable.

The more complex course of substitution reactions may be seen on phosphorus compounds.[127,128] In the reaction of *tert*-butoxy radical with trimethyl phosphine (unoccupied d-orbital) a relatively stable radical intermediate is formed (Equation 57) which was identified directly in the reaction medium.

$$(CH_3)_3CO\cdot + P(CH_3)_3 \rightarrow (CH_3)_3CO\overset{\cdot}{P}(CH_3)_3 \qquad (57)$$

This radical intermediate (RI) does not split out only a methyl radical (Equation 58), but also a *tert*-butyl radical (Equation 59) and

$$RI \rightarrow CH_3\cdot + (CH_3)_3COP(CH_3)_2 \quad (\alpha\text{-scission}) \qquad (58)$$

$$RI \rightarrow (CH_3)_3C\cdot + O{=}P(CH_3)_3 \quad (\beta\text{-scission}) \qquad (59)$$

there are simultaneously formed stable substituted molecules. While α-scission corresponds to the mechanism of substitution of methyl radicals by *tert*-butoxy radicals, β-scission represents rather an addition-fragmentation reaction. Two possible pathways of the reaction course consist in a variable electron structure of a radical adduct. The unpaired electron, however, may be located either on the 3rd or 4th orbital.

The ratio of α/β-scission depends also on the type of attacking radical. At the reactions of radicals having the unpaired electron on sulfur (Scheme 36) there occurs only

$$RS\cdot + PX_3 \rightarrow RS\overset{\cdot}{P}X_3 -\!/\!/\!\rightarrow \alpha\text{-scission}$$
$$\Big|\ \beta\text{-scission}$$
$$\underline{\qquad\qquad} \rightarrow S{=}PX_3 + R\cdot$$

SCHEME 36

β-scission. One-way decomposition of the radical adduct follows from the lower strength of the C-S bond when compared with the C-O bond. The reaction of thiyl radicals with trialkylphosphines and phosphites is very fast and may be used for desulfuration of organic compounds.

Not all substitution reactions of radicals taking place on the central atom of a molecule have an addition-fragmentation character. Instead of transient adduct formation there occurs the electron transfer as the primary step of substitution. Such reactions may be illustrated on oxidation of an alkyl radical by metal ions, where the formed alkyl cation is neutralized by the counter-anion (Equation 60 and 61).

$$R\cdot + M^{(n+1)+} \rightarrow R^+ + M_n^+ \tag{60}$$

$$R^+ + X^- \rightarrow RX \tag{61}$$

The course of the reaction may, however, be alternatively explained by displacement of the ligand (Equation 62) since the final result of the alternative reaction pathway is identical.

$$R\cdot + MX_m \rightarrow RX + MX_{m-1} \tag{62}$$

One should confess here that sometimes it is difficult to distinguish the right mechanism of the substitution.

C. Substitution Mediated by Nonbonding Electrons

The site in the molecule where the substitution reaction may start is, e.g., the peroxidic bond in peroxides. The reaction itself is then called an induced decomposition of peroxide. The term of induced decomposition is, of couse, somewhat more general and includes also other radical reactions which lead to peroxide decomposition. The peroxidic bond may be attacked especially by nucleophile radicals. The mechanism of substitution on the peroxide bond was studied in detail on reaction of benzoyl peroxide with α-hydroxyalkyl, triphenylmethyl, cyclohexenyl, and also with organometallic radicals (Equation 63) which have an unpaired electron located on the atom of metal.

$$R_3Sn\cdot + C_6H_5COO-OCOC_6H_5 \rightarrow R_3SnOCOC_6H_5 + C_6H_5COO\cdot \tag{63}$$

The increase in the nucleophile character of the corresponding radical of the same type decreases the rate of the attack on the peroxidic bond rather significantly. The rate constant of the interaction of, e.g., a perfluortriphenylverdazyl radical with an O-O bond of benzoyl peroxide in benzene is $8 \cdot 10^{-3}$ dm^3 mol^{-1} sec^{-1} at 20°C, whereas the triphenylverdazyl radicals alone which are more nucleophilic react with benzoyl peroxide by 5000 times faster.[129]

A substitution reaction on the oxygen atom of the peroxyacid group (Equation 64) takes place at the decomposition of peroxy-acids in solvents.

$$R\cdot + HOOCOR \rightarrow ROH + RCO_2\cdot \tag{64}$$

The yield of alcohol ROH depends on the nucleophility of the radical as well as on the degree of its delocalization.[130] If the unpaired electron is not delocalized, the reactivity of the radical increases with its nucleophility, correlated with its ionization potential. As a consequence, σ-radicals like phenyl, cyclopropyl, or 1-bicyclo-[2.2.1]-heptyl are unreactive with respect to peroxyacids and give little alcohol (Equation 64). On the other hand, π-alkyl radicals lead to a quasi-quantitative decarboxylation into alcohol. If the unpaired spin is delocalized, the orbital overlap integral between the radical and the peroxidic oxygen becomes an important reactivity factor. A radical seemingly more nucleophilic on the basis of the ionization potential then proves to be less reactive than a nondelocalized alkyl radical. This is the case for the benzyl radical. The electrophilic alkoxy radical does not react on the O-O bond but abstracts the H atom of the peroxyacid group.

Epoxidation of alkenes (Equation 65) may be considered as an intramolecular substitution reaction on the peroxidic bond, too.

$$ROO\cdot + \,\,{>}C{=}C{<} \rightarrow \underset{O-OR}{-\overset{|}{\underset{\cdot}{C}}-\overset{|}{C}-} \rightarrow -\overset{|}{\underset{O}{C}}\!\!-\!\!\overset{|}{C}- + \cdot OR \tag{65}$$

Such reactions of hydroperoxide alkyl radicals are of importance during oxidation of hydrocarbons in the gaseous phase. They give relatively high yields of cyclic ethers.

Also, substitution reactions on sulfur compounds are very frequent.[124] Unambiguous evidence of substitution reaction of 3-chlorbenzyl radicals with dibenzylsulfide (Equation 66)

$$ClC_6H_4CH_2\cdot + (C_6H_5CH_2)_2S \rightarrow ClC_6H_4CH_2SCH_2C_6H_5 + C_6H_5CH_2\cdot \qquad (66)$$

is the formation of 3-chlordibenzylsulfide.

For reaction of the methyl radical with deuteromethylmercaptane (Equation 67) there were even determined the kinetic parameters ($A = 5 \cdot 10^7$ dm^3 mol^{-1} sec^{-1}, $E = 32$ kJ/mol), from which it is obvious that substitution on the sulfur atom of mercaptane will play a significant role in the whole set of other possible reactions.

$$\cdot CH_3 + CD_3SH \rightarrow \cdot CD_3 + CH_3SH \qquad (67)$$

Sulfur compounds also undergo fast substitution reactions with thiyl radicals (Scheme 37).

$$R^2S\cdot + R^1SSR^1 \Leftrightarrow R^1S\!-\!S\!\!\begin{array}{c} \diagup SR^2 \\[4pt] \diagdown R^1 \end{array} \Leftrightarrow R^2SSR^1 + R^1S\cdot$$

SCHEME 37

Unpaired electrons in a sulfur-centered radical show a strong tendency to coordinate with p-electron lone pairs of another sulfur atom.[131] The unpaired electron is shared between the two equivalent S-S bonds. The actual distribution of the electron density will, of course, be affected by the character of substituents R.[132] The unpaired electron is expected to be accomodated in an antibonding σ^*-orbital. Each of the sulfur-sulfur bonds will be thus weakened as compared to a normal S-S σ-bond which provides a rationalization for the establishment of an equilibrium between the adduct radical and its constituents.

Substitution reactions of sulfur radicals explain the release of the mechanical stress in polysulfide rubbers.[124] During the deformation of polysulfidic polymer, linear macromolecules orient in the direction of the applied force, and the energetically favorable initial arrangements is thus disrupted. The forced inner strain is capable of restoring the initial shape of the rubber sample whenever the deformation force is removed. At a large deformation, some macromolecules are broken down, and formed macroradicals change the conformation of environmental strained chains by a substitution reaction on sulfur atoms of the main chain. The macromolecule breaks down there and its fragments link together with another part of the polymer chain. Recombination and splitting off the macromolecules goes on until the most advantageous conformation of polymer chains without the inner strains is gradually restored.

D. Substitution Reactions on Carbon Atoms

Substitution on carbon atoms is mediated by π-electrons as a sequence of two elementary reactions. This may be illustrated on substitution of allyl compounds as well as on phenylation of aromatic compounds.[133] At the reaction of allyl compounds with phenyl radicals (Equation 68) addition takes place there first.

$$C_6H_5\cdot + CH_2\!\!=\!\!CHCH_2X \rightarrow [C_6H_5CH_2\dot{C}HCH_2X]$$

$$\rightarrow C_6H_5CH_2CH\!\!=\!\!CH_2 + X\cdot \qquad (68)$$

The formed radical, which is chemically activated, then decomposes. After elimination of the X group, the sequence of the above reactions may appear as a direct substitution on the carbon atom. The possibility of the course of the substitution reaction depends on the strength of binding of substituent X to the allyl group. Provided tha X is bound via the carbon or oxygen atom, the substitituion by the phenyl radical does not proceed and, consequently, allylbenzene is not observed in the reaction products. Decreasing the strength of the bond between the substituent and carbon of the allyl group leads to the increase of the relative rate of substitution in the order: C-Cl > C-Si > C-S > C-Br. In allyliodide, the iodine-carbon bond is so weak that the phenyl radical, before the addition to the π-bond of the allyl group, eliminates the iodine atom from the molecule and the substitution does not occur.

In benzene phenylation (Scheme 38), the radical adduct does not release the hydrogen atom as it should be expected according to the mechanism of addition-fragmentation reaction. The hydrogen atom is abstracted from the adduct only in its disproportionation reaction with the other radical.

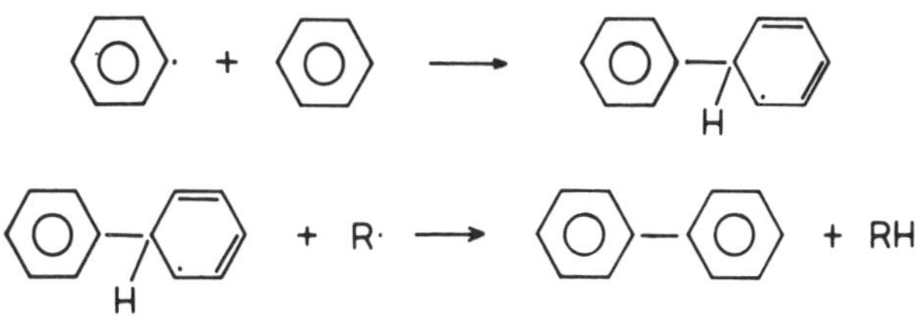

SCHEME 38

A simple radical substitution reaction at a saturated carbon atom (Equation 69) which was suggested almost 50 years ago, has apparently never been observed.[134]

$$CH_3\cdot + CH_4 \rightarrow C_2H_6 + H\cdot \qquad (69)$$

E. Transfer Reactions

The displacement of the side atom from the molecule to the attacking radical occurs after a mutual interaction of the unpaired electron on the radical with the displaced atom (Scheme 39).

$$D\cdot + -CH_2 \longrightarrow [D\cdots HCH] + \longrightarrow HD + -\dot{C}H- \qquad (A)$$

$$-/\!/\longrightarrow \left[\begin{array}{c} H \\ | / \\ D..C \\ \backslash \\ H \end{array} \right]^{\neq} -/\!/\longrightarrow H\cdot + -HCD- \qquad (B)$$

SCHEME 39

Even though the displacement of the hydrogen atom may be explained also by addition-elimination or addition-fragmentation mechanisms (B) taking place on the central atom of molecule, for hydrocarbons it is not probable and contradicts the experimental results. It was, e.g., found that at the reaction of deuterium with hydrocarbons there is only HD and not a molecular hydrogen. Also, all other results indicate that the transfer reaction of hydrogen proceeds according to the pathway A.

The most frequent atoms taking part in the transfer reactions are iodine, bromine, hydrogen, chlorine; less frequent is fluorine. Very reactive is especially hydrogen in aldehydes, hydrogen on α-carbon of alcohols, amines and ethers, hydrogen on carbon in the neighborhood of a double bond (allyl hydrogen), hydrogen of the thiol group, and hydrogen in organometallic compounds. Lower reactivity of fluorine follows from the high strength of the C-F bond. The higher strength of the C-H bond in benzene and some aromatic compounds is also the reason for their lower reactivity in transfer reactions.

It is assumed that the reaction is more facile when three important reaction sites (atom with unpaired electron, transferred atom, and atom of substrate with transferred atom) are colinear or nearly so in the transition state. The bond forming between a radical and transferred atom and decaying bond between atom and molecule moiety remain longer in the transition state when compared with corresponding bonds in reactants and products, the relative prolongation being the same approximately only for thermoneutral reactions.

Provided that there exists a significant difference in the electron affinity of radical and substrate, the reaction proceeds as an electron transfer. The central atom in free radicals for which such a course is valid may be boron, silicon, germanium, tin, lead, phosphorus, chromium, cobalt, vanadium, tungsten, and some other transition metals, the reaction itself being a halogen abstraction from organic ahlides.[135]

1. The Role of the Strength of Broken and Newly Formed Bond

It seems obvious to assume a priori that the facility of the transfer reaction depends on the strength of the broken and newly formed bond between radical and side atom of a molecule. Energy necessary for the breaking of the bond between atom and the rest of the molecule is usually so high that the transfer reaction hardly occurs in the given temperature region unless there is simultaneous formation of a new bond between radical and transferred atom which is capable of compensating for the energy requirements. It may be suggested that the probability of the transfer reaction will increase with the increasing exothermicity of the process. At the reaction of the methyl radical with hydrogen halide (HX) (Equation 70) there is, therefore,

$$\cdot CH_3 + HX \rightarrow CH_4 + X \cdot \tag{70}$$

transferred a hydrogen atom and not a halogen atom, the transfer reaction being fast for HI. For room-temperature reactions of deuterated methyl radical with HI and HBr the rate constants are $4.6 \cdot 10^9$ and $2.8 \cdot 10^9$ dm^3 mol^{-1} sec^{-1} while for the reaction with HCl, the value $4.3 \cdot 10^7$ dm^3 mol^{-1} sec^{-1} is recommended.[136] The activation energies in the reaction (Equation 70) decrease in going from HCl (2.5 kJ/mol) to HBr (1.7 kJ/mol) and to HI (0 kJ/mol). The transition state has the C-H-Cl collinear arrangement, with the C-H and H-Cl distances both extended (~30%) compared with those of CH$_4$ and HCl, respectively. In the reaction of HBr and HI the forming C-H bonds will be more extended.

2. Abstraction of Hydrogen

Provided that we compare activation energy of transfer reaction of several radicals with four selected hydrocarbons and hydrogen[137,138] we see that, indeed, it sinks with the decrease of dissociation energy of the attacked bond (Figure 2). The dependence is valid for individual radicals only, which is a good demonstration of the decisive effect of the energy of newly formed bonds; the stronger a new bond is, the lower the activation energy necessary for the transfer reaction. At the same time, from Figure 2, it is obvious that for the identical pair of radical and transferred atom, but for different substrates, there exists a distinct declination from the dependence of activation energy on the dissociation energy of the broken bond. In our case it concerns the transfer reactions from the hydrogen molecule compared with those from hydrocarbons. From the above reasoning, the dependence between the activation energy

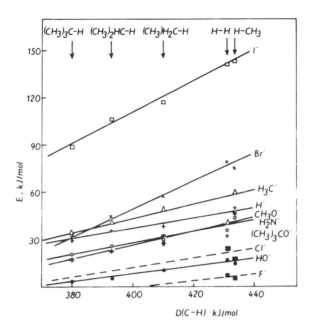

FIGURE 2. Plot of activation energy E of hydrogen transfer on dissociation energy of C-H bond for different radicals. Data are average values from different authors quoted in References 137 to 139.

E and dissociation energy D(R-H) of the broken bond (Equation 71) or reaction heat Q (Equation 72) may

$$E = \alpha \cdot D(R\text{–}H) + \beta \qquad \text{(Evans-Polanyi rule)} \qquad (71)$$

$$E = a + b \cdot Q \qquad \text{(Semenov rule)} \qquad (72)$$

be approximated by the straight line only within the series of compounds of the same type. The constant α, which depends on the kind of attacking radical, is from the interval 0.2 to 0.9. It increases with endothermicity of the transfer reaction. The constant α may simultaneously be taken as a rating of the selectivity of a radical at the abstraction of hydrogen from hydrocarbons; the radicals with higher α are more selective. With respect to a small difference in the values of the frequency factor for the transfer reactions of a given radical with the same type of chemical bond, the rate constants may be correlated with the values of the dissociation energy of the broken bond or with the corresponding C-H stretching frequencies. Such empirical correlations are particularly advantageous for radicals with known rate constants which were verified in several laboratories. This is, e.g., the case of the reaction of hydroxyl radicals[140,141] which play an important role in the chemistry of the atmosphere and in combustion processes. Its rate constants have been determined for a large number of organic compounds and used in numerical modeling. The correlation of log (k/n), where k is the constant in dm^3 mol^{-1} sec^{-1} of the reaction of the hydroxyl radical with organic substrate, n is the number of equivalent reactive C-H bonds in the molecule, and D(C-H) denotes the dissociation energy in kJ/mol of broken bond, is linear. For the temperature 298 K, this dependence may be expressed by Equation 73. The correlation is true for alkanes and their

$$\log (k/n) = 28 - 0.05 \cdot D(C\text{–}H) \qquad (73)$$

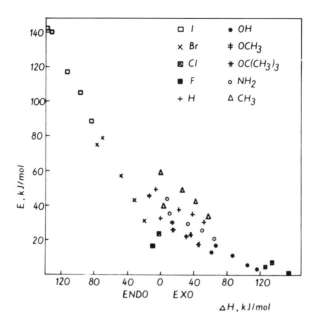

FIGURE 3. Plot of activation energy E of transfer reaction of hydrogen atom on reaction heat ΔH for different radicals and substrates. Data are average values from authors quoted in References 137, 138, 140, 141.

derivatives whose rate constants span five orders of magnitude. The ·OH radicals react with aromatics and alkenes primarily by addition to an unsaturated carbon atom, although abstraction of H occurs to a limited extent, the more important being at higher temperatures. It is noteworthy that aldehydes lie on the same line as the other compounds, despite the much weaker aldehydic C-H bond. This fact indicates that sp^3 or sp^2 hybridization of carbon does not significantly affect the reactivity outside of its influence on the C-H bond energy.

Simple correlations of this kind are useful in signaling unusual behavior or possible experimental error.

Provided that we plot the above data of activation energies of transfer reactions for different radicals against the difference in energy of the newly formed and decaying bond (Figure 3), we see that the scatter of individual points around the central line is considerable, indicating thus that besides the reaction heat also other factors determining the kinetic parameters of the transfer reaction come into play.

The influence of reaction heat on the transfer reaction may be excluded at the intermolecular migrations of the type (Equation 74) where the hydrogen atom is transfered from a molecule

$$R \cdot + RH \rightarrow RH + R \cdot \tag{74}$$

to a radical in the hopping manner and the chemical composition of the system does not change. The identical radicals are systematically regenerated in such a reaction. Even though the reaction is thermoneutral, the activation energy differs rather considerably for different radicals (Table 11). No correlation exists between the strength of the formed or broken bond and the values of activation energy. At the same time, it could be assumed that the reaction will take place more easily, the more weakly the hydrogen atom is bound to the molecular system. The obvious disagreement between this rule and the experiment indicates that the effect of the strength of the exposed bond is masked by a more significant effect of the structure of both the attacking radical and the attacked molecule. Within the obtained results the most obvious tendency is seen in the decrease of the activation energy of the reaction

Table 11
VALUES OF ACTIVATION ENERGY E FOR RELAY LIKE REACTIONS OF HYDROGEN ATOM TRANSFER BY DIFFERENT RADICALS R· AND DISSOCIATION ENERGIES D OF R–H BONDS

R·	E kJ/mol	D (R–H) kJ/mol	Ref.
·CH$_2$C$_6$H$_5$	71	334	137
·CH$_3$	60	435	138
·C$_2$H$_5$	56	410	137
·SiH$_3$	60	376	137
RO$_2$·	57	376	142
·NH$_2$	42	456	138
H·	32	431	138
Cl·	25	427	137
F·	20[a]	564	137
HO·	17[b]	497	143
(CH$_3$)$_3$CO·	11	426	137
t-BuPhO·	4	336	144
Br·	7	362	137
Na·(+ NaCl)	76	408	142

[a] The calculated value.
[b] The estimated value.

when we are going from polyatomic to monoatomic radicals. The lower activation energy for atoms seems to be a consequence of lower requirements on the reorganization of reactants in the activated complex when compared with polyatomic and particularly with carbon centered radicals.

In addition to the number of atoms in the molecule and the radical, the other important factor is the character of the atom carrying the unpaired electron. From alternatives of migration transfer reactions, radicals with an unpaired electron localized on an atom with nonbonding electrons have a lower activation energy. The higher activation energy of the intermolecular migration for benzyl radicals compared with the other alkyl radicals may be conditioned by a higher degree of delocalization of the unpaired electron. The more distinct influence of delocalization of the unpaired electron on the activation energy of the relay-like reaction can be seen from comparison of the activation energies or rate constants of the reaction of peroxy (ROO·) and alkoxy radicals with different X-H bonds. Table 12, constructed from data given in References 145 to 149, demonstrates the effect of the reactant on the rate of hydrogen transfer more generally. The results also show on the different selectivity in the abstraction of hydrogen atoms by different radicals. The lowest difference in the rate constants may be seen in the case of *tert*-butoxy radicals which react very fast. On the contrary, *tert*-butyl peroxy radicals react the slowest, but they are the most selective from the presented examples.

Another fact which ensues from Table 12 is the diffent effect of the structure of the substrate on the rate constant of individual radicals. Methyl and *tert*-butylperoxy radicals, e.g., abstract more easily allyl hydrogen than methylene hydrogen. The opposite is true for *tert*-butoxy radicals. The reasons for this difference consist in the different contribution of reacting partners to the formation of the activated complex. In reactions of methyl and *tert*-butylperoxy radicals these contributions are determined by the dissociation energy of formed and decaying bonds predominantly while polar effects play a more significant role in the case of *tert*-butoxy radicals. At the transfer of hydrogen from other than carbon atoms the

Table 12

RATE CONSTANTS (dm³ mol⁻¹sec⁻¹) OF HYDROGEN ATOM ABSTRACTION FROM MOLECULES EVENTUALLY GROUPS BY METHYL, *TERT*-BUTYLOXY AND *TERT*-BUTYLPEROXY RADICALS[145]

Group or substrate[a]	D(X–H) kJ/mol	CH₃· 338 K 435	(CH₃)₃CO· 313 K 434	(CH₃)₃COO· 303 K 370
–CH₃	410	5	$2.5 \cdot 10^4$	—
>CH₂	393	50	$1.4 \cdot 10^5$	$2.7 \cdot 10^{-4}$
>CH	380	500	$3.8 \cdot 10^5$	$4.8 \cdot 10^{-3}$
>C=CHCH₂R	351	120	$1.4 \cdot 10^4$	$8.4 \cdot 10^{-2}$
C₆H₅CH₃	334	420	$2.4 \cdot 10^5$	$1.2 \cdot 10^{-2}$
C₆H₅CH₂R	328	560	$1.6 \cdot 10^5$	0.10
C₆H₅CH(CH₃)₂	315	1800	8.4	0.16
–COCH₂–	359	850	$8.5 \cdot 10^4$	—
C₆H₅SH	343[b]	$2 \cdot 10^{8c}$	—	$5 \ 10^{3d}$
(C₂H₅)₂OPH	343[b]	—	$5 \cdot 10^{5e}$	—
(C₄H₉)₃SnH	313	$2 \cdot 10^{6c}$	$2 \cdot 10^{8f}$	—

[a]   The site of hydrogen abstraction in molecules is underlined.
[b]   The result from Reference 146.
[c]   Data for primary alkyl radical.
[d]   The result from Reference 147.
[e]   Reference 148.
[f]   Reference 149.

sizes of the reaction center on the hydrogen donor should probably be considered as important as the dissociation energy of the broken bond and the polar factors. This could explain a relatively high rate constant of the hydrogen abstraction from the Sn atom by alkyl and by alkoxy radicals. The larger volume of the hydrogen donor facilitates the formation of the activated complex with reacting radicals. In such circumstances the significant displacement of the surrounding atoms is not necessary. The transfer reaction may similarly be supported by a lower degree of sterical hindrance on the atom with potentially available hydrogen. The effect is the same as in the case of the larger bulkiness of the hydrogen donor. The hydrogen from thiophenol is thus abstracted more easily than from cumene.

3. The Abstraction of Halogens

The preferential course of the transfer reaction with decreasing dissociation energy of the broken C-Cl bond may be illustrated on the Arrhenius parameters of the reaction of methyl radical with chlorfluormethane derivatives.[150] The tendency is not, of course, so distinct as in the case of hydrogen atom transfers. The constant α of the Polanyi equation for the transfer of Cl atom by methyl radical is only 0.1 whereas for the transfer of hydrogen in the same reaction its value is 0.45. This does not follow from the high exothermicity of the chlorine atom transfer, but from the incorporation of the polar effect into the thermochemical dependence.

Comparing rate constants of the transfer reaction of hydrogen and chlorine of chloroform to methyl radical we may find that hydrogen is abstracted about 2 orders faster than chlorine. At the same time, the dissociation energy of C-Cl bond is by 66 kJ/mol lower than that of C-H bond. The abstraction of hydrogen by methyl radicals is, however, by 18 kJ/mol more exothermic than the abstraction of chlorine, and the observed difference in these two transfers could be interpreted also in this way. The *tert*-butyl radical has a considerably lower ratio of the rate constants of hydrogen and chlorine transfer but still higher than 1, even though the reaction with the chlorine atom is more exothermic by 16 kJ/mol than the similar reaction with hydrogen.[151] The relatively slower transfer of chlorine when compared with hydrogen

Table 13
KINETIC PARAMETERS OF TRANSFER REACTION[135,150,152,153] Y· + RX → YX + R· OF RADICALS Y· WITH HALOGEN DERIVATIVES OF METHANE RX AND DISSOCIATION ENERGIES D OF BROKEN AND NEWLY FORMED BOND

Y·	D(R–X) RX	D(Y–X) kJ/mol		A E	k(450°C) dm³ mol⁻¹ sec⁻¹	
Na·	H_3C–F	452	481	82	$4 \cdot 10^{11}$	28
	H_3CCl	351	412	44	$5 \cdot 10^{11}$	$2 \cdot 10^6$
	H_3C–Br	239	370	18	$3 \cdot 10^{11}$	$1 \cdot 10^9$
$(C_2H_5)_3Si\cdot$	H_3C–Cl	351	477	17	$1 \cdot 10^8$	$8 \cdot 10^5$
	ClH_2C–Cl	307	477	9	$2.5 \cdot 10^9$	$2 \cdot 10^8$
	Cl_2HC–Cl	280	477	5	$1.6 \cdot 10^9$	$4 \cdot 10^8$
	Cl_3C–Cl	293	477	3	$1.6 \cdot 10^{10}$	$6 \cdot 10^9$
	$C_6H_5CH_2$–Br	211	393	5	$2.0 \cdot 10^{10}$	$4 \cdot 10^9$
	C_2H_5–I	222	—	4	$2.5 \cdot 10^{10}$	$8 \cdot 10^9$
$\cdot CH_3$	F_3C–Cl	356	351	49	$6 \cdot 10^8$	$5 \cdot 10^2$
	Cl_3C–Cl	293	351	42	$6 \cdot 10^8$	$4 \cdot 10^3$
	F_3C–Br	297	293	39	$4 \cdot 10^8$	$6 \cdot 10^3$
	F_3C–I	222	234	18	$1 \cdot 10^8$	$8 \cdot 10^5$
$\cdot CF_3$	Cl_3C–Cl	293	356	47	$6 \cdot 10^9$	$8 \cdot 10^4$
$\cdot CF(CF_3)_2$	Cl_3C–Cl	293	356	59	$1 \cdot 10^{10}$	$6 \cdot 10^3$
$\cdot CH(CH_3)_2$	Cl_3C–Cl	293	337	43	$1 \cdot 10^8$	$5 \cdot 10^3$

may be due to the repulsion between the unpaired electron of the methyl radicals and the closed electron shell of the chlorine atom.

The transfer of halogen from alkylhalogenides by nucleophile radicals, however, clearly predominates the hydrogen transfer (Table 13). The sodium atom, e.g., abstracts the strongly bound fluorine from alkyl fluorides faster than the hydrogen atom. The faster abstraction of halogen atoms when compared to alkyl radicals was also observed for triethylsilyl radicals and other nucleophile radicals, as, e.g., diethoxyphosphonyl or trialkyltin radicals.[148]

When we compare the rate constants for halogen atom abstractions by the different radicals, we see that the more electronegative the carrier of the radical center, the more slowly the abstraction of halogen takes place.

For the different halogens, the influence of the strength of the broken bond or the reaction heat on the activation energy of the transfer reaction may be contemplated, too. An increase in the preexponential factors for the reaction of $(C_2H_5)_3Si\cdot$ radical is ascribed to the polar effect.[135] The greater the polar contribution, the smaller will be the restriction on the orientation of the $(C_2H_5)_3Si\cdot$ radical with respect to the carbon-halogen bond being broken. In the limiting case of complete electron transfer, the resulting ion pair would not be subject to any restriction in this rotational motion. The gain of two rotational degrees of freedom in the transition state would enhance the preexponential factor by two orders.

4. The Influence of Radical and Substrate Polarity

The important structural factor which affects the transfer reaction is the polarity of both the radical and substrate. The effect of substrate polarity may be demonstrated on approximately thermoneutral transfer reaction of methyl radicals with different polar donors of hydrogen (Table 14). The reaction where the attacked hydrogen atom has the lowest partial charge has the highest activation energy. The increase of the partial charge on hydrogen leads to the decrease of activation energy of the transfer reaction. This may be explained by an electrostatic attraction between the negative partial charge on the carbon (-0.045) of the methyl radical and transferred hydrogen atom. The decrease of the activation energy causes, moreover, the slight endothermicity of the transfer of methyl radicals to trifluoro-

Table 14
ACTIVATION ENERGY OF APPROXIMATELY THERMONEUTRAL TRANSFER REACTIONS OF METHYL RADICALS WITH DIFFERENT HYDROGEN DONORS[136,138] (AH) AND DATA OF CALCULATED VALUES OF PARTIAL POSITIVE CHARGES (PPC) ON HYDROGEN ATOMS OF AH COMPOUNDS[156]

AH	D (A–H)[b] kJ/mol	PPC	E kJ/mol
H_3C-H	437	0.015	61
F_3C-H[a]	447	0.32	57
CH_3O-H	440	0.10	32
$Cl-H$	427	0.16	3

[a] Reference 157.
[b] References 154 and 155.

methane. Besides the attraction, however, there exists also the repulsion between the negative partial charge of fluorine atoms and the central atom of methyl radical.

The effect of the partial charge may be more evidenced by a comparison of the transfer reaction of methyl and trifluoromethyl radicals. The latter radicals already have a positive charge on carbon atoms. Since the bond dissociation energy of the C-H bond in CF_3H is greater than that of the C-H bond in CH_4 by about 10 kJ/mol,[154] it might reasonably be expected that the activation energy of the $\cdot CF_3$ reaction would be lower than that of $\cdot CH_3$ reaction. The actual pattern of behavior is much more complicated (Table 15). $\cdot CF_3$ radicals are more reactive than $\cdot CH_3$ radicals for attack on hydrocarbons and hydrides. The hydrocarbons with lower dissociation energy of C-H bond have lower activation energy but the ratio of the reactivity $\cdot CF_3/\cdot CH_3$ remains constant. For halogenmethanes the activation energy of the reaction of both radicals is already the same, approximately. When compared to methane its value decreases more for the methyl than for the trifluoromethyl radical. The decrease in the activation energy is usually substantiated by the lower dissociation energy of C-H bond in halogen methanes. The lower decrease of activation energy in the case of $\cdot CF_3$ radicals is due to the opposite effect of higher positive partial charges in halogen methanes when compared with hydrocarbons. The reduction of bond dissociation energy is counterbalanced to some extent by increased repulsive polar forces in the transition state. The higher reactivity of methyl radicals compared to trifluoromethyl radicals at the transfer to hydrogen halides and hydrogen sulfide can be rationalized in the same qualitative way of polar effects.

It was already pointed out that the changes of activation energy are also due to different bulkiness of the atom carrying hydrogen and a complexity of the attacked molecule. The fewer the atoms in the molecule, the lower the activation energy corresponds to the transfer reaction of radicals.

Another view on the role of polarity of hydrogen donors may be obtained when comparing the course of the transfer reaction of chlorine and bromine atoms with butane and its 1-substituted derivatives (Table 16). In close proximity of electronegative substituents the expected selectivity of chlorine and bromine atoms is disrupted and the reactivity of hydrogen atoms will decrease especially in the α-position. For reaction sites where the electrostatic repulsion of the electronegative substituents becomes important, the relative rate of hydrogen

Table 15

**ACTIVATION ENERGY E OF HYDROGEN
TRANSFER BY METHYL AND
TRIFLUOROMETHYL RADICALS FROM
DIFFERENT SUBSTRATES (A-H)[136,137,154,157-159] AND
DATA ON SOME PROPERTIES OF RADICALS
AND SUBSTRATES**

Property of radicals (R·) and substrate	CH_3·	CF_3·
D[a] of the bond radical–H, kJ/mol	437	447
Electronegativity (EN)	3.61	5.18
Partial charge (PC) on the carbon of R·	−0.045	0.34

A-H	D (A-H) kJ/mol	EN[b]	PC/H[b]	E kJ/mol	
CH_4	437	3.60	0.015	61	48
H_2	431	3.55	0	47	41
c–C_6H_{12}	398	3.63	0.020	41	26
NH_3	435	3.76	0.049	41	34
SiH_4	375	3.40	−0.040	29	21
$FH_2C–H$	423	3.96	0.105	49	47
$F_2HC–H$	426	4.36	0.207	44	47
$ClH_2C–H$	422	3.84	0.074	49	44
$Cl_2HC–H$	410	4.10	0.141	38	32
$Cl_3C–H$	401	4.38	0.212	28	28
H_2S	343	3.73	0.046	10	18
HCl	427	4.18	0.162	3	22
HBr	363	4.01	0.117	2	12

[a] Dissociation energy.
[b] Calculated data of electronegativity and of partial charge (PC/H) on
hydrogen atoms of A-H according to Reference 156.

Table 16

**RELATIVE RATE OF HYDROGEN
ABSTRACTION BY CHLORINE[160] AND
BROMINE[161] ATOMS FROM DIFFERENT
POSITIONS IN BUTANE AND ITS
SUBSTITUTED DERIVATIVES**

Radical	Substituent X	X – CH_2	–CH_2	–CH_2	–CH_3
Cl·	H	1	4	4	1
	F	0.9	1.7	3.7	1
	Cl	0.7	2.2	4.2	1
	CF_3	0.04	1.2	4.3	1
Br·	H	1	80	80	1
	CF_3	1	7	90	1
	F	9	7	90	1
	CH_3OCO	19	29	73	1

abstraction by electronegative chlorine and bromine atoms decreases.[162] The presence of the chlorine atom reduces the reactivity of the geminal hydrogens in the CH_2Cl group by about a factor of 10 at 298 K. The presence of two chlorine atoms, as in the $CHCl_2$ group, decreases the reactivity of the attacking chlorine atom even further by a factor of about 33 relative to the >CH group in alkanes. The chlorine substituent in the vicinal group is also important. The kinetic isotope effect for H/D abstraction sinks with increasing number of chlorine substitutents in the geminal group in parallel with the trend established for C1-substitution in the adjacent group.

The problem of the polar factor may be treated from the viewpoint of attacking radicals, too. As an example, we can consider the reaction of chlorine and hydrogen atoms with methane. Since the dissociation energies of bonds H-H and H-Cl are comparable, the reaction heat of both reactions will be approximately the same (slightly endothermic reactions). Despite this fact, the activation energies are considerably different. E for the reaction of hydrogen atoms with methane is 50 kJ/mol, while that for chlorine atoms is only 16 kJ/mol. Such a large difference may be explained by the polar structure of the activated complex in which electrostatic attraction of opposite charges on atoms in the activated complex lowers the energy barrier for occurence of the reaction. The polar structure of the activated complex will be formed as a consequence of high electron affinity of chlorine atoms.

The existence of the polar effect demonstrates also the fact that the nature of the abstracted atom depends on the type of attacking radical (Scheme 40). In polyatomic molecules

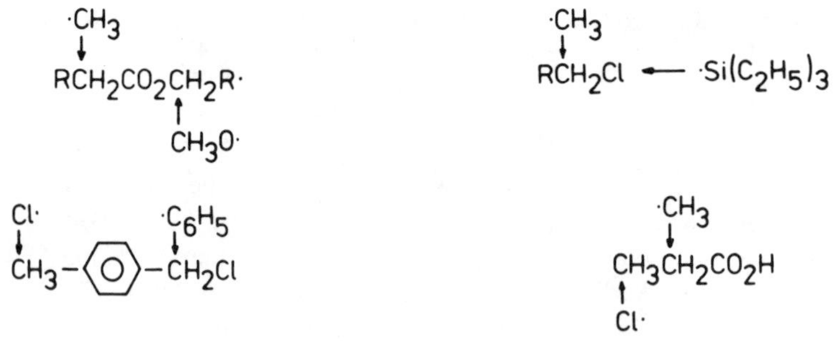

SCHEME 40

there is not generally available only one reactive site but several of them, whose reactivity is ultimately determined also by the reacting radical.

The relative importance of the two kinds of polar structure contributing to the transition state (Equation 75) is

$$R^+X. \; Y^- \leftrightarrow R.X..Y \leftrightarrow {}^-R.XY^+ \tag{75}$$

influenced by the polar character of the substrate and radical. In the polar transition state, the radical may be either the donor or acceptor of the electron. The substituents of both kinds on the substrate or on the radical therefore increase polarization of the transition state.[163]

The reactivity of carbon radicals $(CH_3)_2\dot{C}X$, where $X = OC_6H_5$, $OCOCH_3$, SC_6H_5, C_6H_5 and CN, in the halogen abstraction reaction decreases with decreasing electron-donating power of the radical. The rate constant for the halogen abstraction from CBr_4 is more than 100 times as large as that from CCl_4. This difference does not correspond to the difference of dissociation energies of cleaved atoms but rather agrees with eight times higher electron affinity of CBr_4 when compared with CCl_4. These results show on the participation of the electron-transfer process in the halogen abstraction reactions.[164]

5. The Solvent Effect

Although the effect of solvents in elementary reactions of free radicals is mostly small, in some cases it may be well distinguished. One of such positive examples[165] is the reaction of free and solvated chlorine atoms with 2,3-dimethylbutane. The molecule of this hydrocarbon contains 12 primary and 2 secondary hydrogens. During the chlorination in carbon tetrachloride, the nonsolvated and very reactive chlorine atom attacks up to 60% of methyl groups. Provided that the chlorine atom is solvated by benzene molecules, the selectivity of the reaction increases considerably, and tertiary hydrogens are abstracted almost exclusively. The hydrogen of methyl groups is abstracted only from 10%. With regard to the ratio of primary /p/ and tertiary /t/ hydrogens in the molecule of 2,3-dimethylbutane, the reactivity of the chlorine atom in transfer reaction changes from the original ratio t:p = 4:1 (40/2:60/12) to 54:1 (90/2:10/12). The increase in selectivity due to the solvation is connected with the simultaneous decrease of reactivity of solvated chlorine atom, where the spin of the unpaired electron is partially delocalized on the benzene molecule.

At the transfer reaction of halogens of arylalkyhalogenides with nucleophile and relatively stable pyridinyl radicals (Equation 75a) the unusually great influence of solvent

Going back to the question of frequency factor of a reaction in liquid phase, we may present experiments where the transfer reaction in the liquid phase proceeds with considerably lower (by 4 to 5 orders) frequency factor than are the values generally considered as "normal"

$$(75a)$$

polarity on the course of the reaction may be seen.[166] By the increase of polarity of the reaction medium, the rate constant of the transfer reaction also simultaneously increases. For example, the rate constant of reaction 75 in methyltetrahydrofurane is 2 dm^3 mol^{-1} sec^{-1}, whereas in acetonitrile it is 2.10^4 dm^3 mol^{-1} sec^{-1} at 278 K. The increase of the rate constant by 4 orders brought about by the change of only solvent may be explained by the strongly polarized, almost ionic structure of the activated complex.

6. The Effect of the Aggregation State

The numerous data on the rate constants of transfer reactions in both the gaseous and liquid phase are available by now.[167-170] It follows from the comparison of respective Arrhenius parameters that the data on frequency factors scatter more for the liquid than for the gaseous phase.[171] The variance of frequency factors for different transfer reactions in the gaseous phase is not usually higher than 2 orders, the mean value for polyatomic radicals being $6 \cdot 10^8$ dm^3 mol^{-1} sec^{-1}. On the other hand, in the liquid phase the difference is considerably higher and even reaches 5 to 6 orders. One possible explanation for such a discrepancy may consist in larger experimental error of the measurements in the liquid phase where the temperature interval of the determined rate constants is usually more narrow than that in the gaseous phase. Even though such an interpretation seems to be partially true, it is not satisfactory. A deeper insight could be attained from the analysis of kinetic data for the identical reactions performed in both phases simultaneously. Most of the measurements of transfer reactions were unfortunately made only either in liquid or gaseous phase, separately.

On the other hand, it ensues from the reactions of methyl radicals in the gaseous and liquid phase that the selectivity of the C-H bond attack is approximately the same. This means that in a nonpolar medium the course of transfer reactions will be comparable in both phases.

The more electrophile trichloromethyl radicals have, e.g., almost the same Arrhenius parameters for abstraction of hydrogen from cyclohexane in gaseous /g/ and liquid /l/ phase:

Phase	log A (dm^3 mol^{-1} sec^{-1}	E (kJ/mol)
g	8.8 ± 0.2	43 ± 4
l	9.0 ± 0.3	46 ± 4

This was confirmed by several authors.[172] However, rate constants data for chlorine abstraction by alkyl radicals from carbon tetrachloride in the solution and gas phase are substantially different, those in solution being about 2 orders of magnitude higher.[173] The best explanation of the observed fact consists in the possibility of solvation of the polar transition state of the halogen transfer reaction. The above result is not so rare and shows on higher influence of the medium on transfer reactions of radicals in the liquid phase when compared with those in the gas phase. The molecules of the liquid medium may efficiently support the polarization of the transition state and consequently the course of the reaction.

Going back to the question of frequency factor of a reaction in liquid phase, we may present experiments where the transfer reaction in the liquid phase proceeds with considerably lower (by 4 to 5 orders) frequency factor than are the values generally considered as "normal" for radical H atom abstraction. Low values were, e.g., found at transfer reactions of bis(trifluoromethyl) aminoxyl radicals $(CF_3)_2NO\cdot$. This was attributed to the intermediate formation of a hydrogen bonded free radical-molecule complex which leads to a low activation energy and because of a "tight" transition state to a low frequency factor.[174] A highly restrictive geometry demands a very specific orientation of the $(CF_3)_2NO\cdot$ relative to the C-H bond.

The transfer of hydrogen atoms to radicals was observed also in glassy and crystalline matrices of organic molecules. Methyl radicals abstract, e.g., hydrogen from acetonitrile (Equation 76) even at the temperature 77 K. It is of interest

$$CH_3\cdot + CH_3CN \rightarrow CH_4 + \cdot CH_2CN \tag{76}$$

that the extrapolation of the temperature dependence from an interval of the higher temperature excludes the course of such a reaction in the solid state. The fact that it still proceeds is explained by the tunnel effect.[175] Although the hydrogen atom abstraction proceeds readily under such conditions, deuterium abstraction was not observed. It was established that the deuterium isotope effect expressed as the ratio of the rate constants of transfer reaction with hydrogen and deuterium in the methyl group of acetonitrile k_H/k_D is at least 28,000. This is almost 20 times greater than the maximum isotope effect in the absence of tunneling which is obtained from Equation 77, where $_\Delta E = 4.5$ kJ/mol is the difference in

$$k_H/k_D = \sqrt{2} \cdot \exp(\Delta E/RT) \tag{77}$$

zero-point energy between the C-H and C-D bonds being broken.

The tunnel effect may be often encountered in the study of transfer reactions in the low temperature region. Hydrogen atoms abstract the hydrogen from hydrocarbons by such a mechanism even at 10 K. This was also the main reason for the failure to observe the ESR spectrum of hydrogen atoms during radiolysis of hydrocarbons.[176,177] It is known now that the stabilization of hydrogen atoms in frozen methane requires the temperature of liquid helium, i.e., 4.2 K.

An outstanding example of hydrogen tunneling is the decay of free radicals in organic glasses at 77 K. H-atom transfer rates in the solid matrix are controlled mainly by Van der Waals radii and should not therefore show much variations in different organic matrices.[178]

Kinetics of the reactions in the solid phase depends on the presence of fluctuations in the structure of the glassy or crystalline phase. In the condensed phase each particle interacts with its surroundings which varies with time. Kinetics laws depend on the rate of structural changes of the reacting systems and on the lifetime of the radicals. In the liquid state, the lifetime of the radical is much longer than the time necessary for structural changes of the surroundings. The different arrangements of the surroundings averages the reactivity of the radical and the liquid medium appears to be a homogeneous phase. A different situation arises in the solid state where changes in the physical structure of the medium are slower than the lifetime of the radical. This results in the structural nonequivalency of microregions where the chemical reaction is taking place and deviates the kinetics from classical laws.[179]

The rate constants of transfer reactions in the solid phase are usually lower. The abstraction of hydrogen by methyl radicals in solid polymers has about 2 to 4 orders lower rate constant than the corresponding reaction with the model low molecular compound in the liquid phase.[180] The reduction of the rate constant is due to the lower mobility of polymer chains which retards the formation of the optimum structure of the activated complex of the transfer reaction. The retardation of the reaction brought about by the lower mobility of the polymer reactant comes into consideration only below the melting temperature of the crystalline structure or in the temperature region of the glassy state of a polymer. Provided that the polymer is in the highly elastic state, the difference in the reactivity should be small. In some cases this was also verified experimentally.

Comparing the transfer reaction of methyl and *tert*-butyl radicals with poly(4-*tert*-butyl styrene) it was found that the rate constant for *tert*-butyl radicals is by 2 orders lower approximately, than that for methyl radicals.[181] The difference between rate constants of *tert*-butyl radical with styrene polymer and *tert*-butyl radical with small model molecules is greater than in the case of methyl radicals reactions. We may assume that under extreme conditions of hindered mobility, the radicals of similar bulkiness but differing in reactivity react with approximately the same rate constant. The question of such leveling of reactivity is, however, more the subject of further research than of the current state of knowledge.

Bimolecular displacement of radicals on the central atom of a molecule was investigated on about 15 kinds of atoms of different molecules. On the other hand, the transfer reactions of side atoms of molecules involve a relatively small group of compounds and seem to be better understood at present. The practical aspect of such reactions stems from the fact, that the degree of substitution in the sequence of other elementary reactions is the decisive point of the chemical transformations of many compounds. Their transfer reactions may be of importance even at that time when the composition of the reaction mixture does not change. This is, e.g., the case of relay-like transfer reactions which mediate the chemical diffusion and decay of radical centers in systems of lowered mobility.

VI. REACTIONS OF ELECTRON TRANSFER

Free radicals interacting with the reaction medium may either accept the electron and reduce thus to anions (Equation 78)

$$R\cdot + e^- \rightarrow R^- \tag{78}$$

or donate the electron and oxidize to cations (Equation 79).

$$R\cdot - e^- \rightarrow R^+ \tag{79}$$

At the reaction of radicals, the donor or acceptor of electron may be molecules, ions (very frequently ions of metals), sometimes another free radical or solvated electron. The tendency of a radical to donate or accept electrons is determined by electron-donor or electron-acceptor properties of reacting co-partners. The qualitative picture of electrophility of reacting compounds may be obtained from the data on the electronegativity of its elements.[156] Electron-acceptor properties of a free radical increase by the presence of electronegative atoms in the radical. The atom of an alkaline metal such as, e.g., sodium, which has electronegativity 0.7, easily loses its electron and is capable of reacting vigorously even with such stable but electronegative molecules as H_2O (EN = 4.0), which decompose to an hydroxy anion and a hydrogen atom. Also, radicals with higher electronegativity than sodium have reduction properties. It is known that hydrogen atoms (EN = 3.55) are a strong reducing agent. This is, however, true only for such compounds where more electronegative elements prevail and vice versa. An electron is accepted by halogens, by radicals with unpaired electron on oxygen, sulfur, nitrogen, etc. Such a criterion is, of course, valid only for high difference in values of electronegativity of reactants. At smaller differences the predictability of the facility of the direction of the transfer reaction of an electron is more complicated. The resulting course may be influenced by the overall structure of reacting species.

From this viewpoint, the reduction of radicals by free electrons is unequivocal, since solvated electrons react with electrophile radicals very efficiently and almost at each mutual collision.[182] An anion radical of oxygen is, e.g., formed by the reaction of a hydrated electron with triplet oxygen molecules (Equation 80). The reaction has the same rate constant as the

$$e_{aq}^- + O_2 \xrightarrow{\quad 2 \cdot 10^{10} \text{ dm}^3 \text{ mol}^{-1} \text{ sec}^{-1}, \ 20°C \quad} O_2^{\ominus \cdot} \tag{80}$$

addition of hydrogen atoms to oxygen molecules.

Molecules of oxygen in its triplet ground state are capable of abstracting electrons from many other systems but the rate constant is usually lower. The anion radical of oxygen thus formed is successively reduced or oxidized. In strongly acidic medium (pH < 4.8) it converts quantitatively to $HO_2 \cdot$ radicals, which are considerably weaker donors of electrons.[183,184]

It is worth noting that superoxide anion radicals are metabolites of aerobic microorganisms, plants, and animals, where they are formed by one electron enzymatic reduction of oxygen.[185] The enzyme which makes possible such reduction is, e.g., xanthinoxidase.[186] Substrates for this enzyme, xanthin or acetaldehyde, are oxidized to ureic acid and acetic acid, while the enzyme, accepting two electrons, is reduced. In the presence of molecular oxygen the reduced enzyme can lose electrons by an univalent reaction step leading to $^-O_2^\cdot$ formation (Equation 81) or by a divalent reaction step

$$\text{Enzyme} - H_2 + 2 O_2 \rightarrow \text{Enzyme} + 2 H^+ + 2 \ ^-O_2^\cdot \tag{81}$$

(Equation 82) leading to direct formation of hydrogen peroxide.

$$\text{Enzyme} - H_2 + O_2 \rightarrow \text{Enzyme} + H_2O_2 \tag{82}$$

The predominance of one reaction over the other is controlled by the pH of the medium, by concentration of oxygen, and by the turnover rate of the enzyme.

The transfer of electron between anion and radical is fast when the change of free enthalpy is negative or approximately equal to zero. This is, e.g., the case of reduction of phenoxy radicals by phenoxy anions (Equations 83 and 84).

$$H_5C_6O \cdot + \ ^-OC_6H_5 \leftrightarrow H_5C_6O^- + \cdot OC_6H_5 \tag{83}$$

$$\cdot Q_a + \ ^-Q_b \underset{k_{-1}}{\overset{k_1}{\longleftrightarrow}} \ ^-Q_a + \cdot Q_b \tag{84}$$

At the same time, the rate of electron transfer with participation of neutral phenols is too slow to be observed under pulse radiolysis conditions before radicals decay.

Like other elementary reactions, the reactions of electron transfer are at equilibrium. The equilibrium constant can be determined either from the kinetic data ($K = k_1/k_{-1}$) or from the equilibrium concentrations (Equation 85).

$$K = [Qa^-][Qb \cdot][Qa \cdot]^{-1}[Q_6^-]^{-1} \tag{85}$$

The rate constants of electron transfer depend on the structure of the reactants and in examples given in Table 17 vary in the magnitude of 5 orders. The reactivity of phenoxy radicals decreases with the increasing extent of intramolecular transfer of electrons, i.e., with the increasing number of ionizable hydroxyl groups in the molecule. The reduction of phenoxy radicals may take place not only with anions but also with electron donors.[189]

Table 17
RATE CONSTANTS OF ELECTRON
TRANSFER FOR REACTION OF
DIFFERENT PHENOXYL
RADICALS WITH PHENOLATE
ANIONS[187,188]
$$Q_a + -Q_b \Leftrightarrow -Q_a + \cdot Q_b$$
IN STRONGLY ALKALINE
MEDIUM (pH = 13.5) AT
TEMPERATURE 25°C

Position of OH group on the benzene ring		k_1 $\mathrm{mol^{-1}}$	k_{-1} $\mathrm{dm^3\ sec^{-1}}$
Q_a	Q_b		
1	1	$2 \cdot 10^9$	—
1	1	$2 \cdot 10^8$	$2 \cdot 10^8$
1,3	1,2	$8 \cdot 10^8$	$<10^5$
1,2	1,2,4	$9 \cdot 10^6$	$2 \cdot 10^4$
1,2	1,4	$2 \cdot 10^6$	$9 \cdot 10^5$
1,2,3	1,4	$\sim 10^5$	$\sim 10^5$
1,2,3	1,2,4	$\sim 10^5$	—

Galvinoxyl reacts, e.g., with dimethyl aniline according to Equation 86. Reactive hydroxyl radicals[190]

$$(86)$$

and alkylperoxy radicals[191] abstract the electron from disulfides (Equations 87 and 88). On the other hand, disulfides can also

$$HO\cdot + RSSR \rightarrow \overset{\cdot\oplus}{RSSR} + HO^- \qquad (87)$$

$$RO_2\cdot + RSSR \rightarrow \overset{\cdot\oplus}{RSSR} + RO_2^- \qquad (88)$$

accept electrons as occurs at radiolysis of aqueous solutions (Equation 89). Disulfide anion radical RSSR may thus serve as an

$$e^- + RSSR \rightarrow \overset{\ominus}{R\dot{S}SR} \leftrightarrow RS^- + RS\cdot \qquad (89)$$

example of both the powerful oxidant and reductant. The strong oxidizing effect is ascribed to the thiol radical.[192] The transfer of electrons to disulfides and other sulfur compounds is of the highest interest since sulfur structures play an important role in the chemistry of redox proteins. In metal-containing enzymes, thiol groups frequently act as binding ligands, while in other systems they have the function of separate entities in the electron transport pathway.

The reduction of alkyl radicals may be performed by alkaline metals. Carbon-centered radicals more readily donate the electrons and are thus oxidized. The more electrophilic radicals than alkyl radicals prefer reduction.

For interpretation of results on transfer reactions of electrons the data can be of use taking into account the fact that electron transfer from or to a radical may be conducted electrochemically. The electrode is either the donor (cathodic reduction) or acceptor of electrons (anodic oxidation). An electron from the electrode is transferred to a radical at that time when the energetical level of electrons on the cathode is comparable with that of the semioccupied (SOMO) orbital of the radical. The reverse direction of transfer from radical to electrode requires the same or somewhat higher energy of SOMO of the radical than corresponds to the energy of virtual levels of the anode metal. In dependence on energetic levels of SOMO, some radicals are more easily oxidized, others reduced. As may be seen on the example of verdazyl radicals (Scheme 41),

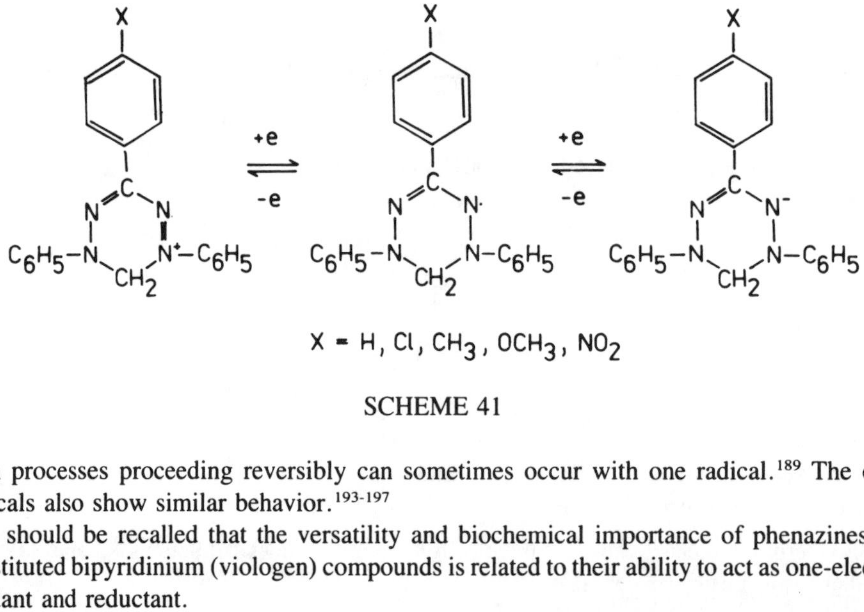

$$X = H, Cl, CH_3, OCH_3, NO_2$$

SCHEME 41

both processes proceeding reversibly can sometimes occur with one radical.[189] The other radicals also show similar behavior.[193-197]

It should be recalled that the versatility and biochemical importance of phenazines and substituted bipyridinium (viologen) compounds is related to their ability to act as one-electron oxidant and reductant.

In many other cases, most of oxidation or reduction polarography waves of radicals are irreversible since the products of the process are unstable in the surrounding medium. The competitive reaction of ions formed by oxidation or reduction of radicals under conditions of polarography measurements takes place faster than the reversible reaction on the mercury drop. Comparison of the results of electrochemical investigation of stable radicals made it possible to estimate their electron-donor properties. According to the value of half-wave oxidation potential, the stable radicals can be arranged into the sequence of increasing electron-donor properties as follows:

Chlorinated triphenylmethyl radicals < 2,4,6-alkylphenoxy radicals < diphenylpicrylhydrazyl < di*tert*-butyl nitroxyl < NO < triphenylmethyl < triphenylverdazyl. According to the above scale, verdazyl radicals are the best donors, whereas chlorinated triphenylmethyl radicals are the best acceptors of electrons.[198] The latter are almost inert with respect to the electrophilic compounds such as halogens and concentrated nitric acid. Verdazyl but also nonchlorinated triphenylmethyl radicals interacting with halogens transmit the electron and transform themselves to verdazyl or triphenylmethyl cations. Verdazyl radicals (V.) are oxidized (Equation 90) by many other compounds such as acids or by

$$2 \text{ V·} + RCOOH \rightarrow VH + V^+ + {}^-OOCR \tag{90}$$

other acceptor molecules.

Structurally different radicals transfer their electrons mutually only when the difference in electron-donor and electron-acceptor properties is sufficiently large. This is, e.g., the case of the reaction of phenoxy radicals with K, or halogen—alkyl or other radicals with verdazyl radicals. The more nucleophile alkyl radicals, such as methyl, cyclohexyl, or benzyl, combine with verdazyl to tetrazenes.

The effect of substituents on oxidizability of alkyl radicals may be demonstrated at the electrochemical oxidation of salts of carboxylic acids. During anodic oxidation of different derivatives of phenyl acetic acid (Scheme 42)

$$XC_6H_4CH_2CO_2^- \xrightarrow{-e, -CO_2} XC_6H_4CH_2· \xrightarrow{-e} XC_6H_4CH^+$$
$$\downarrow \qquad\qquad + \downarrow A^-$$
$$[XC_6H_4CH_2-]_2 \qquad \text{products}$$

SCHEME 42

the radicals having electron-donor substituents lose electrons the most easily.[199] This is also the reason why electrolysis of 4-methoxyphenylacetic acid yields preferably the products of reaction of carbocations with anions of the reaction medium, while that of perfluorophenylacetic acid gives dimerized product, 1,2-diper-fluorophenylethylene.

Oxidizability of alkyl radicals depends on their structure also at the purely chemical mode of generation. The rate constant of electron transfer from alkyl radicals to tetrachloromethane (Equation 91) increases with electron donor properties of

$$R_2\dot{C}OH + CCl_4 \rightarrow R_2CO + H^+ + Cl^- + {}^·CCl_3 \tag{91}$$

the substituents R.[200] Provided that R is hydrogen, the rate constant is at least by two orders lower than it is in the case of R = CH_3 where its value is equalled to 1.10^8 dm^3 mol^{-1} sec^{-1} (293 K).

A reaction of electron transfer is, of course, governed also by the reactivity of the reaction co-partner. The proper example can be the reaction of the substituted 4-nitrobenzenes (Scheme 43) with α-hydroxyalkyl radicals.

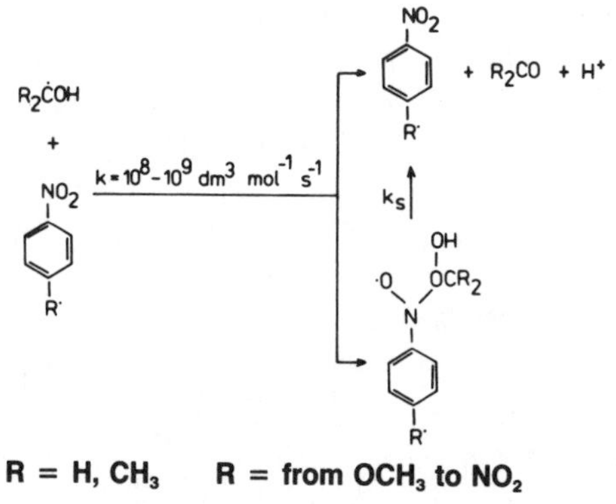

R = H, CH₃   R = from OCH₃ to NO₂

SCHEME 43

At pH = 7 the α-hydroxyalkyl radicals are oxidized quantitatively and give the corresponding ketone or aldehyde and H^+; nitrobenzenes are reduced to radical anions. The mechanism of this redox reaction depends strongly on the substituents of the α-hydroxyalkyl radical (electron donor) and nitrobenzene (electron acceptor).[201] α-Hydroxymethyl radical, $\cdot CH_2OH$ reacts by primary adduct formation and its ks is $<10^2$ sec^{-1}, whereas with $(CH_3)_2COH$ ks is $> 10^6$ sec^{-1}. However, the partial effect of substitution at Cα can be compensated by varying the substituent R· at the phenyl group of nitrobenzene. With $(CH_3)_2COH$, in order to receive ks $< 10^6$ sec^{-1}, R· has to be strongly donating as, e.g., NH_2 group; with $\cdot CH_2OH$, in order to have ks $> 10^2$ sec^{-1}, R· has to be strongly electron removing as, e.g., N_2^+ $(BF_4)^-$ group. With the α-hydroxyethyl radical, $CH_3\dot{C}HOH$, both the addition and the direct electron transfer take place. The ratio of electron transfer increases with the electron withdrawing power of substituent R·. Electron transfer via addition elimination constitutes the case of organic inner-sphere electron transfer (delayed path) as compared to direct (outer-sphere) electron transfer. The outer-sphere mechanism characterizes a reaction in which electron transfer takes place from a reductant to an oxidant without formation of a bond between them, or if any bond forms, it plays no role in the act of electron transfer. Besides this mechanism, two other classes of electron transfer are possible and involve the inner-sphere with and without ligand exchange. An electron transfer mechanism is said to be of inner-sphere if it can be experimentally proved that a bridge through which the electron transfer occurs is established between the oxidant and the reductant that are forming. Therefore, several reactions described as outer-sphere examples could, in fact, possess an inner-sphere character.[202] On the other hand, it should be recalled that the transfer of electrons between reactants can be realized without their contact up to a distance of several nanometers.[203-205] The maximum rate constant of electron transfer is, at the close contact of reactants, between 10^{13} to 10^{14} sec^{-1}.[206] At many reactions the bimolecular rate constant of electron transfer is, therefore, controlled by diffusion of reactants. The transfer of electrons to a radical or from a radical is especially efficient at the inteaction of radicals with metal ions. Although there occurs the transfer of electrons between partners with different electronic configurations, their orbital energies are very close in total energy.[207] These facts indicate the possibility of a transfer from one energy surface to the other, i.e., the transfer of an electron from the metal to the ligand or vice versa. From transition elements, Ag(II), which has a high electrode potential, can mediate the electron transfer smoothly. On the other hand, the complexes of Cu(II) are very effective at selective oxidations of carbon-centered radicals, while the complexes of Fe(III) oxidize strongly nucleophile radicals such as allyl, benzyl, cyclohexadienyl, α-hydroxyalkyl, acyl, etc., but do not react with methyl and primary and secondary alkyl radicals.[208,209] Each transition element has thus specific properties and use in catalysis of radical reactions. The facility of electron transfer with metal ions could thus express the rating of electron donor properties of the radicals with respect to known redox potentials of metals. According to the rate constants of the reaction of a radical with ions of different metals, the quantitative scale of electron-donor or electron-acceptor properties of the radical could then be determined. In reality, such simple attribution of redox properties is not possible since the electron-transfer between metal ion and radical is influenced not only by redox potentials of metals but also by other factors. This may clearly be demonstrated on the high rate of oxidation of alkyl radicals by two-valent copper where redox potential of the pair $Cu^{2+}/Cu+$ is substantially lower than that for other metal ions.[210,211] For determination of redox potentials of free radicals several approaches, therefore, should be used and the results should be compared. One of such complementary methods follows from the electron transfer reaction conducted with a series of organic electron acceptors of known redox potentials.[212] The transfer of electron to acceptor depends on the difference of redox potentials of electron donor and electron acceptor. The more positive redox potential of acceptor when compared to radical, the more easily the reaction proceeds.

Table 18

**THE IONIZATION POTENTIAL (IP) AND
ELECTRON AFFINITY (EA) FOR
RADICALS IN THE GASEOUS PHASE[214-217]**

Radical	IP eV	EA eV	Radical	IP eV	EA eV
$\cdot Na$	5.1	0.5	$\cdot H$	13.6	0.7
$\cdot CH_2C_6H_5$	7.7	0.9	$\cdot OOH$	10.9	3.0
$\cdot OC_6H_5$	8.5	1.2	$\cdot OH$	13.2	1.8
$\cdot NO$	9.3	0.9	O_2	12.2	0.9
$\cdot SH$	9.2	2.3	$\cdot NH_2$	11.4	1.2
$\cdot CH_3$	9.9	1.1	$\cdot CN$	15.1	3.0
$\cdot CF_3$	9.5	1.8	$\cdot I$	10.5	3.3
$\cdot CCl_3$	8.7	2.1	$\cdot Br$	11.8	3.6
$\cdot C_6H_5$	9.9	2.2	$\cdot Cl$	13.2	3.8
$\cdot OOCH$	9.0	3.6	$\cdot F$	17.4	3.6

Metals are not only efficient donors but they may also mediate the transfer of electron between two different radicals. The cations of metals are even able to stimulate the electron transfer between two identical radicals such as, e.g., verdazyls.[213] In Scheme 44, M^{n+}

$$M_m^{n+} + 2 R\cdot \Leftrightarrow [R\cdot-M_{m-2}^{n+}-R] \Leftrightarrow R^+ + [M_{m-1}^{n+}(R^-)]$$

SCHEME 44

denotes the metal ion and m is the number of solvent molecules in the inner coordination sphere. The valency state of the metal does not change at the reaction. Electron transfer takes place in the inner coordination sphere of the intermediate short-lived complex in which the metal ion can act as an electron conducting bridge.

The strength of the attachment of an electron to a radical can be also deduced from the ionization potential defined by the energy necessary for the release of an electron. That is why, the lower the ionization potential of a radical, the more easily the radical gives the electron out to the reaction co-partner. Such a radical has nucleophile properties and vice versa, electrophilic radicals have high ionization potentials. In dependence on the character of the second reactant, the radicals with medium values of ionization potential can either accept or donate electrons. In correspondence with electron transfer, electron affinity defined as the amount of energy released after accepting an electron is an equally important parameter. Taking into account both the values and definitions of ionization potential and electron affinity (Table 18) we could think that radicals spontaneously accept electrons only. This is not so, of course, since the above data are valid for isolated radicals in the gaseous phase only. In the condensed phase, the energetical requirements on the release of electrons from radicals are quite different. Ionization energy of radicals is compensated by solvation energy of ions formed from both reaction partners and, consequently, the oxidation of radicals in the condensed phase can be an exothermic process. At the correlation of the reaction course data with ionization potentials or with electron affinity these facts should, therefore, be considered, and only reactions taking place in the same reaction medium can be compared.

It should be noticed that the ionization potential of a radical is lower than that of the molecule formed by combination of the respective atom and radical. As may be seen from Fig. 4, the electron affinity has the opposite tendency.

The reactions of electron transfer do not involve reduction and oxidation of radicals, only. They are of much hgher impact; the electron transfer with participation of free radicals

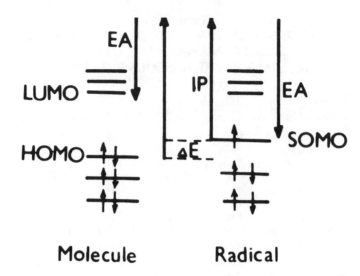

Molecule   **Radical**

FIGURE 4.   Schematic orbital diagram showing the decrease of the ionization potential (IP) and the increase of the electron affinity (EA) of a radical in comparison with a parent molecule, giving a radical by hydrogen abstraction.

represents the linkage between radical and ionic reactions. With regard to fast transfer of electrons in an equilibrium system (Equation 92), each of its components may efficiently

$$R^- \leftrightarrow R\cdot \leftrightarrow R^+ \tag{92}$$

take part in the overall process despite possible low concentration in the mixture of intermediates.[218-220] Depending on the structure of the donor and the acceptor, and on the medium, the general scheme could evolve from a typical ionic to typical radical mechanism with a continuum of possibilities. This is indeed a characteristic of electron-transfer catalyzed reactions and the actual problem is now to evaluate the participation of each contribution.[202]

It appears that the electron transfer does not follow Arrhenius' law. The observed activation energy often sinks when the temperature decreases and exhibits three different behaviors, depending on the temperature range.[221] At very low temperature (T < 100 K), the electron transfer occurs without any activation energy, via a tunneling effect. At intermediate temperatures (100 K < T < 300 K), the activation energy increases with temperature. At high temperature (T > 300 K), the activation energy reaches a maximum value and remains constant.

The electron transfer plays an important role not only in the mechanism of current reactions but in photochemical, electrochemical, radiation, and biochemical processes, too.[222] In the metabolic processes the electrons are transferred between metal sites of enzymes or organic prosthetic groups that are spatially arranged within a single macromolecule or a complex of proteins.[223,224]

VII. DECAY OF FREE RADICALS

In each reaction system where free radicals are formed there inevitably occurs also the process of their decay. Since it terminates the sequence of elementary steps of the radical reaction, it is called termination. Provided that two identical atoms are linked back to a molecule we speak about recombination (Equation 93). The similar reaction of identical polyatomic

$$A\cdot + A\cdot \rightarrow A-A \tag{93}$$

radicals is called dimerization, while the linkage of two different radicals is usually denoted as combination (Equation 94).

$$A\cdot + B\cdot \rightarrow AB \tag{94}$$

The last term is more general.

The particular case of mutual decay of radicals is disproportionation (Equation 95) which involves the transfer of one

$$AC\cdot + AB\cdot \rightarrow ABC + D \tag{95}$$

atom or group from one radical to the other; one radical is thus reduced, the second is oxidized.

Besides combination and disproportionation other alternatives of free radicals termination are also possible. The decay of free radicals as the process opposite to their formation has its antithesis in a particular mechanism. The radical combination corresponds to nonpolar dissociation, while disproportionation to bimolecular reaction with radical intermediates. Taking into consideration the large variety of intermolecular reactions (including those of trimolecular nature) leading to the pair of radicals, other diffeentiation is possible in termination accompanied by fragmentation (Equation 96). In the example of the most studied decay of peroxy

$$AB\cdot + AB\cdot \rightarrow AC + AD + E \tag{96}$$

radicals we may see that 3 particles are formed here from each two radicals. The potential barrier of the termination reaction of radicals in liquid media is very small and flat. The activated complex for such highly exothermic reactions resembles the reactants. In principle, one might expect that the activation energy of the combination of free radicals has a nonzero value when geometrical or electronic rearrangement and the long distance interaction of radicals predominating that of short distance are involved.[225] The contributions of one or more of these effects to activation energy should become more distinct for combination reactions which are less exothermic.

The rate constant of diffusion controlled bimolecular self-reaction of radicals is commonly expressed in terms of the Smoluchowski equation (Equation 97), where D denotes the diffusion

$$2\,k_t = 8\pi D\sigma\rho N \cdot 0.001 \tag{97}$$

coefficient of the radical, σ is the statistical factor (usually 1/4), ρ is the reaction distance, and N is Avogadro's number, respectively. Fairly good agreement between experimental and calculated rate constants has often been found. The rate constants of decay of many radicals in low viscosity solvents are close to those for the gaseous phase, the observed difference being ascribed to experimental errors.

A. Recombination of Atoms

A molecule formed in recombination acquires the energy of the newly formed bond. This energy should be dissipated in the reaction medium, otherwise the molecule dissociates back. The stabilization of a newly formed molecule by emission of vibrational quantum or light is a relatively slow process (10^{-8} sec) when compared with vibration time (10^{-13} sec) of individual atoms in a chemical bond. Excited molecules composed of low numbers of atoms can be efficiently stabilized in collisions with third body particles, which are capable of

Table 19
RATE CONSTANTS k_r OF
RECOMBINATION OF
HALOGENS (X) IN THE
GASEOUS PHASE IN THE
PRESENCE OF ARGON

X	k_r dm^6 mol^{-1} sec^{-1}		
	300 K	1000 K	2000 K
F	$3 \cdot 10^7$	$1 \cdot 10^7$	$4 \cdot 10^6$
Cl	$6 \cdot 10^9$	$5 \cdot 10^8$	$6 \cdot 10^7$
Br	$3 \cdot 10^9$	$4 \cdot 10^8$	$1 \cdot 10^8$
I	$2 \cdot 10^9$	$4 \cdot 10^8$	$3 \cdot 10^8$

accepting some part of the excess energy. Termination reactions of gaseous atoms such as, e.g., hydrogen, have thus a trimolecular character (Equation 98).

$$H\cdot + H\cdot + M \rightarrow H_2 + M\cdot \tag{98}$$

The role of the third body is played here either by other hydrogen atoms, hydrogen molecules, or molecules of some added gas. The participation of the third body in recombination of atoms and radicals composed of a few atoms is indicated by the increased order of the termination reaction. The rate of recombination depends on the concentration and character of the third body. The deactivation efficiency of the third body increases with its bulkiness and, consequently, with the number of atoms composing its molecule (rare gases, water vapors, etc.). The rate constant of termination is by one order lower only at the presence of hydrogen atoms than that in the presence of hydrogen molecules. Provided that molecular hydrogen is replaced by rare gases the rate constant is higher by 30 to 50%.

The rate constant of the recombination of halogen atoms in the gas phase depends on several factors.[226] At 2000 K it increases, e.g., with increasing atomic number (Table 19). This tendency is more distinct when the role of the third body is played by halogen molecules themselves. The higher rate constant of the recombination of heavier halogens is due to their more efficient cross-section which increases the frequency of collisions of two halogen atoms with the halogen molecule. The other reason will be the easier formation of the activated complex from an atom and more bulky halogen molecule.

This may explain also the different rate constants of iodine recombination in different atmospheres. Their value in argon and neon are of the order 10^9 dm^6 mol^{-2} sec^{-1}, while in the presence of NO they reach up to 10^{13} dm^6 mol^{-2} sec^{-1}. The faster recombination is conditioned by lower reaction heat which is partly used up for decomposition of the activated complex. The close contact of the third-body molecule with an atom in the activated complex facilitates the transfer of excess energy from the reaction product. Since a partially thermalized molecule is less sensitive to reversible decomposition, the subsequent deactivation to the ground state may then be slower.

The existence of an activated complex in the reaction system for recombination reactions corresponds well with the negative temperature coefficient of the process. Parallel with the decrease of the rate constant of recombination with the increasing temperature, the equilibrium concentration of activated complexes probably decreases also. Another factor which affects the rate constant of recombination is the strength of a newly formed bond. This follows from the dependence of the rate constant of recombination of different halogen atoms at 300 K; the higher the dissociation energy of the halogen molecule, the higher is the rate constant of recombination. The effect of the above factors is not isolated, but may be more or less distinct when changing reaction conditions.

Table 20

**RATE CONSTANTS OF THE REACTION X· + I· + He
→ XI + He (X DENOTES ALKALINE METAL),
TEMPERATURE COEFFICIENTS Y OF THE RATE
CONSTANT k = c T^{-Y}, RELATIVE VOLUMES $(V_r)_1$,
IONIZATION POTENTIALS (IP) OF ALKALINE
METALS AND ENERGY OF X-I AND X-X BOND**

X	k(1000 K) dm^6 mol^{-2} sec^{-1}	Y	$(V_r)_1$	IP eV	X-I kJ/mol	X-X kJ/mol
Li	5.2·10^9	0.09	1	5.39	358	107
Na	1.1·10^{10}	0.15	1.5	5.19	302	72
K	1.0·10^{11}	0.13	3.1	4.34	325	49
Rb	1.0·10^{11}	0.18	4.2	4.18	321	45
Cs	2.8·10^{10}	0.24	5.4	3.89	314	43
Ref.	227	227	58	214	58	214

The rate constant of the trimolecular reaction of the combination reaction of atoms of alkaline metals and iodine in the gaseous phase is by one to three orders higher than that for the decay of iodine atoms in argon (Tables 19 and 20). The lowest difference in the above rate constants, tenfold value, was observed for lithium atoms, which are approximately of the same size as iodine atoms. The difference of one order at 1000 K may have several reasons. Apart from methodical errors, the higher rate constant for lithium will be probably due to its lower ionization potential when compared with iodine (10.45 eV). As a matter of fact, the transfer of electrons between nonequal atoms which leads ot the reaction product may considerably increase the extent of interaction of both particles. The twofold higher strength of polar Li-I bond compared with I-I bond can contribute to the observed higher rate constant of combination of Li and I atoms, too. Similarly, as was pointed out for a combination of different halogen atoms, the increase of the rate constant in the sequence of alkaline metals consists particularly in the increasing bulkiness of the reacting atoms. Generally, very weak temperature dependence of combination of alkaline metals with iodine slightly increases with increasing sizes of alkaline metal. The negligible effect of temperature on the rate constants indicates the absence of complexes or thermally unstable molecules of alkaline metals and iodine in the deactivation of reaction product.

Two different detailed mechanisms have been proposed on the basis of facts like the above mentioned ones.[228-230] The first one — the energy transfer mechanism — is represented by the reaction Scheme 45.

$$A· + A· \underset{k_{-a}}{\overset{k_a}{\rightleftharpoons}} A_2^*$$

$$A_2^* + M \xrightarrow{k_b} A_2 + M·$$

SCHEME 45

The solution of this kinetical scheme is based on the steady-state approach to the concentration of unstable complex A_2*. In agreement with the experiment, it is possible to show that the reaction rate is proportional to $[A·]^2 [M]$ and the rate constant k_{3rd} can then be approximated by Equation 99.

$$k_{3rd} = k_b k_a / k_{-a} \qquad (99)$$

Because $1/k_{-a}$ is equal to the mean lifetime of A_2^*, this equation illustrates also the fact that some long-lived A_2^* species have a much higher chance of colliding with M. The partial transformation of relative translation energy of atoms to the rotational energy of the collision complex plays an important role here, because it can prolong the lifetime of A_2^* under certain specific conditions.

The second possible mechanism, which is able to reproduce better the influence of quality of the third particle M, is usually referred to as the bound-complex or atom-chaperon mechanism. In this model, recombination is assumed to occur via complexes in which a radical atom is weakly bound to the third-body species. The mechanism involves then the steps represented by (Equations 100 and 101).

$$A\cdot \ + \ M \ + \ M \ \underset{k_{-2}^1}{\overset{k_4^1}{\longleftrightarrow}} \ AM\cdot \ + \ M \tag{100}$$

$$A\cdot \ + \ AM\cdot \ \underset{k_b^1}{\longrightarrow} \ A_2 \ + \ M \tag{101}$$

Similarly, as in the previous case Equation 99 holds, and the equilibrium constant $k\cdot_a/k\cdot_{-a}$ is proportional to the $\exp(D_o/RT)$, where D_o stands for the dissociation energy of the weakly bound $AM\cdot$ species. When $AM\cdot$ is a typical Van der Waals' complex, the rates predicted by both the energy transfer and atom-chaperon mechanism are similar at room temperature. For larger D_o, the atom-chaperon pathway is likely to dominate and the reaction has strong negative temperature dependence.

The damping function of local absorption and dissipation of energy released at recombination of the radicals may be undertaken also by reaction walls.[231] The role of the wall surface of a reaction vessel is more obvious at a reaction of low concentration of molecules, where the probability of the collisions with reactor walls is rather high. This is especially the case of gaseous reactions.

The ratio of radicals decaying on the walls depends on the velocity of radical motion, on the diameter and surface area of the reactor, and on the nature of molecules already adsorbed on the walls.[232]

The properties of the reactor surface are usually expressed by the coefficient of radical deactivation. Its value is by several orders lower for quartz reactors than for reactors having walls coated with Cu, Ag, or Pt. The slowest decay of radicals was observed for polytetrafluoroethylene. The higher rate of decay of radicals on metallic walls corresponds to higher thermal conductivity of these materials as well as to redox properties of metals and the restricted mobility of electrons in nonmetals. The decay of radicals on the solid surface may be preceded by its reaction with adsorbed molecules.[233]

At the onset of chemical research of radical chain reactions, the wall effect was "lapis infernalis" of the patience of scientists, having brought about poor reproducibility of experiments. Even today, reactions taking place on the walls of a reactor create complications which people prefer to avoid by a proper surface treatment of the reactor. This cautious approach, however, prevents the deeper understanding of heterogeneous catalysis in radical reactions.

For reactions in the condensed phase, the deactivation of excited products and the wall effect is of negligible importance in the overall decay of radicals. The mobility of the system, which determines the frequency of collisions of recombining radicals is much more important, here. Even at perfect deactivation of formed molecules, the recombination reaction of atoms does not occur at each collision. The new molecule may be formed only at that time when a pair of reacting atoms will be in the singlet state, i.e., the spins of electrons on atoms should be opposite. Only one quarter of the collisions of hydrogen or halogen atoms can lead to such an orientation of spins. This factor comes into play also in the case of polyatomic radicals.

1. Dimerization of Radicals

The energy of the newly formed bond at the dimerization of polyatomic radicals is dissipated over the vibrational levels of adjacent bonds. Even though there still exists the probability of reverse decomposition of such a molecule into radicals, the intramolecular dissipation of energy prolongs the lifetime of the excited particle and increases thus the posibility of transfer of energy excess to surrounding medium. The effect of the third body is still measurable at the dimerization of methyl radicals in the gaseous phase.[234]

In comparison with the recombination of atoms, the energy-transfer mechanism is supposed to be the only effective one here. It again consists from steps in Scheme 46.

$$R_1\cdot + R_2\cdot \underset{k_{-a}}{\overset{k_a}{\rightleftarrows}} R_1R_2^*$$

$$R_1R_2^* + M \overset{k_b}{\longrightarrow} R_1R_2 + M$$

SCHEME 46

The collision complex $R_1R_2^*$ (contrary to the A_2^* complex) may survive for many vibrational periods, and the collision energy of radicals is supposed to be statistically distributed between vibrational modes of the complex. This image, together with the fact that the recombination process is the reverse of unimolecular dissociation, is the reason why the statistical conception similar to the RRKM theory can be used for the description of the whole process.[229] In agreement with the experimental observation, the energy transfer mechanism gives a trimolecular rate constant for the low pressure limit and a bimolecular one for the high pressure limit (the second one is simply equal to k_a).

Predictions of rate coefficients for recombination reactions led to a relatively good accordance between measured and calculated values.[235]

The rate constant for the dimerization of methyl radicals is about 3.10^{10} dm^3 mol^{-1} sec^{-1} which shows in the high efficiency of the termination. Since the frequency of methyl radical collisions in the gaseous phase is about $1.5\ 10^{10}$ dm^3 mol^- sec^{-1}, one pair of methyl radicals decays at each fifth collision approximately. Taking into account that dimerization is possible only when methyl radicals of opposite spin encounter, then practically every or every second singlet collision leads to the formation of ethane. The high efficiency of dimerization indicates that radicals become properly oriented during their mutual approach some time before the direct collision. In this final stage of approach there is a strong propensity for colliding radicals to orient themselves in the bonding configuration[236] so that the reaction itself is smooth.

Also, combinations of other radicals are very fast. In the absence of steric hindrances, the bimolecular self-termination of small alkyl radicals occurs close to the diffusion-controlled limit with a rate constant on the order of 10^8 to 10^{10} dm^3 mol^{-1} sec^{-1}. It is of interest that the presence of substituents at the radical center stabilizing a radical, such as the vinyl, phenyl (Table 21), or cyano group,[251] does not decrease these values noticeably. The self-termination of, e.g., benzyl radicals in solution also occurs very rapidly at, or near, the magnitude expected for translation diffusion.[252] This fact is conditioned by zero or a small value of activation energy of alkyl radicals decay. The lowering of ground state energy of the radical due to delocalization of an unpaired spin does not influence the reaction rate to an extent exceeding the control by diffusivity. The comparison of the activation energies for recombination with the temperature coefficient of the solution viscosity proves that the decay of radicals is diffusion controlled.

Table 21

RATE CONSTANTS k (dm³ mol⁻¹ sec⁻¹) OF

DIMERIZATION OF RADICALS AT 300 K

Radical	log k	Ref.	Radical	log k	Ref.
CH_3CH_2	10.0[a,b]	237, 238	$(CH_3)_3CO\cdot$	9.1[a]	246
$(CH_3)_2CH\cdot$	9.9[a]; 9.5[b]	239	$C_6H_5O\cdot$	9.0[a]	242
$CH_2\!=\!CHCH_2$	9.9[b]	240	C_6H_5NH	9.0[a]	242
$(CH_3)_3C\cdot$	9.5[a,b]	241	pyridinyl[c]	7.1[a]	247
$C_6H_5CH_2$	8.8[a]; 9.3[b]	242	$(CH_3)_3Si\cdot$	9.0	248
$c\text{-}C_6H_{11}$	8.7[a]	243	$(CH_3)_3Sn\cdot$	9.2	248
$\dot{C}Cl_3$	8.0[a]; 10[b]	244	$(C_6H_5)_3Sn\cdot$	9.2	248
$\dot{C}_2Cl_5$	8.3[a]	245	$\cdot PH_2$	9.8	249

[a]   Liquid.

[b]   Gas.

[c]   4-*Tert*-butyl-1-methylpyridinyl; rate constant depends slightly on alkyl substituent here.[250]

Table 22

MOLECULARITY AND HALF-TIME

OF DECAY REACTION OF

SUBSTITUTED METHYL RADICALS

AT 300 K[253,254]

Radical	Molecularity	$\tau_{1/2}$, sec
$H_3C\cdot$	2	$2\cdot10^{-5}$
$(CH_3)_3C\cdot$	2	$3\cdot10^{-4}$
$((CH_3)_2CH)_3C\cdot$	2	125
$((CH_3)_3C)_3C\cdot$	1	504
$((CH_3)_3CCH_2)_3C\cdot$	1	600
$((CH_3)_3C)_2CF_3SC\cdot$	1	10,800
$((CH_3)_3Si)_3C\cdot$	1	198,000

Note: The half-time $\tau_{1/2}$ for bimolecular decay was estimated for initial concentration of radicals 10^{-5} mol dm⁻³.

Dimerization of the radicals is determined to a large extent by sterical hindrance of a new chemical bond formation. A methyl radical in which hydrogens are substituted by bulky *tert*-butyl groups, cannot dimerize for sterical reasons. This type of radical fragmentates to smaller particles which combine or disproportionate. The decay of the tri-*tert*-butyl methyl radical appears to be the first order reaction; the rate of decay of its fragments is by several orders higher than that of their formation. The bulky substituents in the close vicinity of the radical center hinder the dimerization of other radicals (Table 22), too. Aggravation of dimerization is the kinetic reason for the relative stability of such radicals.

As is seen on the rate constant of termination of pyridinyl radicals, besides steric hindrances, the rate of radical combination is affected also by the degree of delocalization of the unpaired electron. Delocalization should, however, be of a sufficiently high degree to have an influence on the rate constant of dimerization. Provided that in the pyridinyl radical, a hydrogen or an alkyl group is replaced by phenyl, the rate constant sinks by more than three orders at 300 K, which corresponds to the increase of activation energy of dimerization from 8 to 28 kJ/mol.[255] Let us take the example of the dimerization of relatively stable triphenyl methyl radicals which has the rate constant 3.10^2 dm^3 mol^{-1} sec^{-1} at 300 K; its activation energy is 29 kJ/mol. Dimerization occurs so that the methyl carbon of one triphenyl methyl radical having a spin density of 0.5 combines with a phenyl carbon of spin density 0.1 at position 4 of the second radical. Since the linkage of methyl carbons of both radicals is sterically hindered, the reaction involves the site of lower spin density.

The dimer of triphenyl methyl radicals is not stable and in solution it is in equilibrium with radicals. The equilibrium constant (the ratio of rate constant of dimer decomposition and the rate constant of radical combination) of triphenyl methyl radicals has the value 7.10^{-4} mol dm^{-3}. The reversible combination is not a specific case of triphenyl methyl radicals only, but it may be generalized to other radicals.

Nonstable dimers are formed, e.g., from nitroxyl (K = 10^{-3} to 10^{-4} mol dm^{-3}), phenoxy (K = 10^{-8} to 10^{-3} mol dm^{-3}), peroxy (K = 10^{-6} to 10^{-5} mol dm^{-3}), alkyliminoxy (K = 10^{-8} to 10^{-3}), arylaminyl (K = 10^{-8} to 10^{-3} mol dm^{-3}) and other radicals.[256] The equilibrium constants depend on both the radical structure and medium polarity. The increase of medium polarity increases the equilibrium constant; in a polar solvent the concentration of solvated radicals is, therefore, higher. The effect of medium polarity is lowered when the equilibrium constant decreases.

2. Configuration Isomerism in Combination of Polyatomic Radicals

In the combination of radicals, a new chemical bond is formed between atoms with localized unpaired electrons. Provided that the unpaired electron is delocalized, the products of combination involve several isomeric compounds. Phosphonyl radicals can, e.g., dimerize on both phosphorus and oxygen atoms (Scheme 47).

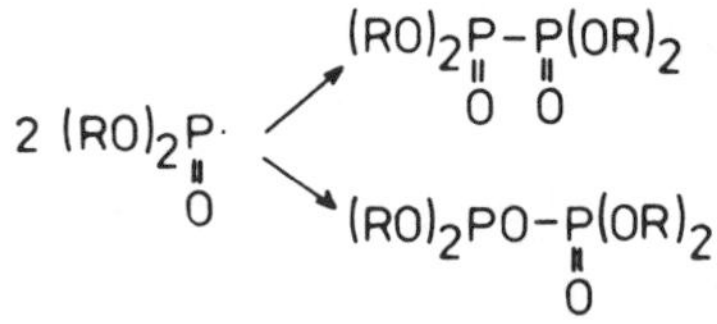

SCHEME 47

The mixture of products is formed also in the dimerization of iminoxyl radicals (R^{1}R^{2}C=NO·) where N-N, N-O, and O-C coupling is possible.[257]

The phenoxy radical combines with the methyl radical to phenyl methyl ether only to a limited extent (8%), the main product being 2- and 4-methyl phenol.[258] This course is due to the delocalization of the unpaired electron in the radical. Apart from the fact that the phenoxy radical is formed by abstraction of hydrogen from the hydroxyl group of phenol. it could be also called ketocyclohexadienyl.

The mechanism for ·CH$_3$ and C$_6$H$_5$O· combination reaction leading to o-cresol can thus be depicted as in Scheme 48.

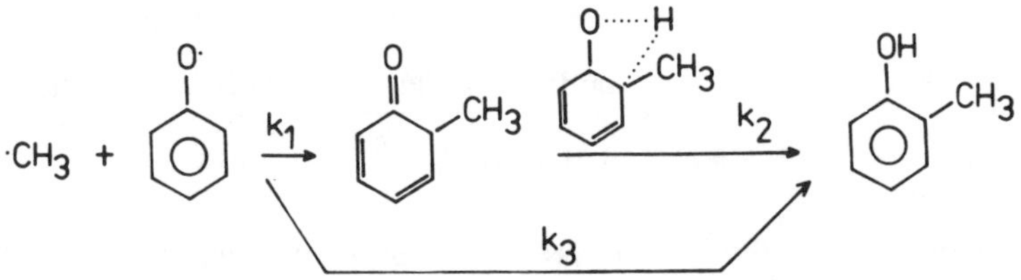

SCHEME 48

The direct formation of *o*- and *p*-cresols is at least as important as the formation of the methylcyclohexadienone intermediate by collisional deactivation. The transition state of the isomerization process which becomes dominant at higher temperatures, involves the transfer of the tertiary hydrogen atom.

Dimerization of two phenoxy radicals leads to a relatively unstable diketoisomer which may either decompose to the original radical or undergo enolization to a final stable product (Scheme 49).

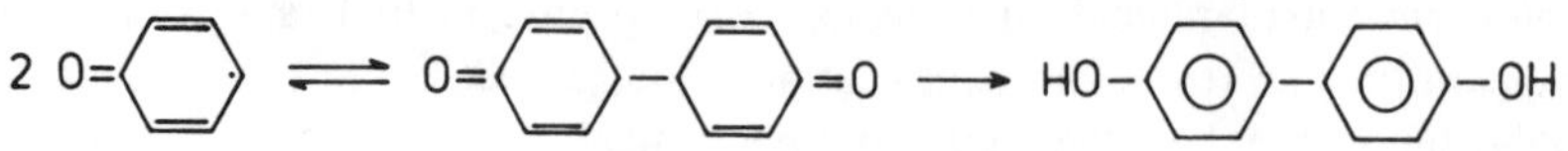

SCHEME 49

Enolization explains the significant effect of medium acidity on the rate of phenoxy radicals decay as well as the relatively large influence of the structure of radicals on the rate constant of termination.[259]

The combination of phenoxy radicals to O-O bond was not observed even at complete substitution of phenyl hydrogens in positions 2, 4 and 6 by alkyls. From the 2,6-*tert*-butyl-4-methyl phenoxy radical an unstable dimer is formed having one keto group, (Scheme 50) which disproportionates or

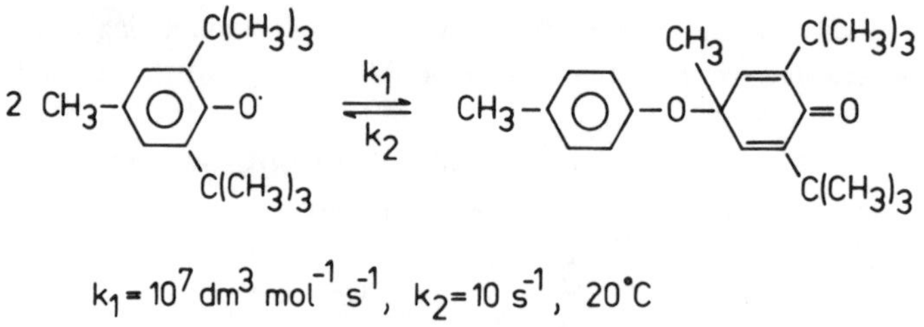

$$k_1 = 10^7 \, dm^3 \, mol^{-1} \, s^{-1}, \quad k_2 = 10 \, s^{-1}, \quad 20°C$$

SCHEME 50

decomposes to its parent radicals.[256]

3. Stereoisomerism of Dimerization Reaction

Substituted alkyl radicals having asymmetric central carbon may dimerize according to Scheme 51 to meso and DL stereoisomers.

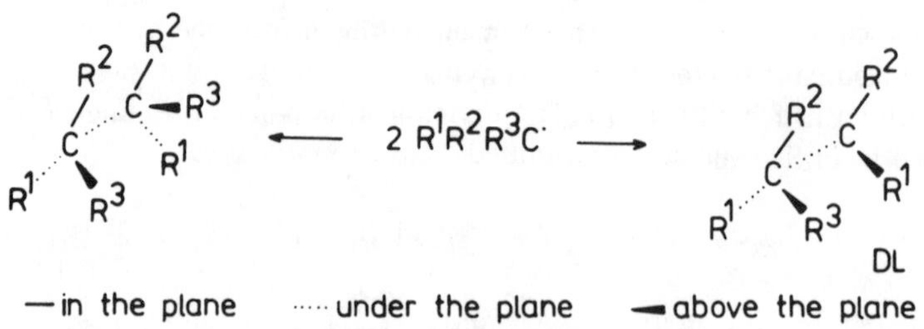

— in the plane   ····· under the plane   ▬ above the plane

SCHEME 51

The ratio of DL and meso-isomers in products depends on substituents R^1, R^2 and R^3, temperature, and the state of aggregation of the reaction medium, and for different radicals lies in the interval 0.4 to 3.[260,261]

Stereoselectivity is influenced by substituents which support the formation of the pyramidal structure of the substituted radical and interact mutually by the attractive or repulsive forces. The predominant structure is often the result of opposite substituents effects.

B. Disproportionation of Radicals

Radicals which form unstable dimers decay by disproportionation, predominantly. At the decay of alkyl nitroxyl radicals, e.g., one radical gives up and the second accepts a hydrogen atom (Scheme 52).

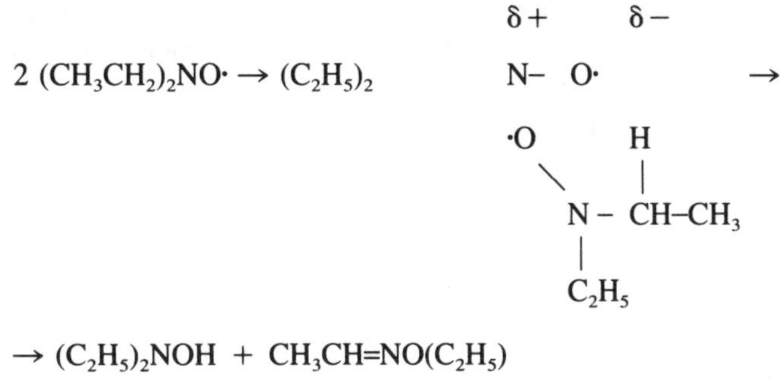

$$2 \ (CH_3CH_2)_2NO\cdot \rightarrow (C_2H_5)_2 \qquad \overset{\delta+}{N-} \ \overset{\delta-}{O\cdot} \qquad\qquad \rightarrow$$

$$\begin{array}{cc} \cdot O & H \\ \diagdown & | \\ N- & CH-CH_3 \\ | \\ C_2H_5 \end{array}$$

$$\rightarrow (C_2H_5)_2NOH + CH_3CH=NO(C_2H_5)$$

SCHEME 52

Displacement of a hydrogen atom from aliphatic substituents of nitroxyl radicals anticipates the formation of the activated complex of disproportionation.[256] The difference in the rate constant of disproportionation of primary dialkyl nitroxyl radicals at 300 K ($k_d = 10^4$ to 10^5 dm^3 mol^{-1} sec^{-1}) and that of diisopropylnitroxide ($k_d = 4$ dm^3 mol^{-1} sec^{-1}) is surprisingly large. At the same time, the strength of the chemical bond of the eliminated hydrogen atom in diisopropylnitroxide is lower. If, in spite of this, the slower disproportionation is observed in the latter case, it indicates the importance of the sterical factor in this reaction. The sterical hindrances in di-*tert*-butyl nitroxide cause this radical not to disproportionate at all. Besides the higher strength of the C-F bond when compared to the C-H bond, the slower disproportionation of perfluoroalkylnitroxyl radicals is caused by the considerably lower strength of the newly formed O-F bond than the O-H bond. The electrostatic repulsion of fluorine, which is much more electronegative than hydrogen also cannot be neglected. The difference is of both a quantitative and qualitative nature. Relating to oxygen which in nitroxyl radicals possesses a partial negative charge, hydrogen atom has a partial positive charge stabilizing the activated complex. On the other hand, in perfluoroalkyl radical, the fluorine atom has a partial negative charge.

As it is, e.g., in the case of alkyl radicals, disproportionation may occur also between radicals forming stable dimers. During disproportionation of α-di-deuterium ethyl radicals (Equation 102), β-hydrogen is transmitted from one radical

$$2 \ CH_3\overset{\cdot}{C}D_2 \rightarrow [\cdot CD_2CH_2 \ \ H \ \ \cdot CD_2CH_3] \rightarrow CH_2=CD_2 + CH_3CD_2H \qquad (102)$$

to another and 1,1-di-deuterium-ethylene and 1,1-di-deuterium-ethane are formed. Comparing this with the transfer reaction of ethyl radicals to ethane, the disproportionation reaction has an activation energy 50 kJ/mol lower, while the frequency factor is 2 orders higher. Disproportionation reactions between two radicals containing C-H bonds can formally be classified as radical H-atom abstraction reactions; their rates are, however, comparable to radical-radical combinations, with little or no energy barriers. Disproportionation reactions are more exothermic than a H-atom abstraction, because an alkene, as opposed to another radical, is formed. Their lower activation energy is due to the lower dissociation energy of the β C-H bond in the alkyl radical when compared with the saturated hydrocarbon molecule and higher reaction exothermicity. The higher value of the frequency factor is an indication of the lesser packing of particles in the activated complex of disproportionation; as though the displaced atom is partially split out. The critical - C·...H - C - distance at which the disproportionation reaction takes place corresponds to the van der Waals contact radius of 0.29 nm, between the carbon of the radical and the β-H atom on the neighboring radical.[262]

Also, atoms other than hydrogen can be displaced at the disproportionation. Similarly as hydrogen, halogen atoms are likely to be transferred at the disporportionation of substituted carbon radicals having halogens in the β-position to the radical site. Unfortunately, little knowledge exists about such reactions at present to make some more definite conclusions.

On the other hand, many papers have been devoted to the displacement of oxygen in disproportionation reactions of inorganic radicals.[263] In spite of it, the mechanism of the disproportionation reactions of halogen monoxide radicals XO· (X = F, Cl, Br and I) remains still highly uncertain. The initial step (Equation 103) may possibly be the formation of transient

$$XO· + XO· \leftrightarrow XOOX^*$$ (103)

XOOX intermediate which can undergo collisional stabilization, rearrangement, and decomposition (Scheme 53).

$$XOOX^* \rightarrow XOOX$$

$$\rightarrow X· + XOO·$$

$$\rightarrow X_2 + O_2$$

SCHEME 53

In should be noted that a number of radicals (O, H, Cl, NO) abstract an O atom from $HO_2·$. This type of disproportionation reaction has Arrhenius parameters characteristic of a loose transition state and probably forms also an HO_2X intermediate followed by a simple bond scission.[264] The radical abstraction of the H atom from $HO_2·$ proceeds by an attack at the hydrogen of the peroxy radical.

C. Competition of Combination and Disproportionation

Polyatomic radicals decay in parallel by different combination pathways; the ratio of individual decay mechanisms depends on the radical structure and on reaction conditions.[265] The ratio of rate constants of disproportionation and combination (k_d/k_c) for numerous radicals is relatively slightly dependent on temperature. The fact that temperature dependence indeed exists indicates that the transition states for disproportionation and combination cannot be exactly the same. It is suggested that the transition state for combination is not greatly polarized while that for disproportionation has a polar character. The alkoxy and halogenated radicals both give high fractions of disproportionated products in comparison with alkyl radicals (Table 23).

Predominance of disproportionation above combination for polar radicals could be understood so that electrostatic repulsion of the radicals having electric dipoles functions against dimerization. The polar structure of radicals, moreover, facilitates the proper orientation at the formation of the activated complex of disproportionation. The role of polar structure in the disproportionation reaction may be noticed even in the case of relatively nonpolar alkyl radicals. Examining numerous measurements of various authors on decay reactions of ethyl

Table 23
THE RATIO OF RATE CONSTANTS OF DISPROPORTIONATION AND COMBINATION (k_d/k_c) IN TERMINATION OF DIFFERENT RADICALS

Radical	k_d/k_c	Ref.	Radical	k_d/k_c	Ref.
·CH$_3$	0[a]	266	C$_6$H$_5$ĊHCH$_2$C$_6$H$_5$	0.11[b]	271
·CH$_2$F	0[a]	267	CH$_3$ĊHCH$_3$	0.66[a]	265
·CDF$_2$	0.11[a,d]	267	CH$_3$ĊHCH$_2$CH$_3$	0.70[a]	265
·CHF$_2$	0.15[a]	267	CH$_3$ĊHCH$_2$CH$_3$	1.0[b]	265
·PH$_2$	0.15[a]	249	Ċ(CH$_3$)$_3$	2.5[d,f]	265
·SiH$_3$	0.70[a]	268	Ċ(CH$_3$)$_3$	5.0[b,f]	265
·CH$_2$CH=CH$_2$	0.008[a]	269	c-Ċ$_5$D$_9$	0.58[b,d]	272
CH$_3$CHCH=CH$_2$	0.014[a]	269	c-Ċ$_5$H$_9$	0.73[b]	272
ĊH$_2$CH(CH$_3$)$_2$	0.076[a]	265	c-Ċ$_6$D$_{11}$	0.38[b]	272
ĊH$_2$CH$_3$	0.14[a,b,e]	270	c-Ċ$_6$H$_{11}$	0.56[b]	272
ĊH$_2$CH$_2$CH$_3$	0.13[a]	265	c-Ċ$_6$H$_{11}$	1.1[c]	273
ĊH$_2$CH$_2$CH$_2$CH$_3$	0.14[a]	265	Ċ$_{14}$H$_{11}$[g]	0.054[b]	274
ĊFHCH$_3$	0.20[a]	265	Ċ$_{14}$H$_{11}$[h]	1.34[b]	274
ĊF$_2$CH$_3$	0.55[a]	265	·OCH$_3$	9.3[a]	265
C$_6$H$_5$ĊHCH(CH$_3$)$_2$	0.037[b]	271	·OCH$_2$CH$_3$	12[a]	265
C$_6$H$_5$Ċ(CH$_3$)$_2$	0.066[b]	265	·NHNH$_2$	4.0	275
C$_6$H$_5$ĊHCH$_3$	0.10[b]	271			

[a] Gaseous phase.
[b] Liquid phase.
[c] Solid phase.
[d] The ratio of $(k/k_c)_H/(k_d/k_c)_D$ lies around 1.4 regardless of radical species and temperature.
[e] The ratio of rate constants was determined by different methods in about 50 papers.
[f] The high temperature/low viscosity limit k_d/k_c should be about $3 + 0.5$, whereas the low temperature/high viscosity limit is estimated as $20 + 10$.[276]
[g] 9,10-Dihydro-9-anthryl.
[h] 9,10-Dihydro-9-phenanthryl.

radicals we may find that values of k_d/k_c determined for the liquid phase lie in the range 0.12 to 0.35. Although the results may be somewhat distorted by both experimental and methodical errors, the overall tendency of an increase of k_d/k_c with the increase of polarity and internal pressure of solvents may be seen.[270] The lower limit of this ratio coincides with the gas phase value. The highest value of k_d/k_c was obtained for the aqueous medium. Solvation in the liquid phase would, therefore, be expected to favor disproportionation. The effect of the polar medium on the k_d/k_c ratio may thus consist in acceleration of disproportionation, deceleration of dimerization, or in a lesser decrease of the disproportionation rate when compared with dimerization.

The ratio of k_d/k_c decreases with increasing viscosity of the reaction medium and outside pressure. The more pronounced effect of viscosity on k_d than on k_c corresponds to higher requirements on mobility of the reactants in disproportionation when compared with combination. As is usually observed, the activation volume of dimerization is negative (about 5 cm^{3}mol^{-1}) and the increasing pressure will support the dimerization and impede disproportionation. Increasing pressure simultaneously increases the viscosity of the system which has a consonant effect on changes of the k_d/k_c ratio.

In the solid phase, the ratio of k_d/k_c of the radical interaction is by about one order higher than that in the liquid phase.[277] Relatively few results are available for the decay of radicals in the solid state and their interpretation should be done cautiously. The radicals in the solid phase can decay in the sequence of fragmentation and combination reactions. Provided that a sufficiently mobile hydrogen atom is ejected from the radical primarily, it may combine with other radicals and the final result will be identical with disproportionation.

According to the generalization that was proposed a good while ago, the k_d/k_c ratio for aliphatic radicals varies statistically with the number of available β-hydrogens for the abstraction leading to disproportionation products. The presence of β-hydrogen itself, therefore, cannot explain the disproportionation even in the case of alkyl radicals.

The convincing objections against this simple and partially valid idea are manifested in examples of disproportionation of unequal radicals. The reaction pathway (a) of Scheme 54 of reaction of difluoromethyl radicals and ethyl radicals is faster than (b), in spite of the fact that from the viewpoint of the number of hydrogens it should be quite opposite.[278]

$$a \quad :CF_2 + C_2H_6$$

$$\cdot CF_2H + \cdot CH_2CH_3$$

$$b \quad CF_2H_2 + C_2H_4$$

SCHEME 54

At the disproportionation of 4-pentene-2-yl radicals (Equation 104)

$$CH_3\dot{C}HCH_2CH=CH_2 \rightarrow a\ CH_3(CH_2)_2CH=CH_2 + b\ CH_2=CHCH_2CH=CH_2 +$$

$$+ c\ CH_3CH=CHCH=CH_2 \qquad (a = b + c) \qquad (104)$$

the predominant product is formed by H abstraction from the methyl group ($b/c = 3.8$) though the statistical viewpoint assumes the ratio $b/c = 1.5$. The b/c ratio should be even lower since the formation of the conjugated diene is expected to be favored energetically.[279] Because this is not so, from this and other examples, it is again seen that disproportionation of many radicals is governed by steric effects. The cross disproportionation of unsaturated alkyl radicals shows that the double bond has an orienting effect on the reaction with a second radical.

The influence of energetical factors on k_d/k_c values may be demonstrated on disproportionation of hydroaromatic alkyl radicals, which is less exothermic (below 200 kJ/mol). In the termination of such radicals, the k_d/k_c ratio increases with increasing enthalpy of disproportionation. Disproportionation of 9, 10-dihydro-9-phenantryl radicals is, e.g., by 88 kJ/mol more exothermic than that of the 9,10-dihydroanthryl radical while the ratio k_d/k_c is 25 fold higher. Similar relations exist also for other alkyl radicals.[274]

The difference in exothermicity of disproportionation can also explain the high difference of k_d/k_c values for cumyl ($C_6H_5C(CH_3)_2$) and *tert*-butyl radical (Table 24). If the disproportionation reaction is more exothermic ($-\Delta H_d > 200$ kJ/mol), there does not, however, appear to be any general relationship between k_d/k_c and reaction enthalpy. In this regime, k_d/k_c values for different hydrocarbon radicals generally fall in a rather narrow range and are apparently controlled by entropic factors. As the reaction exothermicity declines below 200 kJ/mol, k_d/k_c values decline and the rates of disproportionation become sensitive to reaction enthalpy while recombination rates remain insensitive. Nonenthalpic factors are of importance even for this case.

D. Termination of Peroxy Radicals

Fragmentation and termination of peroxy radicals may proceed by several reaction mechanisms (Scheme 55).

Table 24
THE EFFECT OF SUBSTITUENTS (R^1, R^2) ON MECHANISM OF TERMINATION OF PEROXY RADICALS $R^1R^2CHOO\cdot$ IN GASEOUS (g) AND LIQUID (l) PHASE AT 300 K

R^1	R^2	% RO·	Phase	Ref.
H	H	30	g	280
H	CH_3	60	g	281
CH^3	CH_3	60	g	282
C_2H_5	CH_3	4	l	283
$C_6H_5CH_2$	C_6H_5	30	l	283
CH_3CH_2O	CH_3	80	l	283

Note: % RO· denotes the ratio of peroxy radical fragmentation to oxygen and oxyradicals.

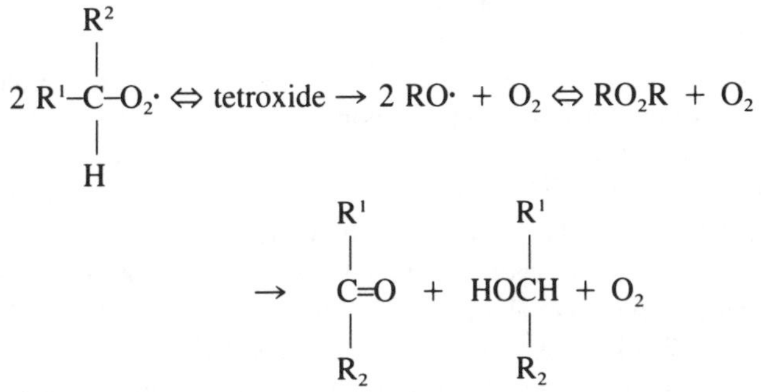

$$2\ R^1\!\!-\!\!\underset{\underset{H}{|}}{\overset{\overset{R^2}{|}}{C}}\!\!-\!\!O_2\!\cdot\ \Leftrightarrow\ \text{tetroxide} \rightarrow 2\ RO\cdot\ +\ O_2\ \Leftrightarrow\ RO_2R\ +\ O_2$$

$$\rightarrow\quad \underset{\underset{R_2}{|}}{\overset{\overset{R^1}{|}}{C}}\!\!=\!\!O\ +\ \underset{\underset{R_2}{|}}{\overset{\overset{R^1}{|}}{HOCH}}\ +\ O_2$$

SCHEME 55

Single step dimerization of peroxy radicals is possible only below temperature 190 K, which is the limit of thermal stability of formed tetroxides. At stepwise termination, there is first eliminated molecular oxygen from two peroxy radicals and formed alkoxy radicals decay by disproportionation and combination. As was confirmed by isotopical analysis of oxygen in labeled radicals, oxygen is split out symmetrically from both peroxy radicals. The ratio of radicals released from the activated complex of the decay reaction depends on the structure of peroxy radicals (Table 24) and on temperature. The mechanism of decay depends also on solvent, state of aggregation of the medium, and on other factors (pressure, radiation, etc.), respectively. This is due to the instability of peroxy radicals and products of combination themselves as well as to a large variety of possible, thermodynamically advantageous chemical transformations of reactive intermediates. We remind the reader that transformation of alkylperoxy radicals to alkoxy radicals does not occur only at self-reaction of two peroxy radicals, but also at fast reaction of methyl and methyl peroxy radicals (Equation 105) or at reaction of *tert*-butyl peroxy and

$$\cdot CH_3\ +\ CH_3OO\cdot\ \rightarrow\ 2\ CH_3O\cdot \tag{105}$$

diphenylamino radicals (Equation 106). The character of the alkyl group in

$$(CH_3)_3COO\cdot\ +\ \cdot N(C_6H_5)_2\ \rightarrow\ (CH_3)_3CO\cdot\ +\ (C_6H_5)_2NO\cdot \tag{106}$$

Table 25
RATE CONSTANTS k$_t$ OF TERMINATION AT 300 K
AND ACTIVATION ENERGY (E) OF DECAY OF
SOME PEROXY RADICALS IN GASEOUS (g) AND
LIQUID (l) PHASE

Radical	Phase	k$_t$ dm^3- mol^{-1}sec^{-1}	E kJ/mol	Ref.
HOO·	g	1 10^9	—	284
CH$_3$OO·	g	2 10^8	*	280
CH$_3$CH$_2$OO·	g	6 10^7	*	285
c-C$_5$H$_9$OO·	l	3 10^7	13	286
(CH$_3$)$_2$CHOO·	g	8 10^5	19	282
(C$_2$H$_5$)CH$_3$CHOO·	l	5 10^6	11	286
(CH$_3$)$_2$COO·	g	3 10^3	32	286
(CH$_3$)$_3$COO·	l	3 10^4	27	287

Note: * Small negative temperature coefficient.

peroxy radicals does not affect the mechanism of the termination reaction only, but also the rate constant and activation energy of termination (Table 25).

The rate constant (in dm^3 mol^{-1} sec^{-1}) of decay for primary peroxy radicals (RCH$_2$OO·) is of the order 10^8 to 10^9, that of secondary (R$_2$CHOO·) 10^6 to 10^7, while that of tertiary peroxy radicals (R$_3$COO·) is only 10^3 to 10^4. The more hydrogen atoms on the carbon in the neighborhood of peroxidic oxygen, the higher the rate constant of peroxy radical fragmentation and disproportionation. The relatively low value of the rate constant of decay of tertiary alkyl peroxy radicals is connected with the absence of easily transferrable hydrogen, low thermal stability of peroxides, and the slower course of fragmentation of these radicals. Since irreversible termination of peroxy radicals is mediated also by secondary radicals, the difference in overall rate constants of their decay is very understandable.

E. Decay of Radicals in the Solid State
Some pecularities of radical decay in the condensed phase are explained by the existence of the so-called cage effect. By the term "cage" we understand the space reserved for the mutual reaction of a radical pair. Provided that two radicals are in such an area, the probability of their reaction with the surrounding molecules is lower when compared with their decay in the cage.

The formulation of the hypothesis of the cage effect (Frank Rabinowitch, 1934) was stimulated by the interpretation of yields of photolysis for different compounds in the gaseous and liquid phase. The comparison of photolysis of equimolar mixture of azobis(methyl) and its deuterium derivative is an illustrative example. In the gaseous phase, three possible dimers of methyl radicals can be formed: CH$_3$CH$_3$, CH$_3$CD$_3$, and CD$_3$CD$_3$ in the statistical ratio 1:2:1. After the decomposition of a parent molecule, the radicals escaping initial pairs are dispersed to the reaction volume and there form randomly new radical pairs and then decay. In the liquid phase of a chemically nonreactive medium up to 75% of a symmetrical product is formed. Molecules of the liquid phase decrease thus the probability of cross-dimerization of unequal radicals and vice versa, they support the linkage inside of the radical pair in the place of its formation. 50% of radicals from initially formed radical pairs approximately succeed in escaping into the surrounding medium in this case. The probability of the escape of radicals from the cage of surrounding molecules of a solvent depends on the reactivity of the "walls" of the cage and on the medium viscosity. For high reactivity

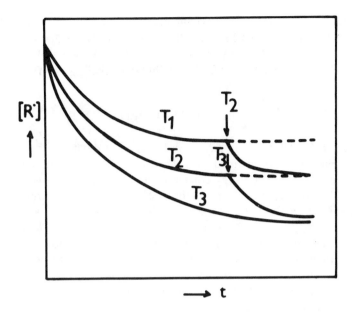

FIGURE 5.   The time changes of concentration of R· radicals in the solid
system at temperatures $T_1 < T_2 < T_3$.

of cage walls only a low ratio of formed radicals can dimerize. This is, e.g., the specific
case of decomposition of azobis(isobutyronitrile) in liquid bromine, where the escape of
radicals from the cage is high. For current solvents, the average value of the radical yield
from the cage is 50%. For solid polymers and crystalline monomers, the probability of the
release of primary radicals from the cage of surrounding molecules is already rather low
(below 10%).[288]

Radical pairs are kinetically nonequivalent which is particularly seen in the solid phase.
By nonequivalency of radical pairs the stepwise decay of radicals in solid systems of the
restricted translational mobility is interpreted. At some temperature when radicals start to
react only a part of the present radicals decays (Figure 5). Radicals which do not recombine
are stable at a given temperature. The further increase of temperature initiates the decay of
a further fraction of radicals, the concentration sink being directly proportional to the change
of the temperature.

The effect of homogeneity of a solid matrix on the course of radical decay may be
investigated at photolysis of monocrystals, polycrystalline samples, and solutions of identical
photosensible compound in a solvent frozen by different rates of cooling.[289] As expected,
the decay of radical pairs in monocrystals is of the first order. In polycrystalline powders
and solid solutions, where the reactivity of radical pairs is determined by the system nonhomogeneity, each type of radical pair decays by first order law, too, but with different rate
constant. The system as a whole then displays the stepwise decay of radicals. The "poly-
chromatism" of the rate constant is conditioned by the different local crystalline field
surrounding the radical pair. A similar kinetic picture may be observed also for homogeneous
matrices where the dependence of the rate constant on time is due to the nonequilibrium
process of the relaxation of radical surroundings.[290] Characteristic stepwise decay of
macroradicals in solid polymers may alternatively be explained by the similar mechanism
of stepwise release of segmental motions of macromolecules with increasing temperature.[291]

The shift to a higher temperature region leads to the increase of activation energy. The
total average activation energy for radical decay is given by the sum of statistically weighed
energy barriers for molecular motions and H atom transfer reactions. In the region of lower

temperature there decay radical pairs with a small distance between reacting partners, predominantly. The increase of activation energy of the decay reaction corresponds to the increasing importance of the process of the physical approach of more distant reaction sites as well as of the chemical approach due to H atom migration in the sequence of elementary transfer reactions.

VIII. COMMON FEATURES OF ELEMENTARY REACTIONS

Analyzing the course of elementary reactions, it was pointed out that the rate of a given reaction is determined by both thermodynamic and kinetic factors. The most substantial effect is due to the difference in energy content of reactants and products, an exothermic reaction being more probable than an endothermic. The height of the energetical barrier to be overcome going from reactants to products depends on reaction heat as well as on kinetic factors, such as polarity and sterical accessibility of the reaction sites. The polar structure of reactants may either increase or decrease the rate of radical reaction. Polar effects are especially seen in thermoneutral reactions where lengthening of partially activated bonds increases the partial charges in the structure of the activated complex. The polar structure of the activated complex may lead to complete transfer of electrons to radical or vice versa and the further course of the reaction has then an ionic character. The transfer of electron comes into consideration in the case of a large difference of electron donor and electron acceptor properties of reacting particles or when the efficient solvation of forming ions is possible.

Sterical hindrance by substituents is most marked at the decay of polyatomic radicals and at the addition of polyatomic radicals to tetrasubstituted alkenes. A similar effect is supposed in substitution reactions of polyatomic groups by polyatomic radicals taking place on the central atom of a molecule. The lesser sensitivity of transfer reactions occurring mostly on side groups of a molecule to sterical hindrance is quite comprehensible. The sterical effect of an opposite sign connected with the release of the tight packing of the substituents after the abstraction of one atom may be more important here. In the transfer reaction, a radical can, therefore, be more easily formed than would be expected from the analogy with reactants of a lower sterical strain.

As was shown in several cases, some radical reactions considered as elementary may appear stepwise after a more thorough investigation. From this aspect, a deeper insight into all elementary reactions is not possible yet because of the lack of proper experimental methods of analysis of shortly living intermediates.

To date, there are only indirect and sometimes rather abstract speculations of the detailed mechanism possible. Consider, e.g., the addition of radical to polyatomic molecule. The more complex radical formed here, may either remain without an apparent structural change for some time or subsequently isomerize or fragmentate. The former case corresponds to the simple addition. Provided that the subsequent isomerization or fragmentation of the polyatomic radical is sufficiently fast when compared with primary addition, we speak about substitution reaction. Isomerization itself can also be represented by the sequence of fragmentation and additions inside of radicals. In such a way all elementary reactions of radicals can formally be described in terms of two fundamental processes, namely, those of addition and fragmentation.

The process of the addition reaction involves the interaction of an unpaired electron with some orbital of a reacting particle. The further fate of products of such a primary interaction depends on the kind of attacked orbital. In the case of a semioccupied orbital, the radical decays by combination reaction. At the attack of the nonbonding orbital, the residual unpaired electron is a carrier of radical character of an addition product. A similar situation occurs also at the typical elementary addition reaction of a molecule with a multiple bond. If an unpaired electron of a radical reacts with a single bond, the addition occurs with the simultaneous rupture of this bond (transfer reaction).

Fragmentation reaction of polyatomic radicals is initiated by spin polarization (or triplet excitation) of the electron pair of the bonding orbital due to the unpaired electron of the radical center. Taking into consideration the destabilization effect of the semioccupied orbital on properly oriented bonding orbitals, the considerably faster transfer reaction within two radicals at the disproportionation, when compared to the reaction of one radical with a saturated molecule, may then be understood. The activated complex, however, decomposes relatively fast even in the case of transfer reaction. At present, the rate of decomposition of a radical adduct can be followed only at some substitution reactions on central atoms of molecules or at fragmentation reactions of more stable radicals.

At the course of chemical transformation of compounds, elementary reactions of radicals take place in a synchronous manner — they compete themselves. The mutual connection and interrelation of elementary reactions and their ratio in the given reaction system depends on radical reactivity as well as on reaction conditions (temperature, reactant concentration, medium polarity, etc.). This will be dealt with in later chapters of this book.

REFERENCES

1. **Burkey, T. J., Griller, D., Sutcliffe, R., and Harding, C. J.**, Inversion barrier of the cyclopentyl radical, *J. Org. Chem.*, 50, 1138, 1985.
2. **Hori, Y., Shimada, S., and Kashiwabara, H.**, ESR Study of free radicals in irradiated thiourea canal complexes. 2. Structure and ring inversion of cyclohexyl and cycloheptyl radicals, *J. Phys. Chem.*, 90, 3037, 1986.
3. **Lloyd, R. V. and Causey, G.**, Electron spin resonance of inversion and conformations in 1-hydroxycyclohexyl radicals, *J. Chem. Soc. Perkin Trans. II*, 1143, 1981.
4. **Walton, J. C.**, Cycloalkylmethyl radicals. 4. Electron spin resonance study of conformation. Equilibria in cyclohexenylmethyl and 4-alkylcyclohexenylmethyl radicals, *J. Chem. Soc. Perkin Trans, II.*, 1641, 1986.
5. **Casarini, D., Grossi, L., and Placucci, G.**, Naphtoyl δ-radicals: elucidation of their conformational behaviour by electron spin resonance, *J. Chem. Soc. Perkin Trans.*, II. 599, 1986.
6. **Beckwith, A. L. J.**, Structure, Reactivity and Rearrangement, in *Free Radical Reactions*, Waters, W. A., Ed., *Organic Chemistry*, Vol. 10, Butterworths, London, 1973, 1.
7. **Bubnov, N. N., Solodnikov, S. P., Prokofjev, A. I., and Kalacnik, M. J.**, Dynamics of forced tautomerism in free radicals, (Russian), *Usp. Khim.*, 47, 1048, 1978.
8. **Kochi, J. K.**, Oxygen radicals, in *Free Radicals II*, Kochi, J. K., Ed., John Wiley & Sons, New York, 1973, 665.
9. **Carter, W. P. L., Darnall, K. R., Lloyd, A. C., Winer, A. M., and Pitts, J. N.**, Evidence for alkoxy radical isomerizations in photooxidation of C_4-C_6 alkanes under simulated atmospheric conditions, *Chem. Phys. Lett.*, 42, 22, 1976.
10. **Niki, H., Maker, P. D., Savage, C. M., and Breitenbach, L. P.**, an FT IR study of the isomerization and O_2 reaction of n-butoxy radicals, *J. Phys. Chem.*, 85, 2698, 1981.
11. **Dobe, S., Berces, T., and Marta, F.**, Gas phase decomposition and isomerization reactions of 2-pentoxy radicals, *Int. J. Chem. Kinetics*, 18, 329, 1986.
12. **Van Sicle, D. E.**, Oxidation of 2,4,6-trimethylpentane, *J. Org. Chem.*, 37, 755, 1972.
13. **Shostenko, A. G., Myshkin, V. E., and Klim, V.**, Kinetics and mechanism of isomerization of alkyl radicals in solution, *React. Kinet. Catal. Lett.*, 10, 311, 1979.
14. **Ibuki, T., Tsuji, A., and Takezaki, Y.**, Isomerization of chemically activated 1-butene-1-yl and 1-butene-4-yl radicals, *J. Phys. Chem.*, 80, 8, 1976.
15. **Mattice, W. L.**, Complex branch formation in low density polyethylene, *Macromolecules*, 16, 487, 1983.
16. **Bovey, F. A., Schilling, F. C., McCrackin, F. L., and Wagner, H. L.**, Short chain and long chain branching in low density polyethylene, *Macromolecules*, 9, 76, 1976.
17. **Valko, L. and Simon, P.**, Extension of BEBO method for the estimation of activation energies of intraradical 1,2-hydrogen shift reactions, *Chem. Phys.*, 99, 447, 1985.

18. **Ingold, K. U.,** Some applications of free radical rearrangements, in *Organic Free Radicals,* Pryor, W. A., Ed., ACS Symposium, Series 69, Washington, D.C., 1978, 187.

19. **Freidlina, R. Ch. and Terentiev, A. B.,** Isomerization of unstable elementoorganic radicals in the liquid phase reactions, (Russian), *Usp. Khim.,* 48, 1548, 1979.

20. **Hjertberg, Th. and Sorvik, E. M.,** Formation of anomalous structures in poly (vinyl chloride) and their influence on the thermal stability, in *Polymer Stabilization and Degradation,* Klemchuk, P. P., Ed., ACS Symposium, Series 280, 1985, 259.

21. **Skell, Ph. S. and Traynham, J. G.,** Radical brominations of alkyl bromides and the nature of β-bromoalkyl radicals, *Acc. Chem. Res.,* 17, 160, 1984.

22. **Maj, S. P., Symmons, M. C. R., and Trousson, P. M. R.,** Bridged bromine radicals. An electron spin resonance study, *J. Chem. Soc., Chem. Commun.,* 561, 1984.

23. **Sakurai, H.,** Group IV B Radicals, in *Free Radicals II,* Kochi, J. K., Ed., John Wiley & Sons, New York, 1973, 741.

24. **Walter, D. W. and McBride, J. M.,** Neophyl rearrangements in crystalline bis(3,3,3-triphenypropanoyl) peroxide, *J. am. Chem. Soc.,* 103, 7069, 1981.

25. **Nonhebel, D. C. and Walton, J. C.,** *Free Radical Chemistry,* University Press, Cambridge, 1974, 503.

26. **Tada, M., Inoue, K., and Okabe, M.,** Aryl rearrangements on the phenolysis of 2-aryl-2-ethoxy-2-phenylethyl cobaloxime, *Bull. Soc. Jpn.,* 56, 1420, 1983.

27. **Lindsay, D. A., Lusztyk, J., and Ingold, K. U.,** Kinetics of the 1,2-migration of carbon centered groups in 2-substituted 2,2-dimethylethyl radicals, *J. Am. Chem. Soc.,* 106, 7087, 1984.

28. **Witt, J. W. and Keller, S. M.,** Free radical rearrangement of a silyl radical via net 1,2-migration of an acetoxy group, *J. Am. Chem. Soc.,* 105, 1395, 1983.

29. **Boothe, T. E., Greene, J. K., Jr., and Shevlin, Ph. B.,** The stereochemistry of free radical eliminations on β-phenyltio radicals, *J. Am. Chem. Soc.,* 98, 951, 1976.

30. **Dowd, P., Shapiro, M., and Kang, J.,** The mechanism of action of vitamin B_{12}, *Tetrahedron,* 40, 3069, 1984.

31. **Halpern, J.,** Mechanistic aspects of coenzyme B_{12}-dependent rearrangements. Organometallics as free radical precursors, *Pure Appl. Chem.,* 55, 1059, 1983.

32. **Gierszak, T., Gawlowski, J., and Niedzielski, J.,** Mutual isomerization of cyclopentyl and 1-penten-5-yl radicals, *Int. J. Chem. Kinetics,* 18, 623, 1986.

33. **Griller, D., Schmidt, P., and Ingold, K. U.,** Cyclization of the 4-cyanobutyl radical, *Can. J. Chem.,* 57, 831, 1979.

34. **Lal, D., Griller, D., Husband, S., and Ingold, K. U.,** Kinetic application of electron paramagnetic resonance spectroscopy cyclization of the 5-hexenyl radical, *J. Am. Chem. Soc.,* 96, 6355, 1974.

35. **Chatgilialoglu, Ch., Woynar, H., and Ingold, K. U.,** Intramolecular reactions of alkenylsilyl radicals, *J. Chem. Soc. Perkin Trans.* II, 555, 1983.

36. **Sarasa, J. P., Igual, J., and Poblet, J. M.,** Electronic effects in the regiochemistry of the alkenylsilyl radical cyclizations, *J. Chem. Soc. Perkin Trans. II,* 861, 1986.

37. **Beckwith, A. L. J. and Schiesser, C. H.,** Regio- and stereo-selectivity of alkenyl radical ring closure: a theoretical study, *Tetrahedron,* 41, 3925, 1985.

38. **Burkhardt, P., Roduner, E., Hochman, J., and Fischer, H.,** Absolute rate constants for radical rearrangements in liquids obtained by muon spin rotation, *J. Phys. Chem.,* 88, 773, 1984.

39. **Park, S. U., Chung, S. K., and Newcomb, M.,** Acceptor, donor and captodative stabilization in transition states of 5-hexene-1-yl radical cyclizations, *J. Am. Chem. Soc.,* 108, 240, 1986.

40. **Julia, M.,** Free radical cyclization, *Pure Appl. Chem.,* 15, 167, 1967.

41. **Stork, G. and Mook, R., Jr.,** Five vs six membered ring formation in the vinyl radical cyclization, *Tetrahedron Let.,* 27, 4529, 1986.

42. **Beckwith, A. L. J. and Ingold, K. U.,** *Rearrangements in Ground State and Excited States,* de Mayo, P., Ed., Academic Press, New York, 1980, 1, 161.

43. **Roberts, Ch. and Walton, J. C.,** Homolytic ring fission reactions of bicyclo (n.1.0) alkanes and bicyclo (n.1.0) alk-2-yl radicals, electron spin resonance study of cycloalkenylmethyl radicals, *J. Chem. Soc. Perkin Trans. II,* 879, 1983.

44. **Ondruschka, B., Zimmerman, G., Remmler, M., Rennecke, D., and Anders, G.,** Reaktionen von Cyclanyl Radikalen in der Gas Phase. Cyclohexyl Radikal aus Azocyclohexane, *J. Prakt. Chem.,* 326, 561, 1984.

45. **Surzur, J. M.,** Intramolecular additions of alkoxyl radical. Reductive application cyclization of olefinic hydroperoxides, *J. Chem. Soc. Perkin Trans. II,* 547, 1983.

46. **Chenera, B., Chuang, C. P., Hart, D. J., and Hsu, L. Y.,** Observations regarding δ- and ε cyano radical cyclization, *J. Org. Chem.,* 50, 5409, 1985.

47. **Held, A. M., Manthorne, K. C., Pacey, Ph. D., and Reinholdt, H. P.,** Individual rate constants of methyl radical reactions in the pyrolysis of dimethyl ether, *Can. J. Chem.,* 55, 4128, 1977.

48. **Cremer, D.,** General and theoretical aspects of the peroxide group, in *The Chemistry of Peroxides,* Patai, S., Ed., John Wiley & Sons, London, 1983, 1.

49. **Gray, P.,** Bond dissociation energies in the phenyl benzoate molecule and in related free radicals *Adv. Chem. Ser.,* 75, 282, 1968.

50. **Gilbert, K. E. and Gajewski, J. J.,** Coal liquefaction model studies. Free radical chain decomposition of diphenylpropane, dibenzyl ether, and phenyl ether via β-scission reaction, *J. Org. Chem.,* 47, 4899, 1982.

51. **Bradley, J. N.,** A general mechanism for the high temperature pyrolysis of alkanes, *Proc. R. Soc. London,* A 337, 199, 1974.

52. **Tsang, W.,** Thermodynamic and kinetic properties of the cyclohexadienyl radical, *J. Phys. Chem.,* 90, 1152, 1986.

53. **Boothe, T. E., Greene, J. L., and Shelvin, Ph. B.,** The stereochemistry of free radical elimination on β-phenylthio radicals, *J. Am. Chem. Soc.,* 98, 951, 1976.

54. **Kerr, J. A. and Trotman-Dickenson, A. F.,** The reactions of alkyl radicals, in *Progress in Reaction Kinetics,* Porter, G., Ed., Pergamon Press, 1961, 108.

55. **Tsang, W.,** The stability of alkyl radicals, *J. Am. Chem. Soc.,* 107, 2872, 1985.

56. **Beckwith, A. L. J., Griller, D., and Lorand, J. P.,** Radical reaction rates in liquids, in *Landolt-Bornstein: Numerical Data and Functional Relationships in Science Technology,* Springer-Verlag, Berlin, 1984.

57. **Nonhebel, D. C. and Walton, J. C.,** *Free Radical Chemistry,* University Press, Cambridge, 1974, 471.

58. **Sanderson, R. T.,** *Chemical Bonds and Bond Energy,* Academic Press, New York, 1976, 161.

59. **Drew, R. M., Kerr, J. A., and Olive, J.,** Relative rate constants of the gas phase decomposition of the s-butoxy radical, *Int. J. Chem. Kinetics,* 17, 167, 1985.

60. **Baignee, A., Howard, J. A., Scaiano, J. C., and Stewart, L. C.,** Absolute rate constants for reactions of cumyloxy radicals in solution, *J. Am. Chem. Soc.,* 105, 6120, 1983.

61. **Kerr, J. A. and Wright, J. P.,** The kinetics of the gas phase decomposition reactions of the trifluoro t-butoxy radicals, *Int. J. Chem. Kinetics,* 16, 1321, 1984.

62. **Jiang, X.-K., Li, X. Y., and Wang, K.-Y.,** CH_3 vs. CF_3. Relative rates of formation from β-scission, *J. Chem. Soc., Chem. Commun.,* 745, 1986.

63. **Ruegge, D. and Fischer, H.,** Rate constants for the decarboxylation of the $+-$butoxycarbonyl radical and concurring radical terminations directly obtained by kinetic ESR, *Int. J. Chem. Kinetics,* 18, 145, 1986.

64. **Nonhebel, D. C. and Walton, J. C.,** *Free Radical Chemistry,* University Press, Cambridge, 1974, 491.

65. **Petraev, E. P., Maslovskaya, L. A., and Shadyro, O. I.,** The effect of oxygen on transformations of α-diol radicals (Russian), *Zh. Org. Khim.,* 19, 2263, 1983.

66. **Ershov, B. G. and Isakova, O. V.,** Formation and thermal transformation of radicals in γ-irradiated cellulose (Russian), *Izv. AN SSSR, Khim, Ser.,* 1276, 1984.

67. **Slagle, I. R., Jong-Yoon Park, M. C., Heaven, M. C., and Gutman, D.,** Kinetics of polyatomic free radicals produced by laser photolysis. 3. Reaction of vinyl radicals with molecular oxygen. *J. Am. Chem. Soc.,* 106, 4356, 1984.

68. **Turro. N. J., Gould, I. R., and Baretz, B. H.,** Absolute rate constants for decarbonylation of phenoxyacetyll and related radicals, *J. Phys. Chem.,* 87, 531, 1983.

69. **Lunazzi, L., Ingold, K. U., and Scaiano, J. C.,** Absolute rate constants for the decarbonylation of the phenylacetyl radical, *J. Phys. Chem.,* 87, 529, 1983.

70. **Kerr, J. A. and Wright, J. P.,** Kinetic study of the gas-phase decomposition of the trifluoroacetyl radical, *J. Chem. Soc., Faraday, Trans.,* 1, 81, 1471, 1985.

71. **Chin-Yu Lin and Lin, M. C.,** Unimolecular decomposition of the phenoxy radical in shock waves, *Int. J. Chem. Kinetics,* 17, 1025, 1985.

72. **Vollenweider, J. K. and Paul, H.,** On the rate of decarbonylation of hydroxyacetyl and other acyl radicals, *Int. J. Chem. Kinetics,* 18, 791, 1986.

73. **Dannenberg, J. J.,** A theoretical study of the decomposition of alkyldiazenyl radicals, *J. Org. Chem.,* 50, 4963, 1985.

74. **Bhattacharya, D. and Willard, J. E.,** Photoeliminiation of H from radicals in CH_4 and Xe matrices, *J. Phys. Chem.,* 86, 962, 1982.

75. **Bentrude, W. G.,** Phosphoranyl radicals: their structure, formation and reactions, *Acc. Chem. Res.,* 15, 117, 1982.

76. **Burkey, T. J., Majewski, M., and Griller, D.,** Heats of formation of radicals and molecules by a photoacoustic technique, *J. Am. Chem. Soc.,* 108, 2218, 1986.

77. **Bloodworth, A. J., Courtneidge, J. L., and Davies, A. G.,** Rate constants for the formation of oxiranes by γ-scission in secondary β-t-butylperoxyalkyl radicals, *J. Chem. Soc. Perkin Trans. II,* 523, 1984.

78. **Olivella, S., Canadell, E., and Poblet, J. M.,** Theoretical study of the addition of vinyl and cyclopropyl radicals to ethylene, *J. Org. Chem.,* 48, 4696, 1983.

79. **Arnaud, R., Donady, J., and Subra, R.,** Theoretical study of radical addition reactions: addition of δ-free radicals to ethylenic and aromatic systems, *Nouveau J. Chem.,* 5, 181, 1981.
80. **Houk, K. N., Paddon-Row, M. N., Spellmeyer, D. C., Rondan, N. C., and Nagase, S.,** Theoretical transition structures for radical additions to alkenes, *J. Org. Chem.,* 51, 2874, 1986.
81. **Griesbaum, K.,** Probleme und Moglichkeiten der Radikalischen Addition von Thiolen und ungesattig Verbindungen, *Angew. Chemie,* 82, 276, 1970.
82. **Yoshihara, M., Nozaki, K., Fujihara, H., and Maeshima, T.,** Asymmetric induction in free radical addition of thiols to olefins, *J. Polym. Sci., Polym. Lett.,* 19, 49, 1981.
83. **Knoll, H.,** Zur Kinetik der Radikalischen Addition an C = O Bindungen in der Gasphase, *Z. Chem.,* 22, 245, 1982.
84. **Abell, P. I.,** Additions to multiple bonds, in *Free Radicals,* Vol. 2, Kochi, J. K., Ed., John Wiley & Sons, New York, 1973, 64.
85. **Johnston, L. J., Scaiano, J. C., and Ingold, K. U.,** Kinetics of cyclopropyl radical reactions, *J. Am. Chem. Soc.,* 106, 4877, 1984.
86. **Tully, F. P., Ravishankara, A. R., Thompson, R. L., Nicowich, J. M., Shah, R. C., Kreutter, N. M., and Wine, P. H.,** Kinetics of the reactions of hydroxyl radical with benzene and toluene, *J. Phys. Chem.,* 85, 2262, 1981.
87. **Pryor, W. A., Lin, T. H., Stanley, J. P., and Henderson, R. W.,** Addition of hydrogen atoms to substituted benzene. Use of the Hammet equation for correlating radical reactions, *J. Am. Chem. Soc.,* 95, 6993, 1973.
88. **Sakurai, H.,** Group IV B radicals, in *Free Radicals,* Vol. 2, Kochi, J. K., Ed., John Wiley & Sons, New York, 1973, 741.
89. **Perkins, M. J.,** Aromatic substitution, in *Free Radicals,* Vol. 2, Kochi, J. K., Ed., New York, 1973, 231.
90. **Veltwich, D. and Asmus, K. D.,** On the reaction of methyl and phenyl radicals with p-benzoquinone in aqueous solution, *J. Chem. Soc. Perkin Trans. II,* 1147, 1982.
91. **Dohmaru, T. and Nagata, Y.,** Kinetics of gas phase addition reactions of trichlorsilyl radicals, *J. Chem. Soc. Faraday Trans. I,* 75, 2617, 1979.
92. **Bat, L. and Mowat, S. I.,** The addition of methyl radicals to hexafluoroacetone, *Int. J. Chem. Kinetics,* 16, 603, 1984.
93. **Kerr, J. A. and Wright, J. P.,** The Kinetic of the addition of alkyl radicals to carbonyl groups. II. The addition of methyl radicals to trifluoracetone in the gas phase, *Int. J. Chem. Kinetics,* 16, 1327, 1984.
94. **Chatgilialoglu, C., Ingold, K. U., and Scaiano, J. C.,** Absolute rate constants for the addition of triethylsilyl radicals to the carbonyl group, *J. Am. Chem. Soc.,* 5119, 1982.
95. **Alberti, A., Chatgilialoglu, C., Peduli, G. F., and Zanirato, P.,** On the addition of silyl radicals to unsaturated carbonyl compounds: regioselectivity of the attack and 1,3-carbon to oxygen silicon migration, *J. Chem. Soc.,* 108, 4993, 1986.
96. **Schmid, P. and Ingold, K. U.,** Rate constants for spin trapping. 1. Primary alkyl radicals, *J. Am. Chem. Soc.,* 100, 2493, 1978.
97. **Zubarev, V. E., Belevskij, V. E., and Bugaienko, L. T.,** The application of spin traps for investigation of mechanism of radical process, (Russian), *Usp. Khim.,* 48, 1361, 1979.
98. **Torssell, K.,** Investigation of radical intermediates in organic reactions by use of nitroso compounds as scavangers, *Tetrahedron,* 26, 2759, 1970.
99. **Lobanova, T. V., Kasaikina, O. T., Povarov, L. S., Shapiro, A. B., and Gagarina, A. B.,** 4-Methyl-2-spirocyclohexyl-2', 3', 3,4-tetrahydrofuran of peroxyradicals, (Russian), *DAN,* 245, 1154, 1979.
100. **Park, J. Y. and Gutman, D.,** Reaction of ethyl, cyclopentyl and cyanomethyl radicals with nitrogen dioxide. *J. Phys. Chem.,* 87, 1844, 1983.
101. **Janzen, E. G., Strouks, H. J., Du Bose, C. M., Poyer, J. L., and McCay, P. B.,** Chemistry and biology of spin-trapping associated with halocarbon metabolism in vitro and in vivo, *Environ. Health Perspect.,* 64, 151, 1985.
102. **Slagle, I. R., Ratajczak, E., Heaven, M. C., Gutman, D., and Wagner, A. F.,** Kinetics of polyatomic free radicals produced by laser photolysis. 4. Study of the equilibrium i-C_3H_7 + O_2 = i-$C_3H_7O_2$ between 592 and 692 K, *J. Am. Chem. Soc.,* 107, 1838, 1985.
103. **Maillard, B., Ingold, K. U., and Scaiano, J. C.,** Rate constants for the reactions of free radicals with oxygen in solution, *J. Am. Chem. Soc.,* 105, 5095, 1983.
104. **Ruiz, R. P. and Bayes, K. D.,** Rates of reaction of propyl radicals with molecular oxygen, *J. Phys. Chem.,* 88, 2592, 1984.
105. **Wu, D. and Bayes, K. D.,** Rate constants for the reactions of isobutyl, neopentyl, and cyclohexyl radicals with molecular oxygen, *Int. J. Chem. Kinetics,* 18, 547, 1986.
106. **Plumb, I. C. and Ryan, K. R.,** Kinetic studies of the reaction of ·C_2H_5 with O_2 at 295 K, *Int. J. Chem. Kinetics,* 13, 1011, 1981.

107. **Saito, K., Ito, R., Kakumoto, T., and Imamura, A.,** The initial process of the oxidation of the methyl radical in reflected shock waves, *J. Phys. Chem.,* 90, 1422, 1986.

108. **Kan, C. S., Calvert, J. G., and Shaw, J. H.,** Oxidation of sulfur dioxide by methylperoxy radicals, *J. Phys. Chem.,* 85, 1126, 1981.

109. **Castleman, A. W. and Tang, I. N.,** Kinetics of the association reaction of SO_2 with the hydroxyl radical, *J. Photochem.,* 6, 349, 1976/1977.

110. **Giese, B.,** Formation of C-C bonds by addition of free radicals to alkenes, *Angew. Chem. Int. Ed. Engl.,* 22, 753, 1983.

111. **Munger, K. and Fischer, H.,** Separation of polar and steric effects on absolute rate constants and Arrhenius parameters for the reaction of tert-butyl radicals with alkenes, *Int. J. Chem. Kinetics,* 17, 809, 1985.

112. **Citterio, A., Minisci, F., and Vismara, E.,** Steric, polar, and resonance effects in reactivity and regioselectivity of aryl radical addition to α, β-unsaturated carbonyl compounds, *J. Org. Chem.,* 47, 81, 1982.

113. **Nonhebel, D. C. and Walton, J. C.,** *Free Radical Chemistry,* University Press, Cambridge, 1974, 278.

114. **Davies, D. I.,** Addition reactions, in *Free Radical Reactions,* Waters, W. A., Ed., *Organic Chemistry,* Vol. 10, Butterworths, London, 1973, 50.

115. **Tedder, J. M. and Walton, J. C.,** Free radical additions to olefins. 11. Addition of bromodifluoromethyl radicals to fluoroethylenes, *J. Chem. Soc., Faraday Trans. I,* 70, 308, 1974.

116. **Fordham, J. W. L.,** Stereoregulated polymerization in the free propagating species, *J. Polym. Sci.,* 39, 321, 1959.

117. **Dijkstra, A. J., Kerr, J. A., and Trotman-Dickenson, A. F.,** Reactions of difluoramino-radicals, *J. Chem. Soc. (A),* 105, 1967.

118. **Levin, Ya. A., Abulchanov, A. G., Nefedov, V. C., Skorobogatova, M. S., and Ivanov, B. E.,** Kinetics and mechanism of phenyl radical addition to unsaturated compounds (Russian), *DAN,* 235, 629, 1977.

119. **Dohmaru, T. and Nagata, Y.,** Kinetics of the gas-phase addition reactions of trichlorsilyl radicals, *J. Chem. Soc. Faraday Trans. I,* 78, 1141, 1982.

120. **Baban, J. A. and Roberts, B. P.,** Polar effects on the addition of alkyl radicals to diethyl vinyl phosphonate, *J. Chem. Soc. Chem. Commun.,* 373, 1979.

121. **Sawaki, Y. and Ogata, Y.,** Reaction of acylperoxy radicals in the photoreactions of α-diketones and oxygen, *J. Org. Chem.,* 49, 3344, 1984.

122. **Sway, M. J. and Waddington, D. J.,** Reactions of oxygenated radicals in the gas phase, *J. Am. Chem. Soc. Perkin Trans. II,* 139, 1983.

123. **Glavas, S. and Heicklen, J.,** Relative reactivity of chlorine atoms with NO, NO_2 and $HCCl_2F$ at room temperature and atmospheric pressure, *J. Photochem.,* 31, 21, 1985.

124. **Ingold, K. U. and Roberts, B. P.,** *Free Radical Substitution Reactions,* Wiley Interscience, New York, 1971.

125. **Griller, D., Ingold, K. U., Patterson, L. K., Scaiano, J. C., and Small, R. D.,** A study of transient phenomena in the reactions of alkoxy radicals with triphenyl phosphine and triphenyl borane, *J. Am. Chem. Soc.,* 101, 3780, 1979.

126. **Howard, J. A. and Tait, J. C.,** Organometallic peroxy radicals, *Can. J. Chem.,* 56, 2163, 1978.

127. **Bentrude, W. G. and Min, T. B.,** Free radical studies of organophosphorus compounds, *J. Am. Chem. Soc.,* 98, 2918, 1976.

128. **Penkovski, V. V.,** Free radicals of phosphorus compounds (Russian), *Usp. Khim.,* 44, 969, 1975.

129. **Ryabokon, I. G., Polumbrik, O. M., and Markovski, L. N.,** Interaction of perfluortriphenylverdazyls with benzoyl peroxide, (Russian), *DAN U.S.S.R.,* 49, 1984.

130. **Lefort, D., Fossey, J., Gruselle, M., and Nedelec, J. Y.,** Reactivity and selectivity of homolytic substitution reactions on the peroxyacid group, *Tetrahedron,* 41, 4237, 1985.

131. **Bonifacic, M. and Asmus, K. D.,** Adduct formation and absolute rate constants in the displacement reactions of thiyl radicals with disulfides, *J. Phys. Chem.,* 88, 6286, 1984.

132. **Gobl, M., Bonifacic, M., and Asmus, K. D.,** Substituents effects on the stability of three electron bonded radicals and radical ions from organic sulfur compounds, *J. Am. Chem. Soc.,* 106, 5984, 1984.

133. **Migita, T., Kosugi, M., Takayama, K., and Nakagawa, I.,** S_H type reactions of substituted allylic compounds, *Tetrahedron,* 29, 51, 1973.

134. **Back, R. A.,** A search for gas phase free radical inversion displacement reaction at a saturated carbon atom, *Can. J. Chem.,* 61, 916, 1983.

135. **Chatgilialoglu, C., Ingold, K. U., and Scaiano, J. C.,** Absolute rate constants for the reaction of triethylsilyl radicals with organic halides, *J. Am. Chem. Soc.,* 104, 5123, 1982.

136. **Donaldson, D. J. and Leone, S. R.,** Absolute rate coefficients for methyl radical reactions by laser photolysis, time-resolved infrared chemiluminescence: $CD_3 + HX \rightarrow CD_3H + X$ (X = Br, I), *J. Phys. Chem.,* 90, 936, 1986.

137. **Gilliom, R. D.,** Activation energies from bond energies, *J. Am. Chem. Soc.,* 99, 8399, 1977.

138. **Lesclaux, R.,** Reactivity and kinetic properties of the NH_2 radical in the gas phase, *Rev. Chem. Intermediates,* 5, 347, 1984.

139. **Moller, W., Mozzhukhin, E., and Wagner, H. G.,** High temperature reactions of Ch_3.1. The reaction $CH_3 + CH_2 \rightarrow Ch_4 + H$, *Ber. Bunsenges. Phys. Chem.,* 90, 854, 1986.

140. **Jolly, G. S., Paraskevopoulos, G., and Singleton, D. L.,** Rates of OH radical reactions. XII. The reactions of OH with c-C_3H_6, c-C_5H_{10} and c-C_7H_{14}. Correlation of hydroxyl rate constants with bond dissociation energies. *Int. J. Chem. Kinetics,* 17, 1, 1985.

141. **Baulch, D. L., Campbell, I. M., and Saunders, S. M.,** Rate constants for the reactions of hydroxyl radicals with propane and ethane, *J. Chem. Soc. Faraday Trans.,* I, 81, 259, 1985.

142. **Tsepalov, V. F., Kharytonova, A. A., and Gladyshev, G. P.,** Experimental values of preexponential factors of rate constants for some liquid and gas phase substitution reactions, (Russian), *DAN,* 225, 152, 1975.

143. **Christiansen, H., Sehested, K., and Corfitzen, H.,** Reactions of hydroxy radicals with hydrogen peroxide at ambient and elevated temperatures, *J. Phys. Chem.,* 86, 1588, 1982.

144. **Buchachenko, A. L. and Vasserman, A. M.,** *Stable Radicals* (Russian), Chimia, Moscow, 1973, 405.

145. **Denisov, E. T.,** Reactivity of polyfunctional compounds in radical reactions (Russian), *Usp. Khim.,* 54, 1466, 1985.

146. **Doba, T. and Ingold, K. U.,** Kinetic application of electron paramagnetic resonance spectroscopy. 42. Some reactions of the bis(trifluoromethyl) aminoxyl radical, *J. Am. Chem. Soc.,* 106, 3958, 1984.

147. **Chenier, J. H. B., Furimsky, E., and Howard, J. A.,** Arrhenius parameters for reaction of the tert.butylperoxy and 2-ethyl-2-propylperoxy radicals with some nonhindered phenols, aromatic amines and thiophenols, *Can. J. Chem.,* 52, 3682, 1974.

148. **Anpo, M., Sutcliffe, R., and Ingold, K. U.,** Kinetic applications of electron paramagnetic resonance spectroscopy. 41. Diethoxyphosphonyl radicals, *J. Am. Chem. Soc.,* 105, 3580, 1983.

149. **Scaiano, J.,** Detection of trialkylstannyl radicals using laser flash photolysis, *J. Am. Chem. Soc.,* 102, 5399, 1980.

150. **Sidebottom, H. and Treacy, J.,** Reaction of methyl radicals with haloalkanes, *Int. J. Chem. Kinetics,* 16, 579, 1984.

151. **Dutsch, H. R. and Fischer, H.,** Rate constants for the reaction of tert.butyl radicals with chloroform in solution, *Int. J. Chem. Kinetics,* 13, 527, 1981.

152. **Husain, D. and Marshall, P.,** Determination of absolute rate data for the reactions of atomic sodium, $Na(3^2S_{+/2})$, with CH_3F, CH_3Cl, CH_3Br, HCl and HBr as a function of temperature by time resolved atomic resonance spectroscopy, *Int. J. Chem. Kinetics,* 18, 83, 1986.

153. **De Vohringer, C. M. and Staricco, E. H.,** Gas phase reaction of heptafluoroisopropyl radicals with carbon tetrachloride, *Int. J. Chem. Kinetics,* 18, 41, 1986.

154. **Martin, J. P. and Paraskevopoulos, G.,** A kinetic study of the reactions with trifluoroethanes. Estimation of C-H bond strength in fluoroalkanes, *Can. J. Chem.,* 61, 861, 1983.

155. **Benson, S. W.,** *Thermochemical Kinetics,* 2nd ed., John Wiley & Sons, New York, 1976, 307.

156. **Sanderson, R. T.,** *Chemical Bonds and Bond Energy,* Academic Press, New York, 1976, 193.

157. **Tedder., J. M.,** Welche Faktoren Bestimmen Reaktivitat und Regioselektivitat bei Radikalischer Substitution und Addition? *Angew. Chem.,* 94, 433, 1982.

158. **Arican, H. and Arthur, N. L.,** Reaction of CF_3 radicals with H_2S, *Int. J. Chem. Kinetics,* 16, 335, 1984.

159. **Macken, K. V. and Sidebotom, H. W.,** The reactions of methyl radicals with chloromethanes, *Int. J. Chem. Kinetics,* 11, 511, 1979.

160. **Trotman-Dickenson, A. F.,** The abstraction of hydrogen atoms by free radicals, in *Advances in Free-Radical Chemistry,* Vol. 1, Williams, G. H., Ed., Logos Press-Academic Press, London, 1965, 1.

161. **Skell, P. S. and Shea, K. J.,** Bridged free radicals, in *Free Radicals,* Vol. 2, Kochi, J. K., Ed., John Wiley & Sons, New York, 1973, 810.

162. **Tschuikov-Roux, E., Faraj, F., and Niedzielski, J.,** Rate constants and kinetic isotope effects for hydrogen (deuterium abstraction by chlorine atoms from the chloromethyl group in CH_2ClCH_2Cl, CD_2ClCD_2Cl, $CH_2ClCHCl_2$ and $CH_2ClCDCl_2$), *Int. J. Chem. Kinetics,* 18, 513, 1986.

163. **Migita, T., Takayama, K., Abe, Y., and Kosugi, M.,** Relative reactivities of substituted phenyl radicals in elementary reactions, *J. Chem. Soc. Perkin Trans. II,* 1137, 1979.

164. **Kuwae, Y., Kamachi, M., Hayashi, K., and Viehe, H. G.,** ESR study of radical halogen abstractions. Participation of electron-transfer mechanism in the radical halogen abstractions, *Bull. Chem. Soc. Jpn.,* 59, 2325, 1986.

165. **Pryor, W. A.,** *Free Radicals,* McGraw-Hill, New York, 1966, 157.

166. **Kosower, E. M.,** Pyridinyl paradigm proves powerful, in *Organic Free Radicals,* Pryor, W. A., Ed., ACS 69, Washington, D.C., 1978, 447.

167. **Fischer, H., Ed.,** *Radical Reaction Rates in Liquids,* Landolt-Bernstein, New Series, Group II, Springer-Verlag, Berlin, 13 a-d, 1983, 1984.

168. **Kondratyev, V. N.,** *Rate Constants of Gaseous Phase Reactions,* (Russian), Publishing House Nauka, Moscow, 1970.

169. **Denisov, E. T.,** *Rate Constants of Homolytic Liquid Phase Reactions,* (Russian), Publishing House Nauka, Moscow, 1971.

170. **Kerr, J. A. and Moss, S. J.,** *Handbook of Bimolecular and Termolecular Gas Reactions,* Vol. I, CRC Press, Boca Raton, Fl., 1981.

171. **Tsepalov, V. F., Kharytonova, A. S., and Gladyshev, G. P.,** Experimental values of the preexponential factors of the rate constants for some liquid and gaseous phase radicals substitution reactions (Russian), *DAN,* 225, 152, 1975.

172. **Alfassi, Z. B. and Mosseri, Sh.,** The kinetics of radiation — induced hydrogen abstraction by CCl_3· and CCl_3O_2· radicals from cyclohexane in the liquid phase, *J. Phys. Chem.,* 88, 3296, 1984.

173. **Hawari, J. A., Davis, S., Engel, P. S., Gilbert, B. C., and Griller, D.,** The free radical reaction between alkanes and carbon tetrachloride, *J. Am. Chem. Soc.,* 107, 4721, 1985.

174. **Doba, T. and Ingold, K. U.,** Kinetic applications of electron paramagnetic resonance spectroscopy. 42. Some reactions of the bis(trifluoromethyl)aminoxyl radical, *J. Am. Chem. Soc.,* 106, 3958, 1984.

175. **Spraque, E. D.,** Deuterium isotope effect on hydrogen atom abstraction by methyl radicals in acetonitrile at 77 K. Evidence for tunneling, *J. Phys. Chem.,* 81, 516, 1977.

176. **Toriyama, K. and Iwasaki, M.,** Tunneling effect on hydrogen abstraction by hydrogen atoms and their trapping in some hydrocarbons irradiated at low temperatures, *J. Phys. Chem.,* 82, 2056, 1978.

177. **Iwasaki, M., Toriyama, K., Muto, H., and Nunome, K.,** Hydrogen abstraction by thermal hydrogen atoms at 10 — 30 K. Electron spin resonance evidence, *Chem. Phys. Lett.,* 56, 494, 1978.

178. **Ingold, K. U.,** Kinetic studies of some free radical reactions, *Pure Appl. Chem.,* 56, 1767, 1984.

179. **Karpukhin, O. N.,** The influence of the reaction medium mobility on the formal kinetic laws of the course of chemical reactions in the condensed phase, (Russian), *Usp. Khim.,* 47, 1119, 1978.

180. **Zaporozskaya, O. A., Yassina, L. L., and Pudov, V. S.,** Kinetics of reactions of low molecular radicals in solid matrices. Methyl radical in polymers, (Russian), *Zh. Phys. Khim.,* 53, 2438, 1979.

181. **Weir, N. A. and Rujimethabhas, M.,** Reactions of radicals with polymers. III. Abstraction reactions of methyl and tert.butyl radicals with poly(p-tert.butyl styrene), *Eur. Polym. J.,* 21, 493, 1985.

182. **Keene, J. P.,** The absorption spectrum and some reaction constants of the hydrated electron, *Radiat. Res.,* 22, 1, 1964.

183. **Morkovnik, A. C. and Okhlobystin, O. Yu.,** Inorganic ion-radicals and their organic reactions (Russian), *Usp. Khim.,* 48, P. 1968, 1979.

184. **Schwarz, H. A. and Bielski, B. H. J.,** Reactions of HO_2· and O_2^- with iodine and bromine and the I_2^- and I. Atom reduction potentials, *J. Phys. Chem.,* 90, 1445, 1986.

185. **Afanasyev, I. B.,** Anionradical O_2^- of oxygen in chemical and biochemical processes (Russian), *Usp. Khim.,* 48, 977, 1979.

186. **Olson, J. S., Ballou, D. P., Palmer, G., and Massey, V.,** The mechanism of action of xanthin oxidase, *J. Biol. Chem.,* 249, 4363, 1974.

187. **Schuler, R. H., Neta, P., Zemel, H., and Fessenden, R. W.,** Conversion of hydroxyphenyl to phenoxyl radicals. A radiolytic study of the reduction of bromophenols in aqueous solution, *J. Am. Chem. Soc.,* 98, 3825, 1976.

188. **Stenkeen, S. and Neta, P.,** Electron transfer rates and equilibria between substituted phenoxide ions and phenoxyl radicals, *J. Phys. Chem.,* 83, 1134, 1979.

189. **Khudjakov, I. V. and Kuzmin, V. A.,** Redox reactions of free radicals (Russian), *Usp. Khim.,* 47, 39, 1978.

190. **Mockel, K., Bonifacic, M., and Asmus, K.-D.,** Formation of positive ions in the reaction of disulfides with hydroxyl radicals in aqueous solution, *J. Phys. Chem.,* 78, 282, 1974.

191. **Monig, J., Gobl, M., and Asmus, K.-D.,** Free radical one-electron versus hydroxyl radical induced oxidation. Reaction of trichloromethyl peroxyl radicals with simple and substituted aliphatic sulfides in aqueous solution, *J. Chem. Soc. Perkin Trans. II,* 647, 1985.

192. **Ahmad, R. and Armstrong, D. A.,** The effect of pH and complexation on redox reactions between RS· radicals and flavins, *Can. J. Chem.,* 62, 171, 1984.

193. **Solon, E. and Bard, A. J.,** The electrochemistry of di-phenylpicrylhydrazyl, *J. Am. Chem. Soc.,* 86, 1926, 1964.

194. **Soos, Z. G., Keller, H. J., Moroni, W., and Nothe, D.,** Cation radical salts of phenazine, *J. Am. Chem. Soc.,* 99, 5040, 1977.

195. **Anderson, R. F. and Patel, K. B.,** Radical cations of some low potentials viologen compounds, *J. Chem. Soc. Faraday Trans. I.,* 80, 2693, 1984.

196. **Levey, G. and Ebbensen, T. W.,** Methyl viologen. Radical reactions with several oxidizing agents, *J. Phys. Chem.,* 87, 826, 1983.

197. **Razumas, V. J., Gudavichus, A. V., and Kulis, Yu.,** Kinetics of peroxidase reduction by viologens (Russian), *Khim. Phys.,* 4, 1398, 1985.

198. **Polumbrik, O. M.,** The success of chemistry of verdazyl radicals (Russian), *Usp. Khim.,* 47, 1444, 1978.
199. **Nonhebel, D. C. and Walton, J.C.,** *Free Radical Chemistry,* University Press, Cambridge, 1974, 193.
200. **Koster, R. and Asmus, K.,-D.,** The reduction of carbon tetrachloride by hydrated electrons, H-atoms and reducing radicals, *Z. Naturforsch.,* 26 b, 1104, 1971.
201. **Jaggannadham, V. and Steenken, S.,** One electron reduction via addition/elimination. An example of an organic inner sphere electron-transfer reaction, *J. Am. Chem. Soc.,* 106, 6542, 1984.
202. **Julliard, M. and Chanon, M.,** Photoelectron-transfer catalysis: its connection with thermal and electrochemical analogues, *Chem. Rev.,* 83, 425, 1983.
203. **Zamaraev, K. I. and Khairutdinov R. F.,** Tunnel transfer of electron to longer distance in chemical reactions (Russian), *Usp. Khim.,* 47, 992, 1978.
204. **Shimada, K. and Szwarc, M.,** How far can electrons be transferred?, *Chem. Phys. Lett.,* 28, 540, 1974.
205. **Hush, N. S.,** Distance dependence of electron transfer rates, *Coord. Chem. Rev.,* 64, 135, 1985.
206. **Marcus, R. A.,** Electron, proton and related transfers, *Faraday Discuss. Chem. Soc.,* 74, 7, 1982.
207. **Fischer-Hjalmers, I., Holmgren, H., and Henriksson-Enflo, A.,** Metals in biology. Iron complexes modelling electron transfer, *Int. J. Quantum Chem., Quantum Biol. Symp.,* 12, 57, 1986.
208. **Minisci, F., Citterio, A., and Giordano, C.,** Electron transfer processes: peroxysulfate a useful and versatile reagent in organic chemistry, *Acc. Chem. Res.,* 16, 27, 1983.
209. **Minisci. F.,** Free radical additions to olefins in presence of redox systems, *Acc. Chem. Res.,* 8, 165, 1975.
210. **Kochi, J. K.,** Mechanisms of organic oxidation and reaction by metal complexes, *Science,* 155, 415, 1967.
211. **Kochi, J. K. and Bacha, J. D.,** Solvent effects on the oxidation of alkyl radicals by lead (IV) and copper (II) complexes, *J. Org. Chem.,* 33, 2746, 1968.
212. **Rao, P. S. and Haydon, E.,** Redox potentials of free radicals, *J. Amer. Chem. Soc.,* 96, 1287-1300, 1974.
213. **Maletin, Y. A. and Strizhakova, N. G.,** Kinetics and mechanism of electron transfer via bridge metal ion, *Int. J. Chem. Kinetics,* 18, 13, 1986.
214. **Vedeneyev, V. I., Gurvich, L. V., Kondratyev, V. N., Medvedev, V. A., and Frankevich, E. L.,** *Energy of Dissociation of Chemical Bonds. Ionization Potentials and Electron Affinity* (Russian), Publishing House Academy of Science U.S.S.R., Moscow, 1962.
215. **Gaines, A. F. and Page, F. M.,** Semiempirical prediction of the electron affinities of gaseous radicals, *Trans. Faraday Soc.,* 62, 3086, 1966.
216. **Gaines, A. F., Kay, J., and Page, F. M.,** Determination of electron affinities, *Trans. Faraday Soc.,* 62, 874, 1966.
217. **Shinohara, H. and Imamura, A.,** The origin of the polarity of a radical in hydrogen abstraction reactions, *Bull. Chem. Soc. Jpn.,* 52, 3265, 1979.
218. **Deuchert, K. and Hunig, S.,** Multistage organic redox systems. A general structural principle, *Angew. Chem. Int. Ed.,* 17, 875, 1978.
219. **Bubnov, N. N., Medvedev, B. Ya., Polyakova, L. A., Bilevich, K. A., and Okhlobystin, O. Yu.,** Free radicals in reactions of organic cations, *Org. Magn. Reson.,* 5, 437, 1973.
220. **Beletskaya, I. P., Rykov, S. V., and Buchachenko, A. L.,** Chemically induced dynamic nuclear polarization in some S_E and S_N reaction, *Org. Magn. Res.,* 5, 595, 1973.
221. **Babonneau, F. and Livage, J.,** Dynamics of electron transfer in mixed valence systems, *New J. Chem.,* 10, 191, 1986.
222. **Cannon, R. D.,** *Electron Transfer Reactions,* Butterworths, London, 1980.
223. **Scott, R. A., Mauk, A. G., and Gray, H. B.,** Experimental approaches to studying biological electron transfer, *J. Chem. Educ.,* 62, 932, 1985.
224. **Gray, H. B.,** Long-large electron-transfer in blue copper proteins, *Chem. Soc. Rev.,* 15, 17, 1986.
225. **Dannenberg, J. J. and Tanaka, K.,** Theoretical studies of radical recombination reactions. Alkyl and azoalkyl radicals, *J. Am. Chem. Soc.,* 107, 671, 1985.
226. **Boyd, A. K. and Burns, G.,** Halogen recombination-dissociation reactions. Current status, *J. Phys. Chem.,* 83, 88, 1973.
227. **Plane, J. M. C. and Husain, D.,** Absolute third-order rate constants for the recombination reactions between alkali-metal and iodine atoms and the measurement for Rb + I + He, *J. Chem. Soc., Faraday Trans.,* 2, 82, 897, 1986.
228. **Bunker, D. L.,** *Theory of Elementary Gas Reaction Rates,* Pergamon Press, Oxford, 1966.
229. **Smith, I. W. M.,** Collisions between free radicals, *Sci. Prog. (Oxford),* 69, 177, 1984.
230. **Smith, I. W. M.,** The role of electronically excited states in recombination reactions, *Int. J. Chem. Kinetics,* 16, 423, 1984.
231. **Jablonski, A.,** Heterogeneous Recombination of Atoms, *J. Chem. Soc. Faraday Trans. I.,* 73, 111, 1977.
232. **Glass, G. P. and Chaturvedi, B. K.,** A novel wall reaction of OH·, *Int. J. Chem. Kinetics,* 14, 153, 1982.

233. **Selamoglu, N., Rossi, M. J., and Golden, D. M.,** Reaction of ·CF₃ radicals on fused silica between 320 and 530 K, *J. Chem. Phys.,* 84, 2400, 1986.

234. **Hole, K. J. and Mulcahy, M. F. R.,** The pyrolysis of biacetyl and the third body effect on the combination of methyl radicals, *J. Phys. Chem.,* 73, 177, 1969.

235. **Greenhill, P. G. and Gilbert, R. G.,** Recombination reactions: variational transition state theory and the Gorin model, *J. Phys. Chem.,* 90, 3104, 1986.

236. **Cohen, L. K. and Pritchard, H. O.,** Oriented collisions in the recombination and disproportionation of free radicals, *Can. J. Chem.,* 63, 2374, 1985.

237. **Golden, D. M., Choo, K. Y., Perona, M. J., and Piszkiewicz, L. W.,** An absolute measurement of the rate constant for ethyl radical combination, *Int. Chem. Kinetics,* 8, 381, 1976.

238. **Griller, D. and Ingold, K. U.,** Rate constants for the bimolecular self-reactions of ethyl, isopropyl, and tert.butyl radicals in solution. A direct comparison, *Int. J. Chem. Kinetics,* 6, 453, 1974.

239. **Golden, D. M., Piszkiewicz, L. W., Perona, M. J., and Beadle, P. C.,** An absolute measurement of the rate constant for isopropyl radical combination, *J. Am. Chem. Soc.,* 96, 1645, 1974.

240. **Throssell, J. J.,** Rates of reaction of allyl radicals. A reassessment, *Int. J. Chem. Kinetics,* 4, 273, 1972.

241. **Bennet, J. E. and Summers, R.,** Electron spin resonance measurements of the termination rate constant for t-bu-tyl radicals in solution, *J. Chem. Soc. Perkin II,* 1504, 1977.

242. **Meiggs, T., Grossweiner, L. I., and Miller, S. I.,** Extinction coefficient and recombination rate of benzyl radicals, *J. Am. Chem. Soc.,* 94, 7981, 1972.

243. **Burkhartt, R. D.,** Radical-radical reactions in different solvents. Propyl, cyclohexyl, and benzyl radicals, *J. Phys. Chem.,* 73, 2703, 1969.

244. **Carlson, D. J., Howard, J. A., and Ingold, K. U.,** Absolute rate constants for the combination of trichloromethyl radicals and for their reaction with t-butyl hypochlorite, *J. Am. Chem. Soc.,* 88, 4726, 1966.

245. **Kiryushin, Y. A.,** The study of kinetics of pentachloroethyl radicals by impulse photolysis (Russian), *Kinet. Catal. Lett.,* 16, 288, 1975.

246. **Wong, S. K.,** An indirect measurement of the absolute rate constant of the self-reaction of tert-butoxy radicals, *Int. J. Chem. Kinetics,* 13, 433, 1981.

247. **Akiyama, K., Kubota, S. T., Ikegami, Y., and Ikenone, T.,** Dimerization rate of the 1-methyl-4-tert-butylpyridinyl radical and the photochemical process of the dimer cleavage. Kinetic and time-resolved ESR studies, *J. Phys. Chem.,* 89, 339, 1985.

248. **Sakurai, H.,** Group IV B Radicals, in *Free Radicals II,* Kochi, J. K., Ed., John Wiley & Sons, New York, 1973, 741.

249. **Zakharin, V. I., Nadtochenko, V. A., Sarkisov, O. M., and Teytelboym, M. A.,** The study of recombination of PH₂ radicals (Russian), *DAN U.S.S.R.,* 263, 127, 1982.

250. **Akiyama, K., Ishi, T., Kubota, S. T., and Ikegami, Y.,** Photolytic generation and subsequent dimerization of 4-alkyl-1-methylpyridinyl radicals in solutions as studied by steady-state and kinetic ESR spectroscopy, *Bull. Chem. Soc. Jpn.,* 58, 3535, 1985.

251. **Korth, H. G., Lammes, P., Sicking, W., and Sustman, R.,** Rate constants for the bimolecular self-reaction of cyano substituted alkyl radicals in solution, *Int. J. Chem. Kinetics,* 15, 267, 1983.

252. **Claridge, R. C. and Fischer, H.,** Self-termination and electronic spectra of substituted benzyl radicals in solutions, *J. Phys. Chem.,* 87, P. 1960, 1983.

253. **Mendenhall, G. D.,** Long-lived free radicals, *Sci. Prog.,* (Oxford), 65, 1, 1978.

254. **Schluter, K. and Berndt, A.,** Tri-neopenthyl, ein Alkyl-radikal mit bemerkenswerten Eigenschaften, *Tetrahedron Lett.,* 11, 929, 1979.

255. **Ikegami, Y.,** Kinetic ESR and CIDEP studies on the monomer-dimer equilibrium systems of pyridinyl radicals., *Rev. Chem. Intermediates,* 7, 91, 1986.

256. **Khudyakov, I. V., Levin, P. P., and Kuzmin, V. A.,** The reversible recombination of radicals (Russian), *Usp. Khim.,* 49, 1990, 1980.

257. **Brokenshire, J. L., Roberts, J. R., and Ingold, K. U.,** Kinetic application of electron paramagnetic resonance spectroscopy. 7. Self-reactions of iminoxy radicals, *J. Am. Chem. Soc.,* 94, 7040, 1972.

258. **James, D. G. L. and Stuart, P. D.,** Kinetic study of the cyclohexadienyl radical, *Trans. Faraday Soc.,* 64, 2735, 1968.

259. **Mahoney, L. R. and Wiener, S. A.,** A mechanistic study of the dimerization of phenoxy radicals, *J. Am. Chem. Soc.,* 94, 585, 1972.

260. **Henning, H. G., Pragst, F., and Fuhrman, J.,** Stereoselective Radicalreactionen, *Z. Chem.,* 24, 1, 1984.

261. **Eichin, K. H., Beckhaus, H. D., Hellman, S., Fritz, H., Peters, E. M., Peters, K., von Schering, H. G., and Ruchardt, Ch.,** 1-Substituierte Neopentyl Radikale und ihre Dimeren, *Chem. Ber.,* 116, 1787, 1983.

262. **Benson, S. W.,** Disproportionation of free radicals, *J. Phys. Chem.,* 89, 4366, 1985.

263. **Sander, S. P. and Watson, R. T.,** Kinetic and mechanism of the disproportionation of BrO· radicals, *J. Phys. Chem.,* 85, 4000, 1981.

264. **Mozurkewich, M.,** Reactions of HO_2· with free radicals, *J. Phys. Chem.,* 90, 2216, 1986.

265. **Gibian, M. J. and Corley, R. C.,** Organic radical-radical reactions. Disproportionation vs. combination, *Chem. Rev.,* 73, 441, 1973.

266. **Cadman, Ph. and Owen, H. L.,** Reactions of fluoromethyl radicals, *J. Chem. Soc., Faraday Trans.,* 77, 3087, 1981.

267. **Nilsson, W. B. and Pritchard, G. O.,** Disproportionation reactions between CF_2H and C_2H_5 radicals in the gas phase, *Int. J. Chem. Kinetics,* 14, 299, 1982.

268. **Reimann, B., Matten, A., Laupert, R., and Potzinger, P.,** Zur Reaktion von Silylradikalen. Das Verhaltnis Disproportionierung/Rekombination, *Ber. Bunsenges. Physik. Chem.,* 81, 500, 1977.

269. **Baulch, D. L., Chouwn, P. K., and Montaugue, D. C.,** Combination and disproportionation reactions of allylic radicals with ethyl, propyl, and t-butyl radicals, *Int. J. Chem. Kinetics,* 11, 1055, 1979.

270. **Bakac, A. and Espenson, J. H.,** Disproportionation and combination of ethyl radicals in aqueous solution, *J. Phys. Chem.,* 90, 325, 1986.

271. **Gibian, M. J. and Corley, R. C.,** Disproportionation vs. combination of secondary α-aryl alkyl radicals from the thermolysis of azo compounds in solutions, *J. Am. Chem. Soc.,* 94, 4178, 1972.

272. **Fujisaki, N. and Gauman, T.,** Self- and cross-disproportionation-to-combination ratios for cyclopentyl and cyclohexyl radicals and their deuterated analogues, *Int. J. Chem. Kinetics,* 14, 1059, 1982.

273. **Bennet, J. E., Gale, L. H., Hayward, E. J., and Mile, B.,** Disproportionation-combination reactions of cyclohexyl radicals at low temperature in the solid state, *J. Chem. Soc., Faraday Trans.,* 69, 1655, 1973.

274. **Manka, M. J. and Stein, S. E.,** Disproportionation-recombination rate ratios for hydroaromatic radicals, *J. Phys. Chem.,* 88, 5914, 1984.

275. **Whittle, E.,** Reactions of free radicals, *MTP Int. Rev. Sci. Phys. Chem.,* 9, 75, 1972.

276. **Schuh, H. and Fischer, H.,** Rate constants for the bimolecular self-reaction of tert-butyl radicals in n-alkane solvents, *Int. J. Chem. Kinetics,* 8, 341, 1976.

277. **Tilman, P., Tilquin, B., and Claes, P.,** Estimation of k_d/k_c ratio alkyl radicals in the solid phase (French), *J. Chim. Phys.,* 79, 629, 1982.

278. **Pritchard, G. O., Johnson, K. A., and Nilsson, W. B.,** Disproportionation reactions between alkyl and fluoroalkyl radicals, *Int. J. Chem. Kinetics,* 17, 327, 1985.

279. **Kelley, R. D. and Klein, R.,** Cross disproportionation of alkyl radicals, *J. Phys. Chem.,* 78, 1586, 1974.

280. **Sander, S. P. and Watson, R. T.,** Temperature dependence of Ch_3O_2 radicals, *J. Phys. Chem.,* 85, 2960, 1981.

281. **Niki, H., Maker, P. D., Savage, C. M., and Brietenbach, L. P.,** Fourier transform infrared studies of the self-reaction of $C_2H_5O_2$ radicals, *J. Phys. Chem.,* 86, 3825, 1982.

282. **Cowley, L. T., Wadington, D. J., and Wolley, A.,** Self-reactions of isopropylperoxy radicals, *J. Chem. Soc., Faraday Trans. 1,* 78, 2535, 1982.

283. **Lindsay, D., Howard, J. A., Horswill, E. C., Iton, L., Ingold, K. U., Cobley, T., and Li, A.,** The bimolecular self-reactions of secondary peroxy radicals. Product studies, *Can. J. Chem.,* 51, 870, 1973.

284. **Kurylo, M. J., Quellette, P. A., and Laufer, A. H.,** Measurements of the pressure dependence of the HO_2 radical self-disproportionation reaction at 298 K, *J. Phys. Chem.,* 90, 437, 1986.

285. **Munk, J., Pagsberg, P., Rataczak, E., and Sillesen, A.,** Spectrokinetic studies of ethyl and ethylperoxy radicals, *J. Phys., Chem.,* 90, 2752, 1986.

286. **Howard, J. A. and Bennett, J. E.,** The self-reaction of sec-alkylperoxy radicals: A kinetic electron spin resonance study, *Can. J. Chem.,* 50, 2374, 1972.

287. **Tavadjan, L. A., Mardoyan, V. A., and Nalbandyan, A. B.,** The study of reactivity of tert-butyl peroxy radicals in liquid phase by ESR method (Russian), *DAN U.S.S.R.,* 259, 1143, 1981.

288. **Smirnov, B. R. and Sukhov, V. D.,** Quantitative verification of the theory of diffusional and chemical escape from the cage from the data on decomposition of radical initiators in liquids and polymers (Russian), *Vysokomol. Soedin.,* 19, 236, 1977.

289. **Yakimchenko, O. Ye., Lebedev, Ya. S.,** Radical pairs in investigation of elementary chemical reactions in solid organic compounds (Russian), *Usp. Khim.,* 47, 1018, 1978.

290. **Tolkachev, V. A.,** The simple non-Markovian model of kinetics of elementary radical reactions in glassy materials (Russian), *DAN U.S.S.R.,* 287, 165, 1986.

291. **Bartos, J. and Tino, J.,** Study of the mechanism of macroradical reactions in solid polymers, *Polymer,* 27, 281, 1986.

Chapter 4

REACTIVITY OF FREE RADICALS

I. INTRODUCTION

The term "radical reactivity" is used very frequently when speaking about radical reactions. Qualitatively, we distinguish among highly reactive, semistable, stable, persistent, and inert radicals. At first sight, it is obvious that such characterization of radical reactivity in transfer or decay reactions is insufficient and that specification of a more objective scale is needed. The existence of a correlation between the structure or some of its parameters and reactivity should be useful since the course of reactions could be predicted upon this basis. The fact that there is not any unambiguous scale of reactivity indicates that the problem is more complex.

Until recently, reactions of one radical with different molecules were analyzed predominantly with an emphasis on the relationship between reactivity and the structure of saturated particles. Only at present, the reactions of different radicals with one kind of molecule are mutually compared. This may throw some light on the relation of radical reactivity and such structural parameters as a type of semioccupied orbital, extent of delocalization of an unpaired electron, radical conformation, sterical shielding of a radical site by substituents, the influence of radical excitation, complexation, etc.

When selecting a larger set of radical reactions with several standard compounds a new problem arises. The sequence of reactivity of radicals with one substrate may be different from that with another substrate. This stems from the more pronounced polar effects in a reaction of one molecule, sterical effects in a reaction of a second molecule, or other mutually different thermodynamic conditions. Even though this is true, there still exists a general tendency in the reactivity of radicals sufficiently far away on the reactivity scale, namely, that stable radicals react slowly with many substrates, while reactive radicals fast. This latter statement may be reversed; the slower the rate of disappearance of a respective radical, the more stable the radical is and vice versa. Provided that the difference in the rates of compared reactions is not so distinct, the more general scale of radical reactivity, including different substrates, does not exist.

From the viewpoint of simple ideas on the easy or difficult course of radical reactions, it may be supposed that reactivity of a radical depends a priori on the extent of interactions of the unpaired spin with neighboring nuclei and electrons of the atom carrying a radical site as well as on the sterical shielding of the semioccupied orbital. It is obvious that radicals, which have minimum inner interactions of unpaired electrons with neighboring atoms which do not hinder the approach of the reactant to the semioccupied orbital, will be reactive.

The problems of radical reactivity following from the generalization of empirical knowledge and experimentally measured quantitative data may be attacked by the more general methods of theoretical chemistry, as well. As will be seen later, each strategy suffers from some limitations.

II. SOME ASPECTS OF THE THEORY OF CHEMICAL REACTIVITY

The qualitative characterization of chemical reactivity requires description of the electronic structure of reactants and products as well as the knowledge whether the transformation in a given direction is probable. The first item was satisfactorily solved already by the classical theory of chemical structure. The second aspect, concerning the probability of some proposed reaction scheme, may be theoretically supported or ruled out by the Wigner-Witmer theorem, according to which the respective transformations are "allowed" or "forbidden" if the total symmetry* of the reaction system is conserved or not.

The best known approach is that of Woodward and Hoffmann who introduced the rules based upon the construction of correlation diagrams for molecular orbitals of reactants and products.[2,3] Reactions are classified here according to the conservation of the symmetry of frontier orbitals which are the most important for the reaction course. Such a demand, however, requires the preservation of all symmetry elements of the reacting system throughout the whole course of the reaction. Many experimentally interesting reactions do not fulfill such a condition because of the insufficient symmetry or either reactants or products.

A series of papers has been published indicating the above fact and showing the possibilities of how to proceed with such limitations.[4-10] This attracted attention to the topological (nodal) properties of wave functions, demonstrating that they have a decisive effect on chemical reactivity.

A. The Principle of Orbital Symmetry Conservation

Assume that the elementary reaction of any molecularity takes place along the assumed reaction coordinate. By the term "reaction coordinate" we understand the sequence of energetically most favorable configurations of the nuclei of the "pseudomolecule" (formed from all reactant atoms) moving from reactant to products. Provided that the original configuration Q_o is changed to configuration Q by a small increment ΔQ (Figure 1), we may analyze it by the sum of displacements corresponding to normal vibrational modes of a given pseudomolecule.[11] The Hamiltonian description of the reacting system may be expanded to Taylor's series around Q_o (Equation 1).

$$H = H_o + (\partial \hat{V}/\partial Q)_o \delta Q + 1/2(\partial^2 \hat{V}/\partial Q^2)_o \delta Q^2 + .. \tag{1}$$

The potential energy V includes the energy of mutual interactions of nuclei and the energy of the interaction of nuclei with electrons. The kinetic energy of electrons and their repulsive interactions may be ignored since it does not depend on nuceli coordinates in the first approximation.

The Hamiltonian of the system should be totally symmetric with respect to any symmetry operation applied to a pseudomolecule. It follows from the group theory, that operators δQ and $(\partial \hat{V}/\partial Q)_o$ should be of the same symmetry, while the operators δQ^2 and $(\partial^2 V/\partial Q^2)$ should be totally symmetric. The energy of the ground electronic state of the reacting system may then be expressed up to the second order of δQ as follows (Equation 2).

$$E = E_o + \delta Q \langle \psi_o : \partial \hat{V}/\partial Q : \psi_o \rangle + 1/2 \delta Q^2 \langle \psi_o : \partial^2 \hat{V}/\partial Q^2 : \psi_o \rangle$$

$$+ \delta Q^2 \sum_{k=1}^{\infty} (E_o - E_k)^{-1} \langle \psi_o : \partial \hat{V}/\partial Q : \psi_k \rangle^2$$

$$= E_o + E^{(1)} + E^{(2)} + E^{(2')} \tag{2}$$

In the first order accuracy in δQ, the disturbed wave function may be written as Equation 3. Even though the relations are

$$\psi = \psi_o + \delta Q \sum_{k=1}^{\infty} (E_o - E_k)^{-1} \langle \psi_o : \partial \hat{V}//\partial Q : \psi_k \rangle \psi_k \tag{3}$$

---

* Orbital symmetry indicates the shape of the valence shell MO and the way in which the phase of the orbital (plus or minus sign) changes from one part of the molecule to another. Each molecule has certain symmetry elements (planes, axes of rotation, center) which place it in one of the point groups. Each point group is characterized by a definite number and kind of symmetry species.[1]

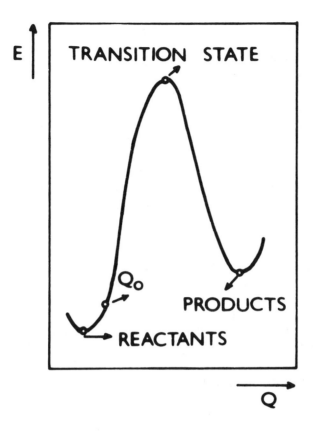

FIGURE 1.   Dependence of energy of the system on the reaction
coordinate.

valid only for small deviations from initial configuration Q_o, they may well be generalized, since Q_o may be chosen at any point of reaction the coordinate which may thus be represented by the sequences of δQ changes.

For a qualitative estimation of individual integrals in Equation 2, the rule from group theory may be used saying that a nonzero value of the integral is obtained only in the case when its integrand is totally symmetric. Since for any extremal point of adiabatic potential, $\partial E/\partial Q = 0$, the linear term $E^{(1)}$ of Equation 2 has to be zero regardless of the symmetry. The situation becomes more complicated when the wave function of the ground electronic state ψ_o is degenerated. With respect to the theory of chemical reactivity, such a case is not of interest; in this text the nondegenerated ground state of φ_o is therefore always assumed. In order that the term $E^{(1)}$ has nonzero values in all the upward and downward parts of reaction coordinate, $(\partial \hat{V}/\partial Q)_o$ as well as δQ must be totally symmetric.

The first selection rule may then be formulated as: except for the extremities of the curve of adiabatic potential all reaction coordinates are totally symmetric. The nonzero value of the integral $<\psi_o|\partial^2\hat{V}/\partial Q^2|\psi_o>$ represents the force constant of the resistance against the motion of nuclei from the initial configuration Q_o along the reaction coordinate, $\psi_o{}^2$ being the electron density.

The last term $E^{(2)}$ of Equation 2 describes the energy alterations brought about by changes of electron distribution when going to new nuclear positions. The summation includes all electronically excited states of the system in the configuration Q_o. The wave function will be changed only when the electron distribution is energetically more favorable for new positions of nuclei.

Since the value of the denominator in the $E^{(2)}$ term is negative, $E^{(2)}$ will be negative too. This corresponds to the decrease of the energy barrier of the respective reaction. For the energy barrier sink it is also important that the integral $<\psi_o | \partial \hat{V}/\partial Q | \psi>$ be nonzero and have as high value as possible. This is true for low laid electronically excited states ψ_k of a proper symmetry.[12] Since the operator $(\partial \hat{V}/\partial Q)$ is symmetrical, the wave function φ_o and φ_k should be of the same symmetry.

The second selection rule may then be formulated as: the acceptable activation energy may be received only at that time when the low laid electronicaly excited states of the reacting system, which have the same symmetry as a ground electronic state, exist. The excited states should lead to such changes of electron density as are consistent with the nuclei motion towards the given reaction course. For monomolecular reactions, another condition was put forward taking into account the contribution of harmonics to force constants, namely, that excitation energy $E_k - E_o$ should decrease throughout the reaction.[13,14] This corresponds to the demand that along the reaction pathway, both the highest occupied and lowest nonoccupied levels are approaching near.

The reactions which satisfy the above two selection rules are called symmetrically allowed. Symmetrically forbidden reactions are those having high activation energy brought about by the absence of proper low electronically excited states.

Equation 2 is accurate up to the second order of δQ. An equal level of accuracy is valid also for symmetry considerations which follow from it. The practical application of the selection rules requires further simplifications.

1.   The assumption that LCAO MO wave functions are eigen-functions of Hamiltonian H_o is the first. Since we focus our attention to symmetric properties of the MO reacting system, usually this does not lead to any mistake. It should be borne in mind, however, that configuration interaction may cause a change of the symmetry of a ground electronic state.

2.   In Equation 2 we consider usually only several lowest excited states instead of an infinite series. It may be shown that the contributions of higher excited states converge fast; as with the increasing $|E_o - E_k|$, the value of the integral $<\psi_o|\partial\hat{V}/\partial Q|\psi_k>$ is quickly decreasing.

3.   In the MO theory the symmetry of state function ψ_o and ψ_k is replaced by the symmetry of molecular orbitals φ_i and φ_j, where φ_i is MO occupied in the ground electronic state while φ_j is MO occupied in the excited state. The transfer of electrons from φ_i to φ_j induces the changes in distribution of electron density of the molecule. The positively charged nuclei move in the direction of increasing electron density. Let's consider such a reaction whose detailed mechanism is not known. During its course, there are changes in the bond characteristics of a pseudomolecule along the reaction coordinate which are in correspondence with the change of MO. Molecular orbitals which correspond to the chemical bond formation or its decay are the most interesting at the analysis of symmetric restrictions.

It should be recalled that molecular orbitals were obtained from the Schrodinger equation and that in the LCAO MO method they were constructed from atomic orbitals of respective atoms, the base of atomic orbitals for both the reactants and products being the same. For the sake of simplicity, we assume that only one bond is changed during the reaction. Let the initial state be expressed by state function ψ_o and the final state by function ψ. The relationship between ψ and ψ_o up to the 1. order accuracy in δQ is determined by Equation 3. The second term of this equation may be considered as a correction of wave function to a new bond situation. According to the symmetry selection rules, only function ψ_o and δ_k having the same symmetry may be combined. In the MO method we define ψ functions as

Slater's determinants consisting of occupied molecular orbitals. If MO φ_i corresponds to the decaying bond and simultaneously, ϑ_j orbital corresponds to the forming bond, we may limit ourselves only to these two MO. A symmetricaly allowed reaction has to have the same symmetry of φ_i and φ_j molecular orbitals of reactants and products and, consequently, the wave functions ψ_o and ψ will correlate, too. The principle of the conservation of orbital symmetry may be thus formulated as:[2,3] the reaction may occur easily when orbital symmetry of reactants and products mutually corresponds, i.e., the symmetry of molecular orbitals is conserved.

At the symmetrically allowed reaction, the electron from the highest occupied molecular orbital (HOMO) of reactants φ_i is transferred to the lowest unoccupied molecular orbital (LUMO) of products φ_j, which have then a positive overlap and thus the same symmetry. Because of the negative value of the last term of Equation 2, the energy of the system decreases and the height of the energetical barrier of the reaction becomes acceptable. The charge transfer HOMO-LUMO will be effective at that time when LUMO is localized on more electronegative atoms. In such a case, the absolute value of the denominator of the last term of Equation 2 is low and the stabilization effect of the electron transfer is large.

Provided that the reacting particle is a radical, its singly occupied molecular orbital (SOMO) plays the role of HOMO or LUMO or both orbitals simultaneously, depending on the relation of SOMO energy to orbital energies of closed shell molecules.[14]

The practical application of the orbital symmetry conservation rule uses the technique of the construction of correlation diagrams linking molecular orbitals of reactants and products. The general approach to their construction is as follows.

Find the molecular orbitals which correspond to forming and decaying bonds. We remind the reader that the bond orbitals of σ and π type, antibonding of σ^* and π^* types, and finally nonbonding orbitals all come into consideration. Determine their approximate scale according to the increasing energy.

For the assumed course of the reaction we choose such elements of symmetry of the reaction system which describe the highest symmetry of the reaction system. If all considered molecular orbitals of reactants and products have the same symmetry with respect to these elements, they may be omitted.

Examine whether corresponding orbitals are bonding or antibonding when moving along the reaction coordinate. In the case of bonding orbitals, the electron density between respective atoms increases (positive overlap) while for antibonding orbitals it decreases (negative overlap).

In the correlation diagram, link orbitals of reactants and products have the same symmetry (the principle of orbital symmetry conservation). The orbital symmetry is conserved if the ground states of reactants and products mutually correspond. Such reactions are called symmetrically allowed, while others are symmetrically forbidden. For the latter kind of reactions, the lines linking twofold occupied and unoccupied or singly occupied and unoccupied orbitals are crossed, the energy of the more occupied orbitals increasing towards the products.

Let us, e.g., take the example of the interaction of two homonuclear biatomic molecules (Equation 4). One should assume that

$$H_2 + I_2 \rightarrow 2\ HI \tag{4}$$

the energetically most favorable will be the side collision of reactants with the formation of the four-center activated complex (Scheme 1).

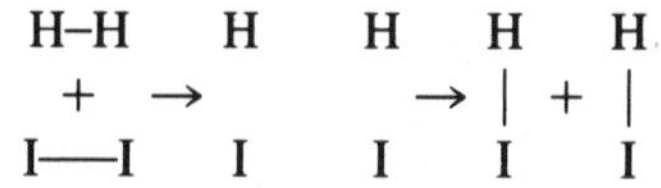

H–H   H   H   H   H
 + →   → | + |
I—I   I   I   I   I

SCHEME 1

The correlation diagram, however, indicates that such a reaction is symmetrically forbidden. Its critical point is crossing over molecular orbitals of a_1 and b_2 symmetry (symmetry group C_2v). The high energy barrier of the above mechanism follows from the high energy of the molecular orbital occupied in activated state (originally it was the σ_g orbital of the H_2 molecule), while the energy of unoccupied b_2 molecular orbital (originally the σ_u orbital of H_2 molecule) strongly decreases (Figure 2). The above reaction mechanism thus seems improbable.

As follows from correlation diagrams (Figures 3 and 4) symmetrically allowed are mechanisms (Equations 5 to 9)

$$I_2 \leftrightarrow 2\ I\cdot \tag{5}$$

$$I\cdot + H_2 \leftrightarrow IH_2 \tag{6}$$

$$IH_2 + I\cdot \rightarrow 2\ HI \tag{7}$$

or

$$I\cdot + H_2 \rightarrow HI + H\cdot \tag{8}$$

$$H\cdot + I_2 \rightarrow HI + I\cdot \tag{9}$$

which are in accordance with experimental results.[15]

Also the direct abstraction of hydrogen from hydrocarbons by a carbonyl group (Equation 10) is forbidden for ground electronic

$$>\!C\!=\!O + HR \rightarrow >\!\dot{C}\ \ddot{O}H + R\cdot \tag{10}$$

states of reactants, since the state S_o symmetrically correlates with the electronically excited state of zwitterionic product Z_1 (Figure 5). If, however, the carbonyl group is excited (n-π^* transition), the formed excited triplet state (^{3}n π^*) correlates symmetrically with the ground biradical state of products (^{3}D σ π).[16]

The orbital symmetry conservation rule was applied to the series of organic and inorganic reactions including catalytic reactions[17-21] and thus became a useful tool of the theory of chemical reactivity.

B. Frontier Orbital Approach

It follows from the perturbation theory that properties of frontier orbitals (HOMO, SOMO, LUMO) decisively affect the reaction course. As was pointed out, the decrease of activation energy is due to the last member of Equation 2. Within a simple MO theory it may be rewritten as Equation 11, where φ_i

$$E^{(2)} = 2\delta Q^2 \sum_i^{occ} \sum_p^{vir} (\epsilon_p - \epsilon_i)^{-1} <\varphi|\partial\hat{V}/\partial Q|\varphi_p>^2 \tag{11}$$

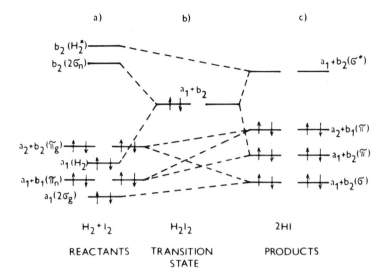

FIGURE 2. Correlation diagram for $H_2 + I_2$ reaction in C_{2v} symmetry. Only the crossing molecular orbitals of the transition state (b) are shown.

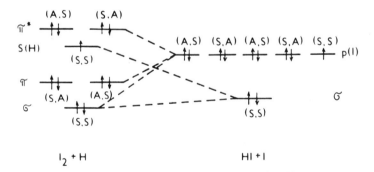

FIGURE 3. Correlation diagram for $I_2 + H \Leftrightarrow HI + I$ elementary reaction in the linear rearrangement. The symmetries of MO - s with respect to two mutually perpendicular symmetry planes are designated in parenthesis.

(φ_p) are occupied (unoccupied) molecular orbitals and ϵ_i (ϵ_p) are corresponding orbital energies. The lowest energy difference is between HOMO and LUMO. This allows one to assume that mainly interactions of these two frontier orbitals contribute to the stabilization of the system.

For bimolecular reactions,[13,14] Equation 11 may be expressed as Equation 12, where dashed indices concern the first and

$$E^{(2)} = 2\delta Q^2 \sum_i^{occ} \sum_{p'}^{vir} (\epsilon_{p'} - \epsilon_i)^{-1} \langle \varphi_i | \partial \hat{V}/\partial Q | \varphi_{p'} \rangle^2$$

$$+ 2\delta Q^2 \sum_{i'}^{occ} \sum_{p}^{vir} (\epsilon_p - \epsilon_{i'})^{-1} \langle \varphi_{i'} | \partial \hat{V}/\partial Q | \varphi_p \rangle^2 \qquad (12)$$

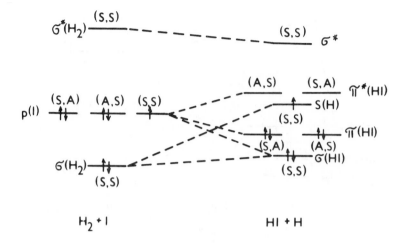

FIGURE 4.   Correlation diagram for $H_2 + I \Leftrightarrow HI + H$ elementary reaction in the linear rearrangement. The symmetries of MO - s with respect to two mutually perpendicular symmetry planes are designated in parenthesis.

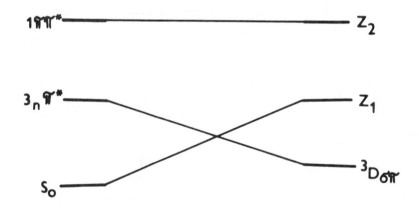

FIGURE 5.   Correlation diagram of photochemical abstraction of hydrogen from hydrocarbons induced by a carbonyl group.

nondashed the second reactant. Since δQ has to be totally symmetric, nonzero terms of Equation 12 will be only those which have φ_i and φ_p or φ_i and φ_p^* of the same symmetry.

Consider the interaction of two electron occupied HOMO of the particle A and unoccupied LUMO of particle B which creates new orbitals of pseudomolecule AB. The interaction leads to the decrease of energy of bonding and to the increase of energy of the antibonding orbital. In the absolute value, the change of energy of the unoccupied orbital is higher than that of the occupied orbital.[12] It is obvious that if the orbitals are twice occupied, their interaction will lead to the increase of the energy of the pseudomolecule AB and to the repulsion.

The condition of low activation energy consists in a high contribution of the delocalization stabilization, i.e., the interaction HOMO-LUMO has to be significant. It is valid that the lower the difference of energy between HOMO and LUMO and the larger overlap of both

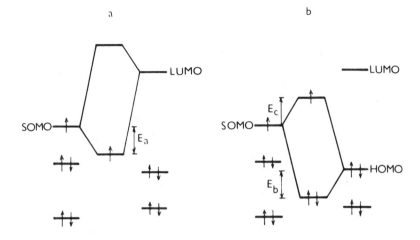

FIGURE 6. The interaction of SOMO of radical with the frontier orbitals of molecule: (a) SOMO-LUMO interaction, (b) SOMO-HOMO interaction.

orbitals, the stronger the interactions HOMO-LUMO exist. In other words, the allowed reactions are defined by the following rules.

1. The orbitals HOMO and LUMO of the reactants are decisive for the reaction course. At the mutual approach of reactants, there occurs the stabilization interaction between HOMO and LUMO. This interaction may be understood as a partial flow of electrons from the filled molecular orbitals to the empty ones.
2. The frontier orbitals should be of a proper symmetry to achieve the most efficient overlap. Only molecular orbitals of the same symmetry can be mixed.
3. The energy difference between HOMO and LUMO should be low (6 eV at maximum). Provided that the point 2 is not fulfilled for any reason, the role of frontier orbitals may be shifted to other orbitals of proper symmetry having the required energy difference.
4. If both orbitals are bonding, then HOMO should correspond to decaying bonds and LUMO to forming bonds. The opposite is true for antibonding orbitals.

Provided that it is not possible to find a pair of orbitals satisfying the above rules, the reaction is forbidden. It should be recalled that the above principles, which are based upon the perturbation theory, are valid only for the initial phases of the reaction. It may be shown, however, that it is satisfactory for the estimation of the reaction course.

The SOMO orbital of a free radical may interact with either HOMO or LUMO of molecule (Figure 6). Its interaction with either LUMO or HOMO leads to the energy decrease by E_a or $2 E_b - E_c$ value, respectively.

Radicals of high energy SOMO will, therefore, react fast with molecules of low energy LUMO (Figure 7) while radicals of low energy SOMO do the same with molecules of high energy HOMO (Figure 7 b). In the former case, radicals are electron donors; in the latter case, they are electron acceptors.

a   b

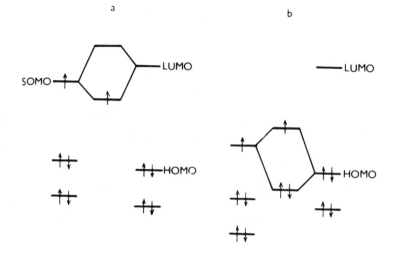

FIGURE 7.   The dominant interaction of frontier orbitals: (a) for radical with high-energy SOMO (nucleophilic radical), (b) for radical with low-energy SOMO (electrophilic radical).

This conclusion may be demonstrated on the alternating copolymerization of vinyl acetate and dimethyl fumarate[9] (Scheme 2). The radical (A), which has an unpaired electron in

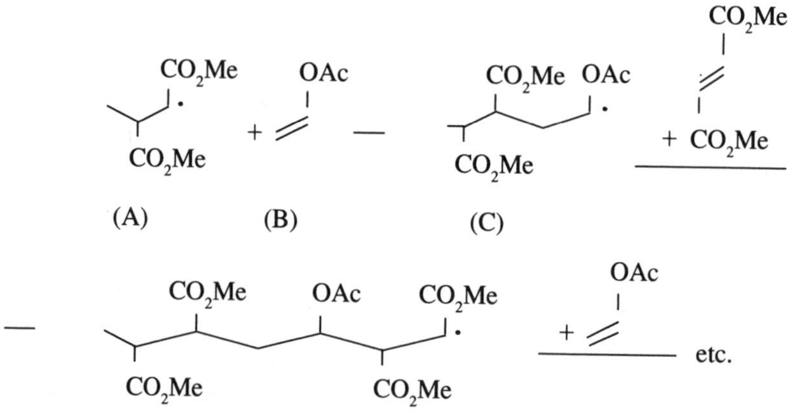

SCHEME 2

the vicinity of the carbonyl group, has low energy SOMO (electrophile radical) and preferably attacks vinyl acetate (B). The formed radical (C) now has in the vicinity of the radical site an oxygen atom and its SOMO will be of high energy (nucleophile radical) and will attack dimethyl fumarate (D). The role of the energy of frontier orbitals for the course of the above copolymerization seems to be quite illustrative.

The effect of the molecule structure on the energy of frontier orbitals may be expressed qualitatively as:

1. Increasing conjugation increases the energy of HOMO and decreases the energy of LUMO, i.e., the energy gap HOMO-LUMO is reduced.
2. Electron acceptor substituents decrease the energy of both HOMO and LUMO, their values getting closer.

3. Substituents having a free electron pair which may participate in the conjugation increase the energy of both HOMO and LUMO.

C. Topological Aspects of Reactivity

The topological approach is based upon the description of specific transformations of bonds from reactants to products.[4,10] The molecules of the reactant and product are represented here by the assembly of chemical bonds, M_R and M_P. The bonds which do not change during the reaction may be omitted.

At the addition of the radical X to the C=C bond (Scheme 3),

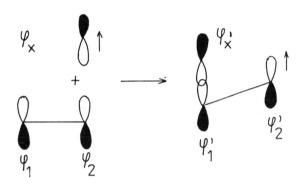

SCHEME 3

the assembly of changing bonds of reactants involves π bond C=C and a free electron of the radical X, while the assembly of changing bonds of products involves σ bond C–X and a free electron on the carbon atom. The system may then be described by determinants (Equations 13 and 14) where π denotes the molecular orbital of

$$\psi_R = |\,\pi\,\tilde{\pi}\,\psi_x\,|\tag{13}$$

$$\psi_P = |\,\sigma_{cx}\,\tilde{\pi}_{cx}\,\psi_z\,|\tag{14}$$

π symmetry occupied by electron with α spin, $\tilde{\pi}$ is the molecular orbital of π symmetry occupied by an electron with β spin, etc. The bonding orbitals may then be expressed by the LCAO approach in Equations 15 to 20. Only the nodal properties (s, P_x, P_y, P_z, etc.)

$$\pi = (\varphi_1 + \varphi_2)\,\alpha\tag{15}$$

$$\pi = (\varphi_1 + \varphi_2)\,\beta\tag{16}$$

$$\sigma_{cx} = (\varphi_1' + \varphi_x')\,\alpha = (\varphi_1 + \varphi_x)\,\alpha\tag{17}$$

$$\sigma_{cx} = (\varphi_1' + \varphi_x')\,\beta = (\varphi_1 + \varphi_x)\,\beta\tag{18}$$

$$\varphi_x = \varphi_x\,\alpha\tag{19}$$

$$\varphi_2 = \varphi_2'\,\alpha = \varphi_2\,\alpha\tag{20}$$

of the above atoms orbitals are of importance here. The chemical reaction may then be qualitatively analyzed in terms of nodal properties of wave function ψ_R and ψ_P of reactants and products, i.e., by the value of the overlap integrals. Provided that nodal properties of ψ_R and ψ_P are mutually different, the overlapping integral is zero, and the reaction is forbidden. For allowed chemical reactions the nodal properties of the wave function of reactants and products are alike and their overlap is different from zero. The overlap of wave functions ψ_R and ψ_P may be expressed by the determinant (Equation 21). The matrix elements of this determinant

$$D = \int \psi_R \, \psi_P \, dV = \underbrace{\begin{vmatrix} a_{11} & a_{12} & \cdots \\ a_{21} & a_{22} & \cdots \\ & & \end{vmatrix}}_{\alpha} \underbrace{\begin{vmatrix} a_{11} & a_{12} & \cdots \\ a_{21} & a_{22} & \cdots \\ & & \end{vmatrix}}_{\beta} \tag{21}$$

are overlap integrals of the respective bonds for α and β spins of electrons, the determinants being constructed separately. (The overlap integrals of α and β electrons are zero since α and β spin functions are orthogonal.) For our case, the elements of the respective determinants are (Equations 22 to 26).

$$a_{11}(\alpha) = \int \pi(\alpha)\sigma_{cx}(\alpha)dV \tag{22}$$

$$a_{12}(\alpha) = \int \pi(\alpha)\varphi_2 dV \tag{23}$$

$$a_{21}(\alpha) = \int \varphi_x(\alpha)\sigma_{cx}(\alpha)dV \tag{24}$$

$$a_{22}(\alpha) = \int \varphi_x(\alpha)\varphi_2(\alpha)dV \tag{25}$$

$$a_{11}(\beta) = \int \widetilde{\sigma}_{cx}(\beta)\widetilde{\pi}(\beta)dV \tag{26}$$

For this estimation, a zero differential overlap approximation may be used (Equation 27) according to which we get Equations 28 to 32.

$$\int \varphi_i \varphi_j dV = \int \delta_{ij} \tag{27}$$

$$a_{11}(\alpha) = \int (\varphi_1 + \varphi_2)\alpha(\varphi_1 + \varphi_x)\alpha dV = 1 \tag{28}$$

$$a_{12}(\alpha) = \int (\varphi_1 + \varphi_2)\alpha\varphi_2\alpha dV = 1 \tag{29}$$

$$a_{21}(\alpha) = \int \varphi\alpha(\varphi_1 + \varphi_x)\alpha dV = 1 \tag{30}$$

$$a_{22}(\alpha) = \int \varphi_x\alpha\varphi_2\alpha dV = 0 \tag{31}$$

$$a_{11}(\beta) = \int (\varphi_1 + \varphi_x)\beta(\varphi_1 + \varphi_2)\beta dV = 1 \tag{32}$$

The value of the overlap determinant then will be Equation 33. The topological approach to chemical reactivity which

$$D = \begin{vmatrix} 1 & 1 \\ 1 & 0 \end{vmatrix}_\alpha \quad |1|_\beta \neq 0 \tag{33}$$

is based upon the same physical principles as the Woodward-Hoffman rule or frontier orbital approach is universal and simple and does not require knowledge of any symmetrical properties of reactants and products, which makes it accessible to nonspecialists in quantum chemistry.

D. Catalysis of Symmetry-Forbidden Reactions

Although orbital symmetry conservation alone is not a sufficient condition for the occurrence of the respective chemical reaction, the symmetry restriction may rule it out. The chemical reaction proceeds spontaneously at that time when its free enthalpy is positive. The practical usage of a given reaction consists in thermodynamic characteristics which imply the reaction rate (kinetics characteristic) and particularly the activation energy. Even simple MO theory accounts surprisingly well for the effect of activation energy on the course of chemical reactions. The necessary condition for low activation energy is conservation of the symmetry of occupied molecular orbitals at the given chemical reaction. Molecular orbitals of reacting compounds correlate with orbitals of products of the same symmetry. Provided that all bonding orbitals of reactants correlate with bonding orbitals of products, the symmetry restrictions are minimal. Such reactions have low activation energy and are symmetrically allowed. If bonding orbitals of reactants correlate with antibonding or nonbonding orbitals of products, the symmetry restrictions are high; the reaction has a high activation energy and is symmetrically forbidden. It should be recalled that only symmetry correlation of molecular orbitals which have different occupation numbers are critical.

The system of molecules tends to react in such a way that it keeps its bond properties. If it is not possible, the process has a high activation energy and the reaction will be improbable. The physical and chemical factors such as excitation, ionization, and catalysis may change the bond correlations for symmetrically forbidden reactions and thus remove the symmetry restrictions.

Let us postulate the existence of the correlation diagram (Figure 8) for reaction $A \rightarrow B$; antibonding orbitals are designated by asterisks, critical correlations are depicted in full lines. The given reaction is symmetrically forbidden since the occupied molecular orbital of reactant $\varphi_{2A}(S)$ correlates with unoccupied antibonding orbital $\varphi_{3B}*(S)$ and unoccupied antibonding orbital $\varphi_{3A}*(A)$ correlates with occupied bonding orbital $\varphi_{2B}(A)$.

The ground state of the system A thus corresponds to the excited state of the system B. Provided that the reaction occurs despite high activation energy, the electron configuration of product B should be $\varphi_{1B}^2(S)\varphi_{2B}^0(A)\varphi_{3P}*^2\varphi_{4B}^0(A)$. Excitation energy of this state then represents a considerable part of the energetical barrier for the reaction.

As was already pointed out, the symmetrical restrictions may be removed by excitation, ionization, or by catalytic effect of some transition metal ions.[17,18]

If system A enters the reaction in the excited state $\varphi_{1A}^2(S)\varphi_{2A}^0(S)\varphi_{3A}*^2(A)\varphi_{4A}*^0(A)$ the transfer of electrons from antibonding to bonding molecular orbital at the transformation $A^{exc} \rightarrow B$ will decrease the energy of the system. The electron population of system A (in its electronically excited state) creates thus the bonding situation in the bonding area of system B. By excitation of system A, the conditions came into existence which allow the transformation of A to B. The excitation removed the symmetry correlations of occupied molecular orbitals (Figure 8).

If compound A participates in the reaction as a two-valent cation A^{2+}, its electronic

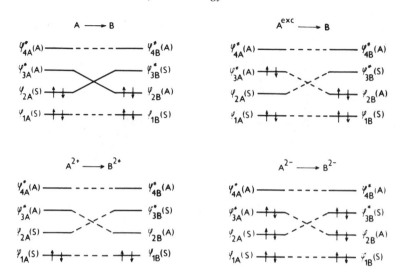

FIGURE 8.   Correlation diagram of reactions: $A \rightarrow B$; $A^{exc} \rightarrow B$; $A^{2+} \rightarrow B^{2+}$; $A^{2-} \rightarrow B^{2-}$.

structure will be $\varphi_{1A}{}^2(S)\varphi_{2A}{}^0(S)\varphi_{3A}{}^{*0}(A)\varphi_{4A}{}^{*0}(A)$. The critical correlations of orbitals $\varphi_{2A}(S)–\varphi_{3B}{}^*(S)$ and $\varphi_{3A}{}^*(A)–\varphi_{2B}{}^{.}(A)$ are removed under such conditions (Figure 8) and the ground electronic state of the system $A^{2+} \rightarrow B^{2+}$ corresponds with the ground electronic state of the system B^{2+}; the energy barrier will decrease.

Compound A which enters the reaction as a two-valent anion has the electronic structure $\varphi_{1A}{}^2(S)\varphi_{2A}{}^2(S)\varphi_{3A}{}^{*2}(A)$. From the correlation diagram (Figure 8) of the reaction $A^{2-} \rightarrow B^{2-}$ it follows that occupation of the antibonding orbital $\varphi_{3A}{}^*$ by an electron pair makes possible a symmetrically forbidden correlation between bonding and originally vacant antibonding molecular orbitals. The ground electronic state of the system A^{2-} corresponds to the ground electronic state of the system B^{2-} and we may again expect a decrease of the energetical barrier.

E. Catalytic Effect of Transition Metal Ions

There are many chemical reactions which are symmetrically forbidden but which occur easily in the presence of some catalyst. This change in reactivity is brought about by redistribution of the electron density and, consequently, by the change in symmetry of occupied molecular orbitals when they are coordinated to a transition metal element. Such reactions are called catalytically allowed. Since these reactions are of high practical significance, interest in studying the changes of the electron structure of reacting compounds coordinated to transition metal steadily increases.[17-21] Molecular orbitals of ligands interact with orbitals of the transition metal of respective symmetry, and the electron density between the transition metal and ligand is redistributed with the formation of a donor-acceptor bond. π-bonding MO of alkene ligand, e.g., affords electrons to free radicals of the transition metal (donor bond) and antibonding π^* MO accepts the electrons from occupied orbitals of the transition metal (acceptor bond) (Figure 9). Of importance is primarily the symmetry of orbitals which are available for ligands from the transition metal element. Both interactions (donor and acceptor) decrease the strength of the bonds holding ligands in the coordination sphere. At the donor interactions, the bond electrons of ligands are partially transferred to orbitals of the transition metal while in acceptor bonds, the antibonding orbital of ligands is partially occupied by valency electrons of the transition element. It reduces the bond order of the ligand and alters its electronic structure in the direction of excitation. Reactions of such a coordinated ligand will be alike as its reactions in the excited electronic state. The extreme case is the filling of orbitals of a transition element by electrons which symmetrically

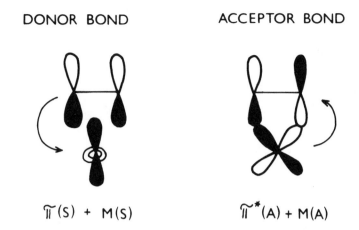

DONOR BOND ACCEPTOR BOND

$\widetilde{\Pi}(S) + M(S)$ $\widetilde{\Pi}^*(A) + M(A)$

FIGURE 9. Scheme of the donor-acceptor interaction of the transition metal-reactant.

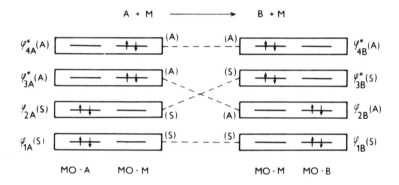

FIGURE 10. Correlation diagram of the reaction A → B in the presence of transition metal.

allow only acceptor interactions of occupied orbitals of the transition element with antibonding π^* orbitals of the ligand. The electron density is here transferred between transition element-ligand; the reactions of the ligand system have the character of reactions of an anion radical (when one electron is transferred) or the system of ligand^{2-} (when the electron pair is transferred). The situation where occupation of transition element orbitals makes possible only donor interactions of ligand to transition metal is an opposite case. The reacting system then has the character of a cation radical or cation ligand^{2+}, respectively.

Let us take the case of the symmetrically forbidden reaction A → B which occurs in the presence of a transition element giving a donor-acceptor bond with A and B coordination (Figure 10). The correlation $\varphi_{2A}(S)–\varphi_{3B}^*(S)$ remains here. The symmetry limitations of the A → B reaction are, however, removed by a redistribution of electron density of ligand orbitals which is brought about by the bond of A and B with the transition element. The transition element affords the symmetrically suitable orbital for the interaction with bonding orbital $\varphi_{2A}(S)$ and thus decreases the electron density here.

The electron density on the orbital $\varphi_{3A}^*(A)$ simultaneously increases, which interacts with the occupied orbital of the transition element. Both the reactant and product form bonds with the transition metal, the coordination of product B being different from that of reactant A. The exchange of an electron pair due to coordination is the consequence of redistribution of valency electrons of the transition metal $(S^0A^2 \rightarrow A^0S^2)$, d orbitals of the transition metal are denoted according to their symmetry. This redistribution is essential for preservation of the coordination bond.

The partial mixing of the excited state of the ligand with the ground state of the coordinated ligand changes its electronic structure towards the excited state. This aspect of the coordination bond is characteristic for catalytically allowed processes.

The increasing electron population of $\varphi_{3A}*$ orbital decreases the bond order in the system A and creates the bond situation in the bond region of the system B (orbital φ_{2B}). The decreasing electron population in the region of φ_{2A} orbital has the similar effect.

The role of the coordination bond at catalytically allowed reactions may be summarized as follows. At the coordination of the ligand system to the transition metal, there is mixed the excited state of the ligand with the ground state of the coordinated ligand. The coordination bond thus creates the conditions for the reaction of the ligand. The orbital symmetry restrictions disappear during coordination of the ligand system by transition element.

III. ACTIVATION ENERGY AS A SCALE OF REACTIVITY

The capability of a radical to enter some reaction may be expressed by a rate constant. In the reaction of radical R· with reaction partner X, the rate of radical disappearance from the system is described by Equation 34.

$$- d[R·]/dt = k[R·][X] \tag{34}$$

The determination of the rate constant k requires knowledge of the reaction rate, i.e., the number of radicals in the unit volume reacted per unit of time, and concentrations of both reaction partners, respectively. From the practical standpoint of the experimenter, changes of radical concentrations may be recorded only to a limited extent. To overcome them, the dependence of the rate constant on temperature extrapolated to conditions exterior to the experimental region is used. The temperature (T) dependence of the rate constant k is usually expressed by the Arrhenius equation (Equation 35) where R

$$k = A \, e^{-E/RT} \tag{35}$$

is the gas constant ($8.31 \, J \, K^{-1} \, mol^{-1}$), A is the preexponential or frequency factor, and E the activation energy. The preexponential factor and activation energy are assumed to be constant for the respective reactions.[22] The values of A and E which thus characterize the whole set of kinetic data and quantify the reactivity may be determined by measuring k at several temperatures.

What is now the relationship between radical reactivity and Arrhenius parameters? The most obvious indication of radical reactivity is the value of the respective activation energy. The difference in activation energies of 150 kJ/mol for two similar reactions, e.g., corresponds to the change of rate constant by 26 orders at 300 K. At the same time, the changes of frequency factors are not higher than 3 orders. The activation energy may be thus an important parameter for the estimation of radical reactivity in some reactions. The obvious tendency is valid, namely, the more reactive the radical, the lower the activation energy of its reaction. It should be remembered, however, that activation energy involves reactivity of both the radical and reactant. Its use as a reactivity rating is therefore justified only within standard or structurally similar substrates.

Provided that overriding the activation barrier can be imagined as a synchronous process of cleavage and a new formation of bonds in the activated complex, the energy profile of the respective bonds can be illustrated on curves of potential energy (Figure 11). We recall that individual curves plot the potential energy along the distance between atoms of the respective bonds. From the mutual interactions of potential energy curves of the respective bonds it may be seen that the rupture of one bond is partially compensated by the energy

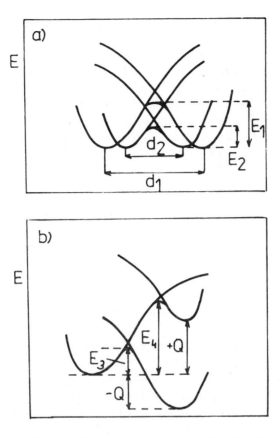

FIGURE 11. Construction of the energy profile of reaction from potential energy curves of forming and breaking bonds. (a) The effect of distance d of reactants in activated complex on activation energies E_1 and E_2 of thermoneutral reactions. (b) The effect of the energetic content of a forming bond on activation energy of exothermic (E_3) and endothermic (E_4) reaction; Q is reaction heat.

release in the formation of the second bond. The interaction of potential energy curves of cleaved and newly formed bonds circumscribes the value of activation energy. The height of the barrier depends on the distance of reactants in the activated complex determined by the structure of both reactants. The tighter the arrangement of particles in the activated complex, the lower the activation energy of reaction (compare E_1 and E_2 on Figure 11).

The height of the barrier is also influenced by the shape of the potential energy curve and by the extent of mutual interactions of reacting particles in the activated complex. Constructing the energy profile of a reaction, we may show that the height of the barrier will be lower for newly formed bonds of higher dissociation energy and vice versa. As a matter of fact, such an energy diagram indicates the increase of activation energy in endothermic and the decrease of activation energy in exothermic reactions. The correspondence of activation energy and reaction heat in competitive reactions is not, however, so simple.[23] It may happen that a less exothermic reaction has a lower activation energy while a more exothermic one has a higher activation energy (Figure 12). Even in this case, the formation of the reaction product is under kinetic control and a reaction of lower activation energy will be faster. On the other hand, provided that the equilibrium constant of the reaction from state A to B will be higher than that to state C, the composition of products of the competitive reaction will depend on time. Prolongation of the reaction increases the ratio

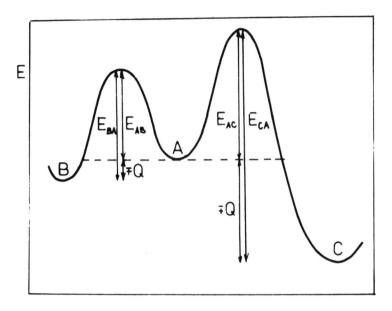

FIGURE 12.   Energy profile of competition reactions B ↔ A ↔ C. Symbols E_{AB} and E_{AC} denote activation energy of direct reactions; E_{BA}, E_{CA} activation energy of reverse reactions; ± Q are reaction heats.

of more stable products corresponding to state C. The overall reaction is said to be under thermodynamic control. In such a way, the situation may come into existence when the activation energy does not apparently determine the reaction course.

IV. STRUCTURE OF ACTIVE AND STABLE RADICALS

The reactivity of a radical is connected with the spin density on the atom of a radical site and with the type of orbital occupied by an unpaired electron. Increasing the number of atoms in a radical and thus the possibilities for intraradical interactions of unpaired electrons decreases radical reactivity. Monoatomic radicals such as atoms of H, Li, and F are therefore very reactive. Comparing atoms with the same type of semioccupied orbital, the reactivity of unpaired electrons decreases with increasing atomic weight. The reactivity, of course, depends not only on the number of atoms in a radical but also on their mutual configuration. While in saturated alkyl hydrocarbon radicals the spin density is concentrated on the α-carbon, in radicals where the radical site is in conjugation with π-bonds, the spin density on this α-carbon is considerably reduced. In a benzyl radical, trivalent carbon was only 70% of the spin density of an unpaired electron whereas the residual 30% is delocalized on the phenyl ring. In accordance with the decrease of spin density on the key atom of the radical, its reactivity decreases. This decrease is not, however, the same in all reactions, the most distinct being in some slower reactions of radical substitution. The other phenyl groups in the phenyl methyl radical lead to the next decrease of the spin density on α-carbon, but not proportionally to the overall number of phenyl groups. In the already mentioned triphenyl methyl radical, the spin density is uniformly distributed among a radical carbon atom and three phenyl groups.[24] Within the framework of one phenyl group, the spin density depends on the position of carbon atoms, the highest being in para and two ortho positions.

The decrease of reactivity is not due to delocalization of the unpaired electron over electron acceptor groups only, but as follows from the existence of several equilibrium concentrations of substituted triphenyl methyl radicals; also, interactions with electron donor moieties of a radical play some role here.[25]

Table 1

DATA ON THE RATIO r OF SUBSTITUENTS AND CENTRAL ATOM
BULKINESS FOR SOME RADICALS

Radical	$\cdot CH_3$	$\cdot C(CH_3)_3$	$\cdot Si(C_6H_5)_3$
r	0.22	3.4	5.4
Radical	$\cdot CCl_3$	$\cdot C(C_2H_5)_3$	$\cdot C(C(CH_3)_3)_3$
r	6.4	7.1	14
Radical	$\cdot C(C_2F_5)_3$	$\cdot C(C_6H_5)_3$	$\cdot CCl_2C_6Cl_5$
r	16	19	21
Radical	$\cdot C(Si(CH_3)_3)_3$	$\cdot CCl(C_6Cl_5)_2$	$\cdot C(C_6Cl_5)_3$
r	22	35	50

Delocalization of an unpaired electron in a radical does not, however, cause absolute stability. The opposite may be sometimes true. By delocalization of the electron, the reaction site involves a larger part of the radical and reactions, which could not be realized from sterical reasons may now easily occur. This is, e.g., the case of dimerization of triphenylmethyl radicals where one radical reacts on methyl carbon while a second does it on 4-carbon of the phenyl ring. The stabilization effect of sterical shielding of the semioccupied orbital is generally valid for each stable radical. Let us take, e.g., again the classical example of the triphenylmethyl radical. Provided that we substitute all hydrogens of this radical by more bulky chlorine atoms, we get a radical of extraordinary stability.[26] Trivalent carbon is protected by nonpermeable perchlorophenyl shields and any formation of a more stable chemical bond is not possible. Such a radical does not dimerize, is stable even in boiling toluene, and does not react with oxygen, chlorine, and bromine. Its decomposition starts at 300°C. Also, perchlorbenzyl radicals are nonreactive, which could be somewhat surprising taking into account the high reactivity of trichlormethyl radicals.

If the central carbon atom of a triphenyl methyl radical is replaced by more bulky silicon, the sterical shielding of three phenyl groups is not sufficient and the radical belongs to the reactive species. It is of interest that hexaphenylsilane is so stable that it withstands boiling in toluene solution at 140°C. In addition, from this fact one may deduce that delocalization of the spin density on three phenyl groups is not as significant in stabilization of triphenyl radicals as the sterical shielding of an unpaired electron.

The preceding ideas accounting for the role of sterical factors in radical reactivity may be supported by quantitative data (Table 1). As a matter of fact, the radical cannot react with some reaction partner if the bulkiness of the surrounding substituents exceeds about 12-fold that of the central atom carrying a radical site. This magic number corresponds to optimum packing of equal balls in a cube lattice; the central ball can have 12 direct neighbors. From this, it is quite obvious that the triphenyl silyl radical is reactive while perchlorobenzyl is not. For similar reasons, namely, due to the formation of branched perfluoralkyl radicals, the persistent part of radicals formed in irradiation of perfluoralkanes may be explained.[27]

The shielding of the semioccupied orbital by surrounding substituents does not prevent, however, intraradical reactions which may release the radical site from encompassing substituents. These reactions are more likely to proceed, the more the unpaired electron is localized on one atom. Increasing temperature and subsequent isomerization and fragmentation reactions may, however, change the reactivity of persistent radicals. Sterically hindered radicals having a delocalized electron are usually stable at a larger temperature interval.

The common tendency of the decrease of radical reactivity with the reduction of spin density on an atom with an unpaired electron is valid also for noncarbon radicals. The hydroxy radical, e.g., in which only very small delocalization of free electrons is possible, belongs to one of the most reactive radicals. In hydrogen peroxy radicals, the unpaired spin is almost entirely located on the π-orbital formed from the 2p-orbital of both oxygen atoms.

On the s-orbital of hydrogen, a very small negative spin density is induced (about 3%). The same degree of delocalization of unpaired electrons exists also in alkylperoxy radicals.[28] The relatively large delocalization of unpaired electrons in $HOO\cdot$ peroxy radicals, when compared with hydroxy radicals, corresponds to a considerable decrease of peroxy radical reactivity particularly in their reaction with molecules.

Substitution of hydrogen of the hydroxy radical by the phenyl group brings about much larger delocalization of unpaired electrons. Spin of the electron is up to 80% delocalized over the phenyl ring. The radical center is shifted to the para and ortho position of the phenyl group and the reactivity is again considerably lower than that of the hydroxy radical. Provided that instead of the phenyl group there is an alkyl, which may interact with free electrons only to a limited extent, we obtain relatively reactive alkoxy radicals. Despite considerable delocalization, the phenoxy radical is not stable and reacts with suitable substrates in para and also in ortho positions of the phenyl ring. Replacement of ortho and para hydrogen by phenyl groups leads to the further delocalization of spin density so that the 2,4,6-triphenyl-phenoxy radical is almost entirely stable in solution and undergoes dimerization only after passage into the solid phase.[29] The sterical stabilization of phenoxy radicals can be achieved when hydrogens on reactive sites are substituted by bulky *tert*-butyl groups. Even though the degree of delocalization in 2,4,6-tri-*tert*-butyl phenoxy radicals does not change practically when compared with nonsubstituted phenoxy radical, the sterical hindrance prevents dimerization in an inert atmosphere for both the solution and solid state of the radical. The *tert*-butyl group particularly protects carbons of an aromatic ring of the highest spin density. The shielding of an oxy group of a phenoxy radical is weaker. This follows from the reactivity of 2,4,6-tri-*tert*-benzyl radicals which dimerize in solution with a high rate constant corresponding practically to the rate of mutual collisions.[30] Replacement of the oxygen or CH_2 group of the above substituted radicals by the NH group which has similar bulkiness leads to the density on nitrogen atom 0.4. Compared with corresponding carbon radicals, the rate constant of decay of nitrogen radicals is by 4 orders lower. On the other hand, it is considerably higher than that of substituted phenoxy radicals. As may be seen on stable galvinoxy radicals (Scheme 4), the most efficient increase of stability of

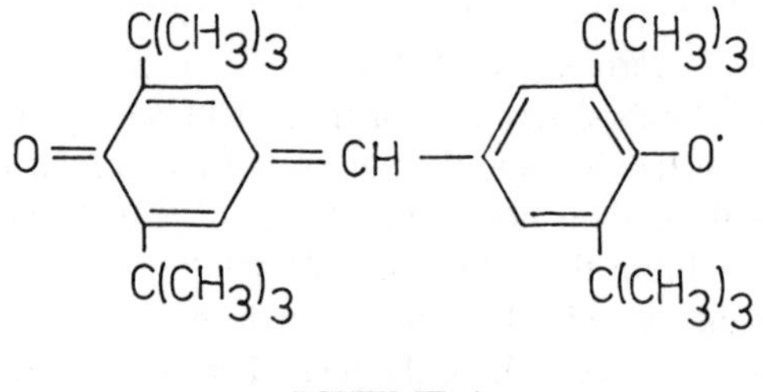

SCHEME 4

phenoxy radicals may be achieved by combination of both delocalization and sterical shielding. The uniform distribution of the unpaired electron in this radical was confirmed by ESR and ENDOR measurements. The unpaired electron appears to be completely delocalized over two aryloxy rings. The π-spin population at the central carbon atom is approximately zero.[31]

A large group of stable and semistable radicals includes nitroxy radicals containing $>N..O$ structural units. An unpaired electron is located here on the π-orbital formed from p_z orbitals of nitrogen and oxygen. The spin density on oxygen is several % higher than that on nitrogen. In both the aromatic and aliphatic nitroxy radicals, the N–O bond is somewhat shorter and therefore stronger when compared with the corresponding single bond in a saturated molecule. The reason for low reactivity of nitroxy radicals consists in the stable electronic configuration around oxygen and nitrogen where a three-electron bond is formed. Nitroxyl radicals, like peroxy radicals, which give tetroxides, dimerize to very unstable products which are wholly dissociated at ambient temperature. This irreversible decay proceeds as disproportionation. Provided that there are no mobile hydrogen atoms in the neighborhood of nitrogen, the nitroxy radicals are stable. This is, e.g., the case of di-*tert*-alkylnitroxides as well as per-fluoroalkylnitroxides. In N-aryl nitroxides, where the electron is partially delocalized, the steric shielding of the reactive position on aryl groups is also necessary to ensure the stability of a radical.

An even higher stabilization effect is achieved when the second nitrogen atom is linked with nitrogen carrying an unpaired electron. This type of radical is represented by the well-known stable radical diphenylpicrylhydrazyl (DPPH) (Scheme 5). The spin density on α-nitrogen is similar to that of the semistable diphenylamine radical. In DPPH, the relatively

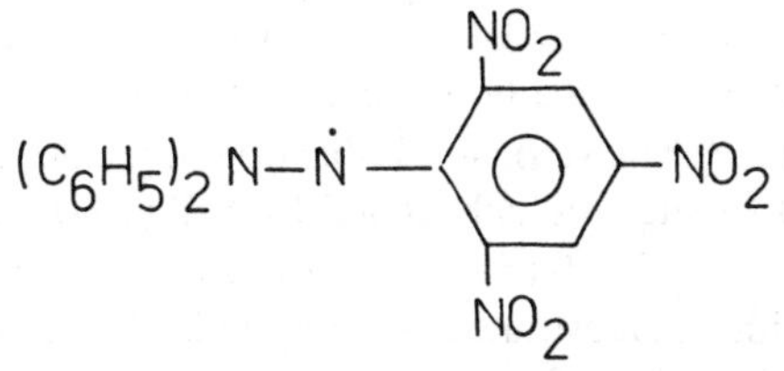

$$(C_6H_5)_2 \, N{-}\dot{N} \longrightarrow \overset{\displaystyle NO_2}{\underset{\displaystyle NO_2}{\bigcirc}} \!\!-NO_2$$

SCHEME 5

high spin density is also on β-nitrogen. The length of bond between these two nitrogens due to the formation of a supplementary fractional bond is lower when compared with saturated diphenylpicrylhydrazine. The same reason explains also the stability of verdazyl and similar radicals.

V. THE EFFECT OF UNPAIRED ELECTRON ORBITAL

At the more detailed classification of radicals according to their structure, the type of orbital of an unpaired electron appears to be an important factor. It follows from the interpretation of ESR spectra of carbon radicals that the hyperfine splitting constant of ^{13}C atom is sensitive to the degree of hybridization of an orbital with an unpaired electron.[32] Increasing the hyperfine splitting constant leads to the increase of the ratio of the s-orbital mixed with the original p-orbital of the carbon atom. This change of orbital character is observed, e.g., in fluoromethyl radicals. From the values of ESR splitting constants we may deduce that the unpaired electron in methyl radicals is located on the p-orbital while in the trifluoromethyl radical it is located on the sp^3 orbital. In the methyl π-radical, the orbital of the unpaired electron is situated perpendicular to the plane of three sp^2-orbitals (Figure 13). The planar arrangement of substituents on carbon with the p-orbital of the unpaired electron is characteristic also for other alkyl π-radicals. On the other hand, the σ-radical has a pyramidal conformation due to localization of an unpaired electron on the sp^3 orbital identical with other three substituents. Also alkoxy RO·, aryloxy R'O·, peroxy ROO·, nitroxy NO·, amidyl –ṄCO, iminyl C=N·, hydrazyl N-N·, phosphinyl -P̈- and other radicals are π-radicals. The very frequent feature of such radicals, the p-orbital of the unpaired electron joined with the system of surrounding π-orbitals, is a reason for their reduced reactivity.

In addition to trifluoromethyl radicals, also phenyl, vinyl, cyclopropyl, and carbonyl radicals such as OĊH, OĊCH$_3$, OĊC$_6$H$_5$ belong to σ-radicals. From noncarbon radicals, iminoxyl radicals with paramagnetic fragment >C=N–O·, phenyldiazyl C$_6$H$_5$–N=N· and radicals of the type (CH$_3$)$_3$X· where X represents atoms of the fourth group of the Periodical Table (X = Si, Ge, Sn), and particularly hydrogen atoms have σ-character, too. In σ-

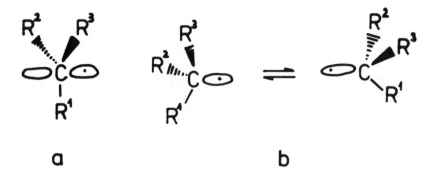

FIGURE 13.   The spatial arrangement of π- and σ-alkyl radicals of type $R^1R^2R^3C\cdot$, (a) planar π-radical, (b) pyramidal arrangement of σ-radical.

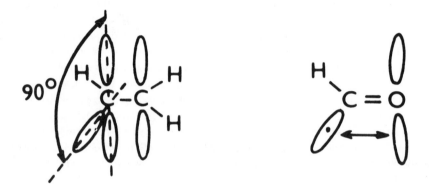

FIGURE 14.   Schematic illustration of semioccupied σ-orbital and π-electrons in vinyl and formyl radicals.

radicals having bonding π-electrons, the σ-orbital of the unpaired electron is located in the nodal plane of π-orbitals (Figure 14). This position of the semioccupied σ orbital of the vinyl radical rules out overlapping with bonding orbitals and the reactivity of the unpaired electron remains unchanged. Similarly in phenyl radicals, the ground state is represented by an unpaired electron on the sp^2 (σ) orbital of carbon. The unpaired electron does not interact directly with the system of π electrons of the phenyl ring. In some σ-radicals as, e.g., in the formyl or other carbonyl radicals, the spin of the unpaired electron can interact with nonbonding electrons of oxygen of the carbonyl group, which leads to its partial stabilization.

Although the difference in energy content of the σ and π radical is not high (about 15 kJ/mol), the reactivity of both kinds of radicals may well be distinguished experimentally. σ-Radicals are generally more reactive than π-radicals. That is why nitroxy radicals are more stable than iminoxy radicals. In nitroxy π-radicals, the unpaired electron is situated on the p-orbital, whose axis is perpendicular to the plane of CNO atoms. The unpaired electron in iminoxy σ-radicals is on the sp-orbital which lies in the plane of these atoms. The radical having the more exposed unpaired electron will induce larger impairment of the electronic structure of the neighboring molecules of a potential reactant and will be, therefore, more reactive. The difference in reactivity of σ and π radicals is more illustrative on chemically identical radicals. The ground state of stable diphenylpicrylhydrazyl has the electronic configuration of a π-radical. Light excitation transforms it to a σ-radical; from the nonbonding electron pair of nitrogen, one electron is displaced to a semioccupied orbital.

The formed σ radical is thereby capable of abstracting hydrogen from reactive hydrocarbons.

Reaction heat or increasing temperature may have similar effects as light. This may be the reason why qualitatively identical reactions of radicals may have a higher activation energy in a lower temperature region when compared with higher temperatures and why identical radicals arising from different chemical reactions differ in reactivity. The problem of radical excitation will be dealt with elsewhere. We recall here only that the radical will be less reactive with respect to surrounding molecules when its spin density will be more delocalized and the semioccupied orbital will be more overlapped with the overall electron system.

The question of how reactivity of unpaired electrons on identical orbitals alters with their distance from the atom of nucleus may be demonstrated on reactions of halogen atoms which have the same number of electrons on the same orbitals of the outer shell. Two electrons are located on the s-orbital and five electrons on three p-orbitals. Since one orbital can contain only two electrons, one of the p-orbitals will be occupied by one unpaired electron. The increase of the distance between atom of nucleus and shell of outer electrons decreases the strength of their binding in an atom and ionization potential unambiguously decreases, too. From this tendency decline fluorine atoms, which have such dense packing of electrons in the L shell that their electrostatic repulsion is also of importance. From halogenide anions this is, therefore, chlorine and not fluorine, which has the highest ionization potential. The effort to fill the electron vacancy on the p-orbital of halogens may be displayed in the energy of the halogen-hydrogen bond which increases significantly going from iodine to fluorine. The reactivity of halogens in transfer reactions with hydrocarbons increases in the same direction. The activation energy of hydrogen abstraction from hydrocarbons, as a rating of halogen reactivity, increases in the opposite direction, from fluorine to iodine, in accordance with the decrease of exothermicity of the transfer reaction. For iodine and bromine atoms, the transfer is already endothermic and the rate of reaction is controlled by the strength of cleaving the carbon-hydrogen bond. In an activated complex, this bond will be almost disrupted and replaced by a new bond halogen-hydrogen.

For fluorine and chlorine atoms the transfer reaction is strongly exothermic and the rate depends on the energy of the broken bond only slightly. The structure of the activated complex is here closer to the structure of the original reactants. The reactivity of a radical is reflected not only in the reaction rate but also in the selectivity of the attack of certain bonds in some part of the molecule or in the mixture of molecules. The rate of abstraction of hydrogen by fluorine or chlorine is rather similar for all hydrocarbons, while in the case of less reactive bromine and iodine it depends strongly on the structure of the reacting molecule. A rule (with a small number of exceptions) follows from this, namely, that the reactive radical is less selective and vice versa. The exceptions exist above all in thermo-neutral reactions where more reactive radicals may be more selective. This opposite behavior can be attributed to the significant effect of polar factors at the activated complex formation.

From the course of hydrogen abstraction by halogens one can deduce also that the reactivity of the same type of semioccupied orbital decreases when increasing its distance from the atom nucleus. This conclusion may be supported by other examples. Silicon radicals abstract hydrogen from hydrocarbons more slowly than alkyl radicals, but still are more reactive, than Ge, Sn, or Pb radicals. Similarly, in the V group of the Periodic Table, nitrogen radicals are more reactive than phosphorus, arsenic, antimony, and bismuth radicals. The same situation exists in the sixth group for radicals of oxygen, sulfur, selenium, and tellurium.

Generalization of the decrease of reactivity with increasing atomic weight of a carrier of the free radical site is, however, valid only for the selected type of transfer reaction. Provided that we consider the abstraction of chlorine atoms from alkyl chlorides as a standard reaction, the sequence of reactivity is changed and silicon radicals become more reactive than carbon radicals. The important role in such kind of reactions is not played only by larger spatial requirements of semioccupied orbital but also by lower electronegativity of the central atom and, accordingly, by increased electron-donor properties of a radical. The same violation of the reactivity sequence of halogens may be observed for dehalogenation of halogen hydrocarbons by halogen atoms as a reaction standard. It is thus confirmed again that delimitation of radical reactivity is not correct without consideration of the respective reaction partner.

VI. STABILIZATION ENERGY OF RADICALS

Quantitative separation of individual reactivity components is not so simple as is needed. The different effects are mutually conditioned and overlapped. We remind the reader now of what is a matter of fact. The shielding of the semioccupied orbital by neighboring atoms does not represent only steric hindrance, but these atoms may also take part in delocalization of unpaired electrons. The surrounding atoms influence the charge distribution and change nucleophility or electrophility of radicals. On the other hand, the mutual coupling and contingency of various factors determining radical reactivity indicate that a different course of reactivity may be described in terms of several physical characteristics of a radical. This approach is used when estimating radical reactivity from its stabilization or destabilization energy attributed to interactions of unpaired electrons with a residual part of a radical. Since it reflects the change of the energetical state of free and molecule-bound radicals, such energy is sometimes called the energy of reorganization.

Energy of reorganization is either released or consumed in the change of electron and geometrical configuration of free radicals related to that in a parent molecule.[33] Energy of reorganization (E_R) of a radical is equal to the difference of the dissociation energy of chemical bond (E_D) and energy of the same bond (E_B) in the parent molecule when all bonds linking atoms of molecules are simultaneously cleaved (Equation 36). The reorganization energy of the first ($E_R(1)$) and

$$E_D = E_B + E_R(1) + E_R(2) \tag{36}$$

second ($E_R(2)$) radical can have an opposite sign. At the same time, the reorganization energy of atoms is zero. The negative values of reorganization energy indicate the stabilization of the radical and vice versa. Since the same radicals may arise from the decomposition of different compounds, the value of E_R should be averaged and the effect of experimental errors on the determination of the reorganization energy of radicals thus diminished.

By analogy, the relative stabilization energy of radicals (E_S) can be expressed by the difference of bond dissociation energies (Equation 37) where S refers to a standard

$$E_S = E_D(S\text{-}X) - E_D(R\text{-}X) \tag{37}$$

moiety of a molecule and R to the radical in question.[34] The data for a variety of X gave fairly constant values of E_B for different kinds of radicals. Besides thermochemical data for determination of the stabilization energy of some radicals, empirical correlations of the stabilization energy and ESR spectroscopic characteristics of spin delocalization were used, too.[35] While relationships of this kind are unlikely to replace thermodynamic measurements of stabilization energies, they are an effective tool for screening the accuracy of the available data.

In the resulting effect of the reorganization energy on radical reactivity, partially opposite tendencies are hidden.[36] This may be illustrated, e.g., on the benzyl radical. It follows from the calculations, that the reorganization energy of this radical has a negative part equal to 70 kJ/mol corresponding to π-electrons and a positive component 16 kJ/mol corresponding to σ-electrons. Accounting for a rather large difference in reorganization energies of π and σ electrons, the reactivity of aromatic polycyclic radicals is predominantly determined by the stabilization energy of π-electrons. The latter is represented by the energy of delocalization of unpaired electrons within an overall system of conjugated π-bonds. From the sequence of reorganization energies of some radicals one may see that reactivity increases when stabilization energy decreases (Table 2). We obtain, thus, the overall pattern of reactivity of structurally different radicals. The above sequence does not fit atoms whose reactivity differs considerably. Apart from this, the stabilization energy does not give unambiguous correlation with reactivity of radicals differing in the central atom of a radical site. On the other hand, the quantification of the reactivity sequence of radicals renders possible comparison and classification of knowledge on radical reactivity and may reveal contingent discrepancies and new relationships in reactions of a homological series of polyatomic radicals.

VII. THE POLAR CHARACTER OF RADICALS

Free radicals, although formally neutral, may have a polar nature and can be classified as electrophilic or nucleophilic according to the electron density on the site of the partner molecule where the reaction starts. Electrophility of a radical may be expressed in terms of electron affinity while nucleophility by the ionization potential of a radical. Radicals having low ionization potentials are nucleophile; those of high electron affinity are electrophile.

The classification of radicals is also possible from the data of radical electronegativity (Table 3). The above approaches to an estimation of polar character allow for detailed classification of radicals only, but the existing demonstration of polar character in some specific reaction depends on subtle structural characteristics of both the radical and its reaction partner as well as on conditions of the reaction course. We know that alkyl radicals have a clear nucleophilic character which is enhanced going from methyl to primary, secondary, and tertiary alkyl radicals. This polar character explains well the increase of the rate constant in the reaction of alkyl radicals with molecular chlorine and other electrophile reagents when the nucleophility of alkyl radicals increases.[38] Carbon-centered radicals with a nitrogen and oxygen atom in the α position have a distinct nucleophilic character.[39] Despite the fact that the electronegativity of nitrogen and oxygen is higher than that of carbon, α-alkoxy alkyl and α-amidoalkyl (CONC<)π-radicals are more nucleophilic than the corresponding unsubstituted alkyl radicals. On the contrary, for the sequence of acyl radicals which are of the σ-type an obvious reduction of nucleophility course with aminocarbonyl (>NĊO) and alkoxy carbonyl (ROĊO) radicals compared with acyl radicals (RĊO) alone appears. The increased nucleophilic character is generally associated with an increased stability and oxidizability of the radical.[40]

The influence of the polar character on the reactions of the benzyl radical is very marked and not less than the influence of the stabilization energy of the radical.[41] The benzyl radical is more nucleophilic than the unsubstituted alkyl radical and the benzoyl radicals, while the acetyl radical is more nucleophilic than the ethyl radical. The reactivity of the benzyl radical is more strongly affected than that of the benzoyl radical by the presence of substituents. Different sensitivity to polar effects can be attributed again to the fact that the benzyl radical is of the π-type and the benzoyl is of the σ-type.

The phenyl radical, perhaps the most studied organic σ-radical, has a small but not nucleophilic character.[42]

From a large set of electrophilic radicals, amino and hydroxy radicals (where the effect of protonation or deprotonation on radical reactivity is obvious) are worth noticing. While nonprotonated ·NH$_2$ radicals attack phenol 5 times faster than benzene sulfonic acid, the reverse is true for the protonated radical. Not only selectivity is changed by protonation,

Table 2
REORGANIZATION ENERGY OF SOME RADICALS[33-37]

Radical	kJ/mol	Radical	kJ/mol
·OC₆H₅	−121	·NO₂, ·SC₆H₅	−59
(structure)	−118	·CH₂C₆H₅, ·OCCH₃	−54
·CH₂(CH=CH)₂CH=CH₂	−102	CH₃ĊHCH=CH₂	−52
·C(C₆H₅)₃	−100	O₂SCH₃	−50
		·OCOC₃H₇	−46
(structure)	−98	c–CH₂ĊHCH₂, ·CH₂CH=CH₂	−42
·OOH	−92	·CH₂C≡CH, ·CH₂COCH₃	−36
		·NF₂	−25
		Ċ(CH₃)₃, c–Ċ₅H₉, c–Ċ₆H₁₁	−21
·CH₂CH=CHCH=CH₂	−88	·CH₂C(CH₃)₂Si(CH₃)₃	−20
(structure)	−84	HOĊHCH₃, ·NO, ·OCOC₂H₅	−17
C₆H₅ṄCH₃	−83	·CH₂Si(CH₃)₃	−10
(structure)	−80	·CH₂OH, ·CH(CH₃)₂, ·CCl₃	−8
		·CH₂C₂H₅, ·C₂H₅	−4
(structure)	−78	·OCH(CH₃)₂, ·SCH₃	0
c–CH=CHĊHCH=CH	−73	·SH, ·OCH₃, ·OC₂H₅	4
		·CH₃	8
(structure)	−71	·OCOCH₃	25
		·OF	29
(structure)	−67	·OH, ·NH₂	42
		·CH=CH₂	50
(structure)	−63	(structure)	63

Table 3
THE VALUES OF IONIZATION POTENTIAL (IP), ELECTRON AFFINITY (EA), AND ELECTRONEGATIVITY (EN) OF RADICALS

Radical	IP eV	EA eV	EN[a]	Radical	IP eV	EA eV	EN[a]
·Na	5.1	0.5	0.70	·NO	9.3	0.9	4.84
·CH₂C₆H₅	7.2	0.9	3.64	·NH₂	11.4	1.2	3.84
·C(CH₃)₃	7.4	1.0	3.65	·OOH	10.9	3.0	4.59
·CH₃	9.9	1.1	3.61	·OOCH	9.0	3.6	4.37
·C₆H₅	9.9	2.2	3.68	·OH	13.2	1.8	4.30
·CCl₃	8.7	2.1	4.62	·CN	15.1	3.0	4.12
·CF₃	10.1	1.8	5.18	·I	10.5	3.3	3.84
·H	13.6	0.7	3.55	·Br	11.8	3.6	4.53
·SH	9.2	2.3	3.82	·Cl	13.2	3.8	4.93
·OC₆H₅	8.5	1.2	4.38	·F	17.4	3.6	5.75

[a]   Relative values estimated according to Reference 33, p. 192.

but also radical reactivity. The more electrophilic $\cdot NH_3^+$ radical reacts more slowly than $\cdot NH_2$. The fact that the more electronegative radical is less reactive cannot be generalized, however.[43] For the reaction of a hydroxy radical and an oxygen anion radical with benzene, the contrary is seen; the more electrophile hydroxy radical is considerably more reactive than the oxygen anion radical.[44]

The polar effect can be explained in terms of the frontier orbital theory by the interaction of the singly occupied molecular orbital (SOMO) of the radical and the lowest unoccupied molecular orbital (LUMO) of the reactant.[9] A dominant SOMO/LUMO interaction exists among nucleophilic radicals and the electrophilic reactant, while a dominant SOMO/HOMO interaction exists among electrophilic radicals and nucleophilic reactants. The energy effect associated with SOMO/LUMO and SOMO/HOMO interaction is proportional to the square of the overlap integral SOMO/LUMO or SOMO/HOMO and inversely proportional to the energy gap between these two orbitals before interaction. The overlap integral of SOMO interaction is determined particularly by the spin density on the radical center. Comparing reactions of different radicals with some substrate, the energy gap can be correlated either with ionization potential or with electron affinity.

The comparison of the O–O bond attack in peroxy acids by different radicals R· (Equation 38) shows the existence of a

$$R\cdot + HO–OCO' \rightarrow ROH + R'CO_2^\cdot \tag{38}$$

correlation between the ionization potentials of alkyl-localized free radicals and their reactivity.[45] Nucleophilic radicals with a low ionization potential easily transfer HO from peroxy acids. On the other hand, σ-radicals with the unpaired electron in a hybridized orbital with a high ratio of s component, such as phenyl or 2-phenylcyclopropyl, react slowly or not at all upon the peroxide oxygen.

For localized alkyl radicals, having a spin density on the carbon close to 1, a constant overlap integral along the series can be assumed, which consequently explains their reactivity difference mainly by the gap difference. The importance of the overlap control can be determined by comparing the reactivity of radicals with different spin density. From this viewpoint, benzyl type radicals are interesting. The spin density on the benzylic carbon is about 0.65. As ionization potentials of the benzyl radical and tertiary alkyl radicals are alike, they should have similar reactivities from the point of view of energy gap control. In reality, benzyl type radicals are less reactive as primary alkyl radicals and much less reactive than tertiary ones in HO transfer reactions from peroxy acids. Comparing the series of benzylic radicals where the spin density at the reactive carbon is of the same order of magnitude, energy gap control becomes again the determining factor.

The complexity of polar effect estimation for the lower difference of electron-donor properties of radical and reacting substrates is due to the fact that both the electron donor and the electron acceptor properties of radical and substrate are put across simultaneously. A comparison of the reactivity of different radicals requires thus knowledge of electron affinity and ionization potential of a radical and, of course, also of spin density on the reaction site of substrate.

VIII. DYNAMIC ASPECTS OF RADICAL REACTIONS

Radical reactions in the gas phase are usually represented by a complex kinetic scheme of different more or less important elementary chemical reactions with corresponding rate constants. The mutual ratio of these rate constants determines the overall picture of the chemical mechanism and is influenced by the circumstances of an elementary chemical act — the collision of two species inducing all subsequent processes and leading to the rearrangement of chemical bonds between atoms. It is necessary to bear in mind that the rate constant represents only an average of initial and final states through the individual state-to-state rate constants for specific translational, vibrational, and rotational states of reactants and products. Present-day experimental techniques allow one to obtain information about such state-resolved characteristics of chemical reactions of small molecules and to extend some of the experience obtained also to large polyatomic systems.

One interesting example is the temperature dependence of the rate constant. On a molecular level, it is represented by the dependence of a reaction cross-section on the relative translational energy of two colliding molecules. The typical character of this curve can be described as follows. It starts at an energy corresponding to the energy threshold of a given reaction channel (approximately equal to the activation energy), proceeds through a maximum, and approaches zero at energies when other channels (with more dissociative character and larger energetical demands) are opened. This means that enhanced translational energy can, up to a certain limit, increase the rate of a chemical reaction. However, in many cases, the same energy can be used much more efficiently if we use it on the enhancement of vibrational energy of reactants and to increase their vibrational temperature. Theory supported by experimental results assumes that the most effective energy input is influenced by the location of the barrier on the electronic potential energy surface.[46] For the early barrier, translational energy is more effective than vibrational. The reverse situation is in the case of a late barrier. As an example, the reaction of H_2 with the ·OH radical is enhanced by a factor of 120 with H_2 (v = 1). On the other hand, the rate of reaction of Br· with CH_3I and CF_3I is more sensitive to stimulated reagent translation than to reagent vibration.[47] The increased rotation motion of reagents has an apparently inhibiting effect. However, it also happens that rotational energy is more efficient in promoting some reactions than the translational energy. From reasons mentioned, hot radicals with increased translational, vibrational, or rotational energy, which is far from its equilibrium value corresponding to a given temperature, show extremely high reactivity for some specific reactions.

Another question is how the input reaction energy will be distributed among the products. As an illustrative example (Equation 39),

$$H· + D_2 \rightarrow HD + D· \tag{39}$$

the practically thermoneutral reaction (ΔH_0 = 4 kJ/mol) can be used.[48] Here, from available energy (1.3 eV), about 70% appears in product translation, about 20% in HD rotation, and 10% in vibrational energy above the HD zero point vibration. The character of the vibrational and rotational energy distribution among individual quantum numbers can give us valuable information about the complexity of the collision process. Its practical value will be demonstrated in the chapter concerning chemical lasers.

The temperature dependence of the rate constant can be also negative (the rate constant decreases with increased temperature). This is the case of some radical-radical reactions which are usually without potential barrier along the reaction pathway. Then the rate-determining moment seems to be the complex formation of radical pairs due to the long-range attraction between the radicals. Increasing temperature has a negative influence on this process because of the enhanced translational and rotational motion of molecules.

In close connection with this problem, is the centrifugal barrier. If the collison complex has a large rotational quantum number, the centrifugal forces act against the mutual approach of two radicals. This effect can be described in the form of an effective potential energy surface with inclusion of rotational energy. Then we can find a small barrier also for potentials where the electronic energy has no barrier. The position and the height of such a centrifugal barrier depend on the relative translation energy of both molecules before collision and on the impact parameter which characterizes their proximity during the collision. For a zero impact parameter (central collisions) the centrifugal forces are also zero. However, at large translational energy and impact parameters this effect prevents any reaction also for large attractive forces between radicals. It can be shown that for the reactions having only a centrifugal barrier the rate constant will usually be independent of the temperature, as is often found for some ion-molecule reactions.[49]

The barriers on potential energy surfaces of radical reactions are significantly smaller than these for reactions between molecules with closed electronic shells. This fact also causes other more subtle factors to play an important role in influencing the total rate constant. It is known that quantum tunneling through the barrier is important for the reactions of H· atom transfer at low temperatures. Theoretical analysis based on the variational transition state theory, for the reaction of hydroxyl radicals (Equation 40)

$$\cdot OH + H_2 \rightarrow H_2O + H\cdot \tag{40}$$

shows the large importance of tunneling for temperatures below 600 k.[50] The rate constant of this reaction is increased by a factor of 17 after inclusion of the tunneling effect even at a temperature of 298 K.

Besides the energy barrier, there are also other features of electronic potential energy surfaces that can influence the rate of chemical reactions. Because radical reactions are usually joined with nonzero total spin number their entrance channel is degenerated and the reaction path splits. As a result, the obtained products can be in different electronic states at sufficient initial energy and the excited ones can lead to other products. During the reaction path of radical reactions, the typical and frequent case is also the existence of a local minima which corresponds to the formation of bound adducts. Then it is necessary to take into account possibilities of stabilization, or redissociation of such complex and kinetical schemes becomes complicated. Also, the total number of reaction channels for a given pair of reacting molecules is of major importance here. This feature, known as the steric factor, must be considered since the center with an unpaired electron on a radical can attack different sites of the second molecule. A simple example (Equations 41 and 42) shows that the result of such action can be different.

$$H\cdot + \cdot CH_3 \rightarrow CH_4(v) \tag{41}$$

$$H\cdot + \cdot CH_3 \rightarrow :CH_2 + H_2 \tag{42}$$

Today's experimental techniques which are capable of orienting the pair of molecules before their collision, allow one to obtain direct information about the steric factor. The result of such elegant experiments is the dependence of the rate constant on the mutual orientation of molecules.[51,52] One of the most complete studies has been performed for the reaction of ·NO with O_3. In this chemiluminiscent reaction (Equation 43), the

$$\cdot NO + O_3 \rightarrow \cdot NO_2 + O_2 + h\nu \tag{43}$$

beam of ·NO molecules is oriented by an electrical field. The intensity of the emitted radiation shows then strong dependence on the orientation of ·NO molecules relative to the direction of their flight. From these experiments it was concluded that a reaction proceeds effectively through two channels with two different transition states. In the first case, the ·NO radical attacks the central O atom of the O_3 to form ·NO$_2$ and recoils backward (Scheme 6). In the second probable configuration (Scheme 7), the ·NO molecule strikes in a broadside orientation the O_3 molecule and abstracts an end O atom and recoils sideways.

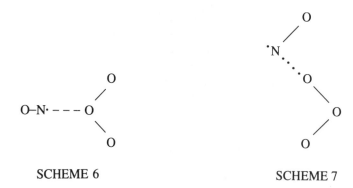

SCHEME 6   SCHEME 7

Another direct possibility of studying the properties of transition-state geometrical arrangements is promised by spectroscopy of the transition state.[53,54] The chemiluminiscent reaction (Equation 44) provides here the radiation with the

$$F\cdot + Na_2 \rightarrow NaF + Na\cdot + h\nu \tag{44}$$

spectrum where the sodium D line has wings on both sides. These low intensive wings correspond (due to the convenient arrangement of potential energy surfaces for the ground and excited states) to the emission from higher electronic energy levels of the unstable ground state (transition state) configuration Na_2F. These examples indicate that many imaginative and exciting experiments concerning very short-lived species can be expected in the near future.

Simple dynamic ideas can help us also at more complex reactions.[55,56] A systematic study of reactions of atomic oxygen with saturated hydrocarbons RH (Equation 45) gives us a

$$\cdot O\cdot (^3P) + RH \rightarrow \cdot OH + R\cdot \tag{45}$$

detailed picture of hydrogen abstraction (R· are primary, secondary, and tertiary alkyl radicals). Molecular beam/laser-induced fluorescence experiments (taking off the vibrational, rotational, and fine structure levels distributions of product ·OH radicals) in combination with quasiclassical trajectory calculations, provide surprisingly much information about the dynamics of the whole process. From the main conclusions it seems important that activation energies and pre-exponential factors per one C–H bond in the Arrhenius form of the rate constant depends significantly on the type of C–H bond (primary, secondary, or tertiary), but are otherwise independent of the exact nature or complexity of the hydrocarbon reactant. Thus, rates for large complex hydrocarbons can be predicted as the sum of rates for the individual C–H bonds. This indicates that an oxygen atom interacts strongly with individual C–H bonds, rather than with the entire hydrocarbon molecule. Even more, a reaction only occurs when ·O·(^3P) is collinear to a C–H bond or the deviations from collinearity are smaller than 15°. This picture explains the relatively small pre-exponential factor for such reactions.

In most of the experiments mentioned, detailed knowledge about the elementary chemical act is obtained under the conditions in which the reagents are formed and encountered in crossed molecular beams. This arrangement allows the creation of defined initial conditions for the chemical reaction (relative translational energy of reacting partners, their internal energy distribution, their mutual orientation, etc.). The chemical reaction proceeds here as a single collision and the study of properties of products is not disturbed by subsequent collision as it is in the case of experiments in bulk. Measurements of scattering angles of products and their velocity distribution can then be interpreted in terms of the mean lifetime of the reaction complex, of the vibrational-rotational energy distribution of products, and of the other dynamic characteristics of chemical reactions. The source of radicals or ion radicals which are frequently used as reagents in molecular beams is usually pyrolysis of some precursors. In the case of $\cdot CH_3$ beams, e.g., it is $(CH_3)_2Hg$, $(CH_3)_2$ Cd, $(CH_3)_2Zn$, azomethane or di-*tert*-butyl peroxide. The velocities of beams are usually supersonic (e.g., about 2100 m sec^{-1} used for the $\cdot CH_3$ beam[57]), which is necessary for reaching the large scale of chemically interesting collision energies. Experience obtained at usually studied atom + diatomic molecule collisions are useful now for the interpretation of the relatively complex mechanism of such molecular beam reactions as in Equation 46, where three possible reaction channels mutually

$$\cdot CH_4^+ + CH_4 \rightarrow CH_5^+ + \cdot CH_3 \qquad (46)$$

compete, namely, H^+-transfer, H-transfer, and intermediate complex formation.[58] The theoretical interpretation of this reaction in terms of the shape of the potential energy surface is very pretentious, but nevertheless it is able to provide a very detailed picture of the elementary chemical act and to give us deep insight into the elementary principles of chemical reactions.[59]

IX. EXCITED RADICALS

The energy of polyatomic particles which may be potentially used in chemical reactions is divided into electron, vibrational, rotational, and translational parts, while for atoms only electron and translational components come into consideration. Even though the individual kinds of energy of corresponding modes of particles motions are mutually interrelated, under certain conditions they may be investigated separately. This is due to the 1000-times-higher rate of electron displacement compared with the rate of vibrations of bound atoms or with even slower rotational oscillations of atom groups. The transformation of the respective mode of motion in polyatomic particles occurs from the higher energetic states to the lower. Excited states exist only transiently until equilibrium distribution of respective forms of energy is attained; it depends on the structure of a radical and conditions of its existence. Since redistribution of excitation energy is very fast, the lifetime of any excited radical does not exceed several microseconds.

Taking into consideration the different energy content of respective motions in radicals, there are various alternatives of excited state formation.

Energy from 0.1 to 15 kJ/mol is capable of exciting rotational states. Vibrational transitions are observed at energy quanta 1.5 to 70 kJ/mol, while electron excitation requires 5 to 500 kJ/mol. The lower limit of the above values may be attributed to spin orbital excitation, whereas the upper limit is connected with the change of electron distribution in a radical. Not only is electron distribution changed by excitation, but also the configuration of atoms. Excitation of electrons, and rotational and vibrational motion occurs in quanta and the energetical level increases stepwise. Only the rate of translation motion of atoms and radicals increases continuously. In radical reactions, the particles of energy content from 0.1 to 17 eV are efficient. At higher energetic collisions, an atom loses part or the whole electronic shell.

The small amount of knowledge on the excitation of radicals at the level of rotational motion of groups of atoms in a radical is due to the only slight effect of such excitation on the rate of radical reactions. Vibrationally excited radicals having an excess of translational energy are called hot radicals. The common feature of excited radicals, the increased reactivity, is brought about by the contribution of excitation energy to activation energy of individual elementary reactions. This explains not only the increase of the rate of excited radicals reactions, but also the course of such reactions which do not occur in the case of thermalized nonexcited radicals.

X. ELECTRON EXCITATION

Electron orbitals of a ground electron state of radicals are occupied according to an increasing energy scale. In electronically excited states, a radical occupies orbitals of higher energy and some of its lower energetical level remain semioccupied or even empty. In addition to the change of configuration of electrons due to their displacement from one orbital to the second, also excitation caused by the change of electron spin orientation with respect to orbital magnetic moment of electron is possible. The orbital motion of an electron around the atomic nucleus is brought about by the magnetic field which affects the spin of an unpaired electron; the energy of spin-orbital interaction is higher, when a higher number of electrons influence the spin of an electron by their orbital motion. This kind of excitation may be seen particularly in heavier atoms which have stronger spin-orbital interaction. The effect of spin-orbital excitation on reactivity may be illustrated on halogen atoms.[60] The difference in energy content of halogen atoms in both the ground state (denoted as $^2P_{3/2}$) and excited state ($^2P_{1/2}$) increases when going from fluorine to iodine:

Atom	F	Cl	Br	I
E, kJ/mol	4.8	10.5	44.0	90.7

The same order is held for the difference in reactivity of the excited and ground state of respective atoms. At the transfer reaction of iodine atoms with a molecule of chlorine (Equation 47), the excited iodine atom has a rate

$$I\cdot + Cl_2 \rightarrow ICl + Cl\cdot \tag{47}$$

constant about 5 orders higher than its ground state at room temperature. The higher rate constant is due to the lower activation energy of the reaction conditioned by the fact that the reaction is exothermic while that of the nonexcited iodine atom endothermic.

The rate of reaction of halogen atoms depends, of course, also on the ratio of formed excited atoms. From this viewpoint, the kinetics of reactions can be significantly affected by the contribution of excited fluorine (6.7% mol) and chlorine (0.7%) atoms. At the same temperature (300 K), the participation of excited bromine atoms is only $1\cdot10^{-6}$ mol% and iodine even less ($7\cdot10^{-15}$ mol%). The transfer of halogen atom from $^2P_{1/2}$ state to its ground state is accompanied by a relatively slow emission of long wavelength light.

Excited halogen	F	Cl	Br	I
Average radiation time, sec	830	83	1.1	0.1
Wavelength, nm	24,752	11,351	2714	1315

Photoexcitation of radicals connected with the change of the electron configuration occurs by a similar mechanism as that of similar molecules. Electron transitions of π-σ* and σ-π* type lead to radical fragmentation, while transitions π-π* and n-π* bring about hydrogen atom transfer and radical ionization.[61] High quantum yields of chemical transformation are typical for the electronic excitation of radicals. The increased sensitivity of radicals to ultraviolet light is connected with the decrease of strength of some bonds in a radical. The shift of excitation radiation to lower wavelengths for a radical R· compared with a molecule RH follows from lower energetical requirements on the first electronical transition. The lowest excited state of the radical has a doublet spin multiplicity. Intersystem crossing from the lowest excited doublet state to the quartet manifold is very unlikely, since the energy level of the lowest quartet state will be usually above the lowest excited doublet state. By contrast, in closed-shell systems, intersystem crossing to the triplet manifold and back to the ground state is a quite frequent process.

The detailed study of light excitation (320 nm) of similar arylmethyl radicals reveals that the fate of the radical depends considerably on its structure.[62] The absorption spectra of the ground states of all studied radicals $(C_6H_5)_2\dot{C}R$, where R = H, C_6H_5, C_3H_5, CH_3 are very similar. The laser light excites the radicals into their third excited state. Radicals then relax by the transition to the first excited state. The chemical reactivity of radicals in this excited state is significantly different. Diphenylmethyl radicals are practically nonreactive. On the contrary, radicals having phenyl groups in propeller-like conformation izomerize very efficiently in cyclization reaction forming a C–C bond between phenyl groups. The solvent dependent photochemical reaction of diphenyl methyl radicals (R=H) is, however, qualitatively different and starts from a higher excited state as a result of consecutive biphotone excitation of the radical. As may be seen, the slight structural changes in the radical produce a pronounced difference in reactivity of irradiated states.

The formation of electronically excited radicals as products of a chemical reaction has not been clearly demonstrated until now, but it is likely. There exists a relatively large and long-term discussion on the possibility of the π and σ character of succinimidyl radicals which has not yet been solved satisfactorily.[63]

XI. HOT RADICALS

Particles with vibrational and rotational motion of atom groups higher than the average value of their thermal energy are formed in exothermic reactions or at the radicals generation. The energy liberated in the bond breaking will often be an outlet through the internal degrees of freedom of the fragment. The production of radicals with intrinsic temperatures from 10 to several thousands Kelvin is, therefore, a reasonable expectation. The way how to obtain thermalized free radicals is to cool them by expansion after production.[64]

During photochemical cleavage of a bond, the internal energy of the hot radical may vary together with the changes in photon energy. For example, the gas phase photolysis of butene isomers (Equation 48) leads to the formation of the $\cdot C_3H_5$

$$C_4H_5 + h\nu \rightarrow \cdot C_3H_5 + \cdot CH_3 \tag{48}$$

radicals which may contain a part of the excess energy of the incident photon. The rate constant of fragmentation of $\cdot C_3H_5$ radicals to alkene and hydrogen atom depends, therefore, on the wavelength of photolyzing light; the higher the energy of incident photons, the higher the rate constant of $\cdot C_3H_5$ radicals fragmentation.[65] The dependence of rate constants of fragmentation of $\cdot C_3H_5$ radicals on the butene structure indicates that primary photoexcited butene molecules are not energetically randomized before their fragmentation processes.

During chemical activation, i.e., in exothermic reactions vibrationally and rotationally excited states are formed, particularly in the process of a new bond formation. Addition of a hydrogen atom to alkenes leads to alkyl radicals having 170 kJ/mol of energy excess. Even higher excitation of the radical occurs when adding a hot hydrogen atom to alkenes or alkines, which contributes by more 50 to 120 kJ/mol to the radical warming up.[66]

The excess of released energy is partially consumed on the increase of rotation of atomic groups and on vibrations among atoms of a radical. At the same time, the realization of rotational excitation is considerably faster than the dissipation of an excess of vibrational energy. Even though the vibrationally excited state may deactivate by emission of infrared light, this latter process is negligible when compared with collisional deactivation. The rate of thermal equilibrium establishment is given by the probability of transformation of vibrational and rotational excitation to translational energy of neighboring molecules occuring in the time lag of the collision. On the other hand, the probability of transfer of one vibrational quantum (20 to 40 kJ/mol) is higher in efficient collision than that of simultaneous transfer of two quanta. The probability of vibrational deactivation during the contact of a radical with another particle increases when increasing the excitation level of the radical. The lifetimes of various vibrational levels of the lowest excited electronic state of cyclopentadienyl decreases as the energy of vibrational level is increased. The lifetimes are the longest in the vibrationless levels, about 70 nsec, and decrease, reaching about 20 nsec for levels with a little over 1000 cm^{-1} vibrational energy content.[67]

Relaxation mechanisms of vibrational energy in gaseous and condensed phases are alike. In the condensed phase and, specifically, in liquids, the number of collisions per time-unit is, however, higher by several orders and deactivation is thus faster. Taking into account these facts, the characteristic properties of hot radicals can display, particularly under conditions of lower probability of collisions with surrounding molecules, i.e., at low pressures in the gaseous phase.

Less distinct dependence of the reaction rate on temperature and lower selectivity in all elementary reactions of hot radicals were observed. Even at the high energy content of hot radicals, their chemical reactions are not completely unselective, since the selectivity mechanism based upon the entropy changes in respective reactions is still put across. Direct information on the energetic content of hot radicals may be obtained from spectroscopic data. The spectra of radicals show that rotational and vibrational energies exceed the medium thermal energy by several thousands of degrees Kelvin.

XII. TRANSLATIONALLY EXCITED ATOMS

The reactivity of an atom depends also on the rate of its motion, i.e., on its kinetic energy. The higher kinetic energy is reflected at the collision with other reacting particles so that the efficiency of chemical transformation increases. At a certain critical kinetic energy of a moving atom, each collision may be chemically effective.

Hot atoms of kinetic energy from 1 to 2 eV arise at photolysis of molecules. Atoms having considerably higher excess of kinetic energy are formed in nuclear reactions.[68] Hot tritium is usually prepared by the bombardment of ^{3}He isotope by neutrons. The initial kinetic energy of tritium forming in a nuclear reaction is 0.19 MeV. A hot atom of tritium enters a radical reaction only when its motion in the medium is slowed down and the kinetic energy decreases below 15 eV. For most of reacting substrates, from 40 to 90% of hot atoms react before their thermalization, i.e., before their kinetic energy falls below 0.1 eV.

The increased reactivity of hot atoms is displayed by the increased variety of possible reaction pathways. While a thermalized tritium atom reacts with methane only in transfer reaction (Equation 49), the hot atom enters also substitution (Equation 50)

$$T\cdot \ + \ CH_4 \rightarrow TH \ + \ \cdot CH_3 \qquad (49)$$

$$T\cdot \ + \ CH_4 \rightarrow CH_3T \ + \ H\cdot \qquad (50)$$

or substitution-fragmentation reaction (Equation 51). The more

$$T\cdot \ + \ CH_4 \rightarrow \cdot CH_2T \ + \ H_2 \qquad (51)$$

complex molecules enter other elementary reactions not known for thermalized atoms. The slight dependence of the reaction of a hot atom on temperature is due to a neglible contribution of thermal energy of the medium to the rection of the translationally excited atom.

Molecules and atoms of iodine represent an effective trap for many thermalized radicals, including hydrogen or tritium atoms which are captured before their reaction with a less reactive substrate. The mechanism of radical capture consists in a very fast reaction of a radical with the molecule or atom of iodine which eliminates the reactive radical from the system. Capturing hot radicals is considerably less efficient since they succeed in reacting with less reactive substrate before they get into contact with the deactivation site.

The addition of an inert moderator, such as rare gases which do not enter reactions chemically, lowers the probability of encountering both hot and thermalized atoms with the reacting compound to the same level. On the other hand, the collisions of hot atoms with the moderator lead to their thermalization and their overall reaction rate sinks more than corresponds to the dilution of reactants, only.

XIII. REACTIVITY OF SOLVATED RADICALS

Free radicals in a reaction system may associate reaction partner molecules of the non-reacting medium, reaction sites of the catalyst, and form secondary (intermolecular) bonds. Provided that physical interaction between each radical and the reacting substrate supports the formation of the activated complex, the reaction rate and reactivity of the radical increases. Interactions of a radical with surrounding solvent molecules may bring about the decrease of radical reactivity due to higher delocalization of unpaired electrons within the formed complex. In addition to higher delocalization, the mobility of radical associate decreases, too. This may contribute to deceleration of fast reactions, namely, of those which are controlled by diffusion of reaction partners.

The effect of solvation on reactivity may be shown on the chlorination of branched alkanes in different solvents.[69] A nonsolvated chlorine atom reacts about 4 times faster with tertiary carbon hydrogen than with that on the methyl group of hydrocarbon. Solvation of chlorine atoms by CS_2 leads to the considerable reduction of a reaction rate; chlorine attacks almost exclusively tertiary hydrogens. Aromatic solvents function similarly. Besides electrophile halogens, the solvation effect was also observed for oxy radicals.

Simultaneously with changes of electronic structure of a radical site due to its association with molecules of reaction medium, the reactivity of radicals may be decreased also by sterical hindrance in a solvated radical. This may be the case of the retarding effect of water on oxidation of some hydrocarbons where peroxy radicals are hydrated by water molecules.[70] The solvation of free radicals by medium molecules also includes the change of reactivity by complexation of radicals with metallic ions. Peroxy radicals decay considerably more slowly when bound in a complex with ions of some metals.[71-73]

The delocalization of spin density may occur also during adsorption of radicals on a solid surface. A methyl radical trapped on the surface of glass changes the characteristics of the ESR spectrum, which indicates that the spin density of the unpaired electron is partially localized on the p-orbital of siloxane oxygen. Adsorption and delocalization of the electron brings about the increased stability of the methyl radical. The course of decay of adsorbed radicals shows in the unequal strength of adsorption and on the different degree of delo-calization, depending on the character of the respective trap. The reactivity of identical radicals adsorbed on the surface of solid phase may, therefore, differ.

The transient stabilization of a radical facilitates the preferable course of slower elementary reactions otherwise suppressed due to the fast decay of radicals. The formation of a complex or adduct of a radical with a catalyst changes, moreover, the electronic structure of the radical, which may enhance the course of such reactions which do not occur with noncomplexed radicals.[74]

XIV. THE STABILIZATION OF REACTIVE FREE RADICALS

The reactivity of radicals is given not only by their structure but also by the physical state of the medium where free radicals were generated. Reactive radicals decay during mutual collisions only in a medium which allows for their free motion. Regulation of the lifetime of free radicals may be achieved by the change of conditions of the mobility of particles in the medium (matrix) and by their isolation. A simple solution to this problem is achieved by hindering the free motion of free radicals by a decrease in temperature and by trapping them in the rigid frozen or glassy medium.[75]

The freezing of a gaseous mixture of radicals and inert molecules leads to the parallel decrease of mobility of both the radicals and surrounding molecules. Before the rigid condensed phase is formed, a large part of atoms and radicals succeeds in recombining to neutral molecules. This decay is due to the relatively long time it takes for the system to cool from the gas temperature to the temperature of the cooling medium and also due to the fact that heat released in the radical recombination induces the local melting of the stabilizing matrix and the decay of further parts of already trapped radicals. This is, moreover, enhanced by the nonhomogeneity of radical distribution in the matrix. If the other radical is trapped in the close vicinity of some radical, this couple may decay, too. Taking this into account and also the fact that each particle is surrounded by 8 to 12 contact neighbors, the maximum concentration of radicals in the condensed phase could reach about 10%. By the technique of rapid freezing, a maximum 0.1% of stabilized radicals was obtained in the solidified phase of inert molecules and reaction products.

Another problem of the stabilization of free radicals arises when compounds, precursors of free radicals, are frozen first, and free radicals are generated subsequently by irradiation. The cage effect is enforced here and only those pairs of free radicals remain stabilized which acquire higher energy than is necessary for chemical bond cleavage. The excess of energy is used up for getting radicals away from the cage which prevents their decay. The efficient alternative of the increase of the concentration of stabilized radicals is a proper choice of photosensitive free radical precursor. Photochemical decomposition of diacylperoxides (Equation 52)

$$R(CO)O-O(OC)R + h\nu \rightarrow R\cdot + 2\ CO_2 + R\cdot \tag{52}$$

yields two molecules of chemically inert CO_2 which impedes the decay reaction of radicals in the matrix of solid argon up to the temperature 30 K. The ratio between a number of molecules of matrix and radicals or their precursors is about 100:1, at minimum.

The particular kind of organic matrix is represented by macromolecular compounds which, due to their high viscosity and decreased mobility of particles, can stabilize free radicals, as well. Radicals can be formed here either directly from the macromolecular or from the low molecular free radical precursor.[76] The stability of free radicals in a macromolecular medium may be increased by pressure.[77] Also, crystalline or porous materials hindering self-decay may be used as stabilizing matrices. Polycyclic hydrocarbon adamantane of melting temperature 541 K is, e.g., one of many examples.[78,79] Reactive radicals formed from dissolved precursors decay at room temperature, while those of the matrix even at temperatures higher than 370 K. From porous matrices, inorganic molecular sieves (as, e.g., zeolites) stabilize many reactive radicals, too.[80,81]

The effort to prolong the lifetime of free radicals was aimed at measuring their properties and structure directly at sufficiently high concentrations. Another goal was to prepare highly energetical fuels containing concentrated systems of free radicals. Even though the suggested result of this latter case was not achieved, it considerably accelerated research, and now the limits of possibilities of the stabilization of free radicals are fairly well known.

There remains one interesting aspect of the reactivity of radicals involving the effect of a magnetic field. Until recently, there were no founded opinions concerning the possible mechanism of the influence of the magnetic field on the course of chemical reactions. The attitude towards realized experiments and to their interpretation was mostly skeptical. This was mainly due to the fact that the splitting of the energy of electron and nucleus spins to two sublevels, corresponding to parallel and antiparallel orientation of a spin with regard to the magnetic field, is negligibly small when compared with the thermal energy of molecules. This might create an impression that research on the effect of the magnetic field on chemical reactions is useless.

Now, there exists a relatively well-elaborated theory on the magnetic field effect on the course of radical reactions.[82] The effect becomes obvious when radical pairs appear in the reaction system. It was already pointed out that particularly at the recombination of radicals the system is sensitive to the spin state of a radical pair. Radicals of opposite electron spins recombine while those with consonant spins do not. If, during the lifetime of radical pairs, their multiplicity is changed, i.e., there occur singlet-triplet transitions, the ratio of recombining radicals may be altered. This is just a pathway through which the effect of the magnetic field may become distinct in the sequence of elementary events of some radical transformation. The rate of singlet-triplet transitions is proportional to the difference of precession motion of spin moments of radicals in radical pairs and will grow linearly with the magnetic field intensity.

The resulting effect may be either an acceleration or a decrease of the reaction rate. If the reaction depends on the ratio of radicals escaping the radical pairs, the magnetic field accelerates the successive reaction step; if, however, the reaction product depends on the ratio of primary radicals recombining in the cage, the effect is opposite.

Although the acceleration or deceleration is not very large, the magnetic field effect has immense value in the research of radical, ion radical, and photochemical reactions. It may, e.g., serve as a diagnostic tool for mechanistic purposes of the development of experimental methods.[83-85]

REFERENCES

1. **Cotton, F. A.,** *Chemical Application of Group Theory,* 2nd ed., Wiley-Interscience, New York, 1971.
2. **Woodward, R. B. and Hoffmann, R.,** *The Conservation of Orbital Symmetry,* Verlag Chemie, Weinheim, 1970.
3. **Woodward, R. B. and Hoffmann, R.,** The conservation of orbital symmetry, *Angew. Chem. Int. Ed.,* 8, 781, 1969.
4. **Pearson, R. G.,** *Symmetry Rules for Chemical Reactions. Orbital Topology and Elementary Processes,* John Wiley & Sons, New York, 1976.
5. **Dewar, M. J. S.,** Aromaticity and pericyclic reactions, *Angew. Chem. Int. Ed.,* 10, 761, 1971.
6. **Trindle, L.,** Mapping analysis of concerted reactions, *J. Am. Chem. Soc.,* 92, 3251, 1970.
7. **Zimmermann, H. E.,** MO following. Molecular orbital counterpart of electron pushing, *Acc. Chem. Res.,* 5, 393, 1972.
8. **Goddard, W. A.,** Selection rules for chemical reactions using the orbital phase continuity principle, *J. Am. Chem. Soc.,* 94, 793, 1972.
9. **Fleming, I.,** *Frontier Orbitals and Organic Chemical Reactions,* John Wiley & Sons, New York, 1976.
10. **Ponec, R.,** Topological aspects of chemical reactivity, I-VI, *Coll. Czechoslov. Chem. Commun.,* 49, 455, 1984; 50, 559, 577, 1121, 1985; 51, 1834, 1986.
11. **George, T. F. and Ross, J.,** Analysis of symmetry in chemical reactions, *J. Chem. Phys.,* 55, 3851, 1971.
12. **Metiu, H., Ross, J., and Silbey, R.,** On Symmetry properties of reaction coordinates, *J. Chem. Phys.,* 61, 3200, 1974.

13. **Salem, L.,** Intermolecular orbital theory of the interaction between conjugated systems. I. General theory, *J. Am. Chem. Soc.,* 90, 543, 1968.
14. **Salem, L.,** Orbital interactions and reaction paths, *Chem. Brit.,* 449, 1969.
15. **Sullivan, J. H.,** Mechanism of bimolecular hydrogen-iodine reaction, *J. Chem. Phys.,* 46, 73, 1967.
16. **Dauben, W., Salem, L., and Turro, N. J.,** *Acc. Chem. Res.,* 8, 41, 1975.
17. **Mango, F. D. and Schachtschneider, J. H.,** in *Transition Metals in Homogeneous Catalysis,* Schrauzer, G. N., Ed., Marcel Dekker, New York, 1971.
18. **Mango, F. D.,** Conservation of coordinate bonding and catalysis of allowed and forbidden reactions, *Fortschr. Chem. Forsch.,* 41, 39, 1973.
19. **Du Boios, D. L. and Hoffman, R.,** Diazenido, dinitrogen, and related complexes, *Nouv. J. Chem.,* 1, 479, 1977.
20. **Boca, R.,** Dioxygen activation in transition metal complexes in the light of molecular orbital calculations, *Coord. Chem. Rev.,* 50, 1, 1983.
21. **Pelikan, P. and Boca, R.,** Geometric and electronic factors of dinitrogen activation on transition metal complexes, *Coord. Chem. Rev.,* 55, 55, 1984.
22. **Pacey, P. D.,** Changing conceptions of activation energy, *J. Chem. Educ.,* 58, 612, 1981.
23. **Snadden, R. B.,** A New perspective on kinetics and thermodynamic control of reactions, *J. Chem. Educ.,* 62, 653, 1985.
24. **Buchachenko, A. L. and Vasserman, A. M.,** *Stable Radicals* (Russian), Publishing House Khimia, Moscow, 1973.
25. **Neumann, W. P., Uzick, W., and Zarkadis, A.,** Sterically hindered free radicals. 14. Substituent dependent stabilization of para-substituted triphenyl methyl radicals, *J. Am. Chem. Soc.,* 108, 3762, 1986.
26. **Ballester, M.,** Inert free radicals: a unique trivalent carbon species, *Acc. Chem. Res.,* 18, 380, 1985.
27. **Allaiarov, S. R., Demidov, S. V., Kiryukhin, D. P., Mikhaylov, I. A., and Barkalov, I. M.,** Formation of stable radicals in radiolysis of perfluoralkanes (Russian), *DAN U.S.S.R.,* 274, 91, 1984.
28. **Howard, J. A.,** The Electron spin resonance spectrum of $(CH_3)_3COO \cdot$, *Can. J. Chem.,* 57, 253, 1979.
29. **Nonhebel, D. C. and Walton, J. C.,** *Free Radical Chemistry,* University Press, Cambridge, 1974, 115.
30. **Bullock, A. T.,** Electron spin resonance spectroscopy and free radicals reactions, in *Reaction Mechanism, Annu. Rep. B,* Chem. Soc., London, 1975, 88.
31. **Kirste, B., Harrer, W., Kurreck, H., Schubert, K., Banner, H., and Gierke, W.,** ^1H and ^{13}C ENDOR investigation of sterically hindered galvinoxy radicals, *J. Am. Chem. Soc.,* 103, 6280, 1981.
32. **Griller, D. and Ingold, K. U.,** Electron paramagnetic resonance and the art of physical organic chemistry, *Acc. Chem. Res.,* 13, 193, 1980.
33. **Sanderson, R. T.,** *Chemical Bonds and Bond Energy,* 2nd ed., Acad. Press, London, 1976, 160.
34. **Griller, D.,** Free-Radical reactions, in *Reaction Mechanism, Annu. Rep. B,* Royal Soc. Chem., London, 1984, 69.
35. **De Nicholas, A. M. and Arnold, D. R.,** The Comparison of electron spin resonance hfc's and the stabilization energy of π-radicals due to spin delocalization, *Can. J. Chem.,* 64, 270, 1986.
36. **Stein, S. E. and Golden, D. M.,** Resonance stabilization energies in polycyclic hydrocarbon radicals, *J. Org. Chem.,* 42, 839, 1977.
37. **Auner, N., Walsh, R., and Westrup, J.,** Kinetic determination of the bond dissociation energy D $(Me_3SiCMe_2CH_2-H)$ and the magnitude of the stabilization energy in a β-silicon-substituted alkyl radical, *J. Chem. Soc., Chem. Commun.,* 207, 1986.
38. **Timonen, R. S. and Gutman, D.,** Kinetic of the reactions of methyl, ethyl, isopropyl and tert. butyl radicals with molecular chlorine, *J. Phys. Chem.,* 90, 2987, 1986.
39. **Citterio, A., Gentile, A., Minisci, F., Serravalle, M., and Ventura, S.,** Polar effects in free radical reactions, *J. Org. Chem.,* 49, 3364, 1984.
40. **Minisci, F., Citterio, A., and Vismara, E.,** Polar effects in free radical reactions, *Tetrahedron,* 41, 4157, 1985.
41. **Clerici, A., Minisci, F., and Porta, O.,** Nucleophilic character of alkyl radicals. 11. The benzyl radical, *Tetrahedron,* 29, 2775, 1973.
42. **Clerici, A., Minisci, F., and Porta, O.,** The polar character of the phenyl radical, *Gazzeta Chim. Ital.,* 103, 171, 1973.
43. **Tomat, R. and Rigo, A.,** Reactivities of aromatic compounds towards the amino radical, *J. Electroanal. Chem.,* 63, 329, 1975.
44. **Taniguchi, H. and Schuler, A. H.,** Ionization of the hydroxy cyclohexadienyl radical in concentrated KOH: a measure of the activity of ^-OH in highly basic media, *J. Phys. Chem.,* 89, 335, 1985.
45. **Fossey, J., Lefort, D., Massoudi, M., Nedelec, J. Y., and Sorba, J.,** Free radical towards peroxyacids in terms of orbital interactions. Competition between energy gap control and overlap control, *J. Chem. Soc., Perkin Trans. II.,* 781, 1986.

46. **Levine, R. D. and Bernstein, R. B.,** *Molecular Reaction Dynamics,* Oxford University Press, New York, 1974, 94.

47. **Leone, S. R.,** State-resolved molecular reaction dynamics, *Annu. Rev. Phys. Chem.,* 35, 109, 1984.

48. **Gerrity, D. P. and Valentini, J. J.,** Experimental study of the dynamics of the H + D → HD + D reaction at collision energies of 0.55 and 1.30 eV, *J. Chem. Phys.,* 81, 1298, 1984.

49. **Levine, R. D. and Bernstein, R. B.,** *Molecular Reaction Dynamics,* Oxford University Press, New York, 1974, 45.

50. **Isaacson, A. D. and Truhlar, D. G.,** Polyatomic canonical variational theory for reaction rates. Separable-mode formalism with application to OH + H_2 → H_2O + H, *J. Chem. Phys.,* 76, 1380, 1982.

51. **Van den Ende, D. and Stolte, S.,** Reactive scattering with a state selected beam; the influence of the internal state and orientation of NO upon the chemiluminiscent reactivity with O_3, *Chem. Phys. Lett.,* 76, 13, 1980.

52. **Van den Ende, D., Stolte, S., Cross, J. B., Kwei, G. H., and Valentini, J. J.,** Evidence for two different transition states in the reaction of NO + O_3 → NO_2 + O_2, *J. Chem. Phys.,* 77, 2206, 1982.

53. **Arrowsmith, P., Bartoszek, F. E., Bly, S. H. P., Carrington, T., Charters, P. E., and Polanyi, J. C.,** Chemiluminiscence during the course of a reactive encounter; F + Na_2 → $FNaNa^{+}$* → NaF + Na*, *J. Chem. Phys.,* 73, 5895, 1980.

54. **Arrowsmith, P., Bly, S. H. P., Charters, P. E., and Polanyi, J. C.,** Spectroscopy of the transition state. II. F + Na_2 → $FNaNa^{+}$* → NaF + Na*, *J. Chem. Phys.,* 79, 283, 1983.

55. **Andresen, P. and Luntz, A. C.,** The Chemical dynamics of the reactions of $O(^3P)$ with saturated hydrocarbons. I., Experiment, *J. Chem. Phys.,* 72, 5842, 1980.

56. **Luntz, A. C. and Andresen, P.,** The Chemical dynamics of the reactions of $O(^3P)$ with saturated hydrocarbons. II., Theoretical model, *J. Chem. Phys.,* 72, 5851, 1980.

57. **Hoffmann, S. M. A., Smith, D. J., Williams, J. H., and Grice, R.,** Reactive scattering of a supersonic methyl beam, *Chem. Phys. Lett.,* 113, 425, 1985.

58. **Herman, Z., Friedrich, B., Koyano, I., Tanaka, K., and Kato, T.,** Ion-molecule collision dynamics via beam scattering and state-selection of reactants, lecture at Microsymposium on Elementary Processes and Chemical Reactivity, Liblice, Czechoslovakia, 1985.

59. **Kamiya, K. and Morokuma, K.,** Potential energy surface and reaction mechanism for the ion-molecule reaction: CH_4 + CH_4^{+} → $\cdot CH_3$ + CH_5^{+}, *Chem. Phys. Lett.,* 123, 331, 1986.

60. **Husain, D. and Donovan, R. J.,** Electronically excited halogen atoms, *Advan. Photochem.,* 8, 1, 1971.

61. **Bogatyreva, A. I. and Buchachenko, A. L.,** Reactions of electronically excited radicals and quenching of excited states of radicals (Russian), *Usp. Khim.,* 44, 2171, 1975.

62. **Bromberg, A., Schmidt, K. H., and Meisel, D.,** Photolysis and photochemistry of arylmethyl radicals in liquids, *J. Am. Chem. Soc.,* 107, 83, 1985.

63. **Shell, P. S., Luning, U., McBain, D. S., and Tanko, J. M.,** Ground and excited state succinimidyl in chain reactions. A reexamination, *J. Am. Chem. Soc.,* 108, 121, 1986.

64. **Heaven, M., DiMauro, L., and Miller, T. A.,** Laser induced fluorescence spectra of free-jet cooled organic free radicals, *Chem. Phys. Lett.,* 95, 347, 1983.

65. **Collin, G. J. and Wieckowski, A.,** Excited radicals in the gas phase photolysis of butene isomers at different photon energies: energy distribution in reaction products, *Can. J. Chem.,* 56, 2630, 1978.

66. **Gierczak, T., Gawlowski, J., and Niedzielski, J.,** Decomposition of highly excited butenyl and pentenyl radicals, *J. Photochem.,* 33, 145, 1986.

67. **DiMauro, L. F., Heaven, M., and Miller, T. A.,** Lifetimes of the lowest excited states of the cyclopentadienyl and the monomethylcyclopentadienyl radicals, *Chem. Phys. Lett.,* 124, 489, 1986.

68. **Wolfgang, R.,** The hot atom chemistry of gas-phase systems, *Prog. React. Kinet.,* 3, 97, 1965.

69. **Huyser, E. S.,** Solvent effect in free radical reactions, in *Advances in Free Radical Chemistry I,* Williams, G. H., Ed., New York, 1965, 77.

70. **Emanuel, N. M., Zaikov, G. E., and Maizus, Z. K.,** *The Role of Medium in Radical Chain Reactions of Oxidation of Organic Compounds,* Nauka Moscow, 1973.

71. **Tkac, A., Vesely, R., Omelka, L., and Prikryl, R.,** Complex bounded and continuously generated peroxy and alkoxy radicals, *Coll. Czechoslov. Chem. Commun.,* 40, 117, 1975.

72. **Clopath, P. and Zelewsky, A.,** Metal complexes of free radicals, *Helv. Chem. Acta,* 55, 52, 1972.

73. **Meisel, D., Czapski, G., and Samuni, A.,** Hydroperoxyl radical reactions. I. Electron paramagnetic resonance study of the complexation of $HO_2\cdot$ with some metal ions, *J. Am. Chem. Soc.,* 95, 4148, 1973.

74. **Tanaka, H., Saki, I., and Ota, T.,** Observation by ESR of an acrylate radical conformationally locked by complexation with $SnCl_4$, *J. Am. Chem. Soc.,* 108, 2208, 1986.

75. **Minkoff, G. J.,** *Frozen Free Radicals,* InterScience New York, 1960.

76. **Carlsson, D. J., Dobbin, C. J. B., and Wiles, D. M.,** Reactivity of polypropylene peroxy radicals in the solid state, *Macromolecules,* 18, 1791, 1985.

77. **Szocs, F., Placek, J., and Borsig, E.,** Generation of free radicals in polymer matrix by thermal decomposition of benzoyl peroxide at high pressure, *Polym. Lett.,* 9, 753, 1972.
78. **Tegowshi, A. T. and Pratt, D. W.,** Free radical decay in adamantane, *J. Am. Chem. Soc.,* 106, 64, 1984.
79. **Dismukes, G. Ch. and Willard, J. E.,** Radiolytic and photolytic production and decay of radicals in adamantane and solutions of 2-methyltetrahydrofurane, 2-methyltetrahydrothiophene and tetrahydrothiophene in adamantane, *J. Phys. Chem.,* 80, 1435, 1976.
80. **Rychly, J. and Lazar, M.,** ESR spectra of some oligomer radicals on synthetic zeolites, *Polym. Lett.,* 7, 843, 1969.
81. **Raghunathan, P. and Sur, S. K.,** EPR study of the chloroperoxy radical C1OO, stabilized in the zeolite host matrix of cancrinite, *J. Am. Chem. Soc.,* 106, 8014, 1984.
82. **Salikhov, K. M., Molin, Yu., N., Sagdeev, R. Z., and Buchachenko, A. L.,** *Spin Polarization and Magnetic Effects in Radical Reactions,* Acad. Kiado, Budapest, 1984.
83. **Adrian, F. J.,** Future prospects for chemically induced polarization in free radical reactions, *Rev. Chem. Intermediates,* 7, 173, 1986.
84. **McLauchlan, A.,** Flash photolysis electron spin resonance, *Chem. Britain,* 825, 1985.
85. **Ulrich, T., Steiner, V. E., and Schlenker, W.,** Control of photo electron transfer induced production by micelar cages, heavy atom substituents and magnetic fields, *Tetrahedron,* 42, 6131, 1986.

Chapter 5

NONCHAIN RADICAL REACTIONS

I. INTRODUCTION

Complex radical reactions having a zero stoichiometrical coefficient of cyclic regeneration of free radicals are of nonchain character, even though they are composed of several elementary steps which mutually compete. The reaction products are formed during the termination of radicals. The sum (P) of products formed at combination and disproportionation of polyatomic radicals depends on the number n of different types of free radicals present in the system so that $P = 2.5 n + 0.5 n^2$. This relation is valid assuming two kinds of products are formed at the disproportionation of one type of radicals. Since most decay reactions of free radicals have low selectivity, the nonchain reaction usually leads to a broad spectrum of various reaction products. This may explain why only a limited number of nonchain radical reactions are used in spite of the many potential possibilities.

Nonchain reactions are mostly used at the cross-linking of linear macromolecules or more generally at the synthesis of larger molecules than correspond to a parent radical. The increase of molecule dimensions by dimerization is suitable for such radicals which decay predominantly by combination. The application of this procedure requires minimizing the probability of side competition reactions of radicals with molecules of the reaction medium. Relatively high local concentrations of radicals formed at the electrolysis of carboxylic acids salts are the most convenient for such a purpose.[1] Electrolysis of carboxylic acids gives, e.g., dimers of acid alkyl chains even in 80% yield. Dimerization of alkyl radicals is preceded by electrochemical oxidation of carboxy anions and by decarboxylation of formed radicals. For synthetic reasons, the combination reactions of unequal radicals are sometimes used at the simultaneous electrolysis of two carboxylic acids. There are, however, three products of radical combination formed together with those of radical disproportionation.

The proper choice of experimental conditions may bring about also efficient dimerization at other modes of free radical generation. This is the case of free radical coupling which occurs at the decomposition of diacylperoxides (Equations 1 and 2).

$$R(CO)O-O(CO)R \rightarrow \cdot R + \cdot R + 2\ CO_2 \tag{1}$$

$$\cdot R + \cdot R \rightarrow R-R \tag{2}$$

At the thermolysis of diacylperoxide solutions, the yield of alkyl radicals dimerization is relatively low, whereas at their photochemical decomposition in the solid state, the yield of asymmetrical dimers is higher than that in the electrochemical methods.[2]

Polyrecombination is another example of nonchain manifold repeating dimerization.[3] The compounds entering polyrecombination have to contain at least two reactive hydrogen atoms. From the parent monomer, the free radicals are formed by the transfer reaction (Equation 3) and a dimer is formed in their combination reaction (Equation 4). The dimer keeps two reactive

$$(CH_3)_2CH-\!\!\!\left\langle\right\rangle\!\!\!-CH(CH_3)_2 + R\cdot \rightarrow RH + (CH_3)_2\dot{C}-\!\!\!\left\langle\right\rangle\!\!\!-CH(CH_3)_2 \tag{3}$$

$$2(CH_3)_2\dot{C}-\!\!\!\left\langle\right\rangle\!\!\!-CH(CH_3)_2 \rightarrow$$

$$\rightarrow (CH_3)_2CH-\!\!\!\left\langle\right\rangle\!\!\!-C(CH_3)_2C(CH_3)_2\!\!\!\left\langle\right\rangle\!\!\!CH(CH_3)_2 \tag{4}$$

hydrogens and becomes a new larger monomer. In its subsequent reactions, the polymer molecules are gradually built up. For the formation of a macromolecular product, 1 mol of peroxide, as the radical source, is consumed per each mol of structural units of a polymer.

Combination reactions following an electron-transfer step are also known.[4] For example, the electron transfer from metallic sodium to 1,1-diphenylethylene results in the monomeric radical anion. This immediately combines to the dimeric dianion (Equation 5) which cannot polymerize further because of steric

$$
\begin{array}{cccc}
C_6H_5 & C_6H_5 & C_6H_5 & C;6H_5 \\
| & | & | & | \\
2C=CH_2 + 2e^- \rightarrow 2 \ominus C\!-\!C\!\cdot\!H_2 \rightarrow & \ominus CH_2CH_2\!-\!C\ominus \\
| & | & | & | \\
C_6H_5 & C_6H_5 & C_6H_5 & C_6H_5
\end{array}
\qquad (5)
$$

reasons. When we choose the substance containing vinylidene group twice in its molecule (Scheme 1) intramolecular and

$$
\begin{array}{cc}
C_6H_5 & C_6H_5 \\
| & | \\
H_2C{=}C{-}C_6H_4{-}C{=}CH_2
\end{array}
$$

SCHEME 1

intermolecular combinations lead to cyclic compounds and to polymers, respectively.

By the mechanism of stepwise combination of alkyl radicals, the macromolecular compounds are formed from saturated aliphatic hydrocarbons under conditions of plasma discharge.[5]

II. CROSS-LINKING OF MACROMOLECULES

The dimerization reaction of macroradicals is used for modification of properties of polyethylene and of silicon rubbers. Also here, for one cross-link, one molecule of precursor of the reactive radical pair is necessary. The wide application which this nonchain reaction has found follows from the fact that a relatively small number of cross-link bridges between macromolecules change the properties of a polymer significantly. The cross-linking of the polymer consists in the generation of macroradicals by transfer reaction of low molecular radicals or by irradiation.[6] We recall here that the cross-linking of polyethylene belongs to the economically most profitable result of research in radiation chemistry.[7]

It was believed in the early days of radiation research that the energy of photons, higher by 5 orders of magnitude than that of chemical bonds, may initiate new types of chemical reactions. Irradiation of organic compounds and polymers, however, has shown that γ rays generate radicals, ions, free electrons, and excited states of functional groups in the systems, the reactions of which may well interpret all principal peculiarities of radiation-induced reactions.

In polyethylene irradiated in a vacuum, the secondary alkyl macroradicals, $-CH_2\overset{\cdot}{C}HCH_2-$, were identified, above all. Transiently formed hydrogen atoms and primary alkyl radicals $-CH_2\cdot$ abstract the hydrogen from surrounding macromolecules above 50 K, and secondary alkyl radicals are predominantly formed. The other types of less reactive radicals may be detected elevating either temperatures or the irradiation dose. At a temperature of about 300 K, allyl $-CH_2\overset{\cdot}{C}HCH{=}CHCH_2-$ and at higher temperatures, polyene $-\overset{\cdot}{C}H(CH{=}CH)_nCH_2-$ radicals are only observable.

By the combination of secondary alkyl radicals cross-links in polyethylene are formed with radiation yield about 1 at temperatures below 230 K. The theoretical radiation yield should be, however, twice of that actually observed. The efficiency of radiation cross-linking may be increased by the presence of acetylene or by a temperature increase above the temperature region of the melting of polyethylene crystallites.[8] The effect of acetylene may be explained by its addition to alkyl radicals and by the formation of allyl radicals which decay with a considerably higher ratio of combination to disproportionation than secondary alkyl radicals. Allyl radicals, as side groups of macromolecules, have, moreover, a higher degree of motional freedom. The effect of mobility of side groups in polyethylene is mostly pronounced below the melting point of polymer crystallites.

We may thus expect that monomers with several double bonds will have an increased yield of cross-links in an irradiated polymer.[9] Besides the mediation of the contact among reactive radicals, one supplementary fact should be taken into account here, namely, that unreacted double bonds may remain in the chain of the polymer. These double bonds may react with macroradicals, too, and thus increase the efficiency of cross-linking. By irradiation of polyethylene, *trans* vinylene bonds appear simultaneously with hydrogen and cross-link formation, probably due to disproportionation and dehydrogenation of secondary alkyl radicals. The end vinyl and vinylidene double bonds, which in polyethylene are considered as anomalous structural units of polymerization, however, decay. This indicates that some cross-links may be formed by a chain mechanism even in pure polyethylene.[10]

Quantitative comparison of the cross-linking efficiency for mers of different structural units cannot be performed with absolute reliability since it is affected by the difference in physical properties of a polymer as well as by anomalous structural units in the polymer chain. Apart from this, from the large number of published papers the comparison of approximate efficiencies of some linear polymers may be attempted.[11,12] The effect of different kinds of unsaturated bonds suggested above may thus be confronted with cross-linking efficiency of polydienes. While natural rubber and 1,4-polybutadiene cross-link with the same efficiency as polyethylene, the structural units of 1,2-polybutadiene are by 100-fold more effective. This is again due to a chain cross-linking reaction.

Since the substituted macroradicals undergo more fragmentation and, moreover, they are also more likely to disproportionate than recombine, displacement of hydrogen in the chain of vinyl polymers by alkyls reduces the cross-linking efficiency. Accordingly, the efficiency of cross-linking for vinylidene polymers is lower.

Polymers having aromatic groups cross-link with about 100-fold lower radiation yield than polyethylene. The retardation effect of aromatic groups consists in the lower rate of radiolysis of the polymer and, in addition, of hydrogen atoms to aromatic groups. When hydrogen is, e.g., added to the phenyl group of polystyrene (Equation 6), the formed cyclohexadienyl radical undergoes disproportionation with the alkyl macroradicals to the original polymer molecules.

$$H\cdot + \text{—CH}_2\text{—CH—} \rightarrow \text{—CH}_2\text{—CH—}$$

$$\tag{6}$$

As initiators of chemical cross-linking, organic peroxides are mostly used. They dissociate homolytically in polymer medium to oxy radicals which abstract hydrogen from surrounding macromolecules. The subsequent reactions of macroradicals are the same, as has already been shown. When compared with radiation, the chemical formation of macroradicals is more selective. At the cross-linking of polydimethylsiloxane by benzoylperoxide, there are, e.g., formed dimethylene cross-link bridges, while at radiation initiation methylene and Si-Si cross-links also appear. In the case of other methods of initiation, the differentiation of respective cross-links is rather complicated since detailed analysis is still lacking.

The number of cross-links formed in a macromolecule related to the amount of decomposed peroxide depends not only on the structure of a polymer chain but also on the source of primary radicals.[13] When compared with radiation cross-linking, at the peroxide initiation, dissociation of the main chain bonds plays a minor role, although it cannot be absolutely excluded. For the cross-linking course, it is important that the transfer reactions of oxyradicals succeed in competing with fragmentation reactions of primary radicals. In such a case, two macroradicals will appear in close proximity and may form one cross-link by recombination. The less advantageous situation arises when oxyradicals fragment. Since carbon-centered radicals are less active in transfer reaction than oxyradicals, the probability of their decay increases.

In the first papers devoted to polyethylene cross-linking the authors usually stated that the number of cross-links determined from the polymer solubility is approximately equal to the number of decomposed molecules of peroxide. At the dimerization of pentadecane and of other low molecular alkanes with *tert*-butyl perbenzoate, efficiency of only about 50% was found. One third of primary radicals are consumed in the formation of double bonds in alkane, the next 15% of low molecular radicals from peroxide self-decay, and about 5% combine with alkyl radicals of alkane.[14] Secondary radicals in alkane did not fragment to a measurable extent, which means that this reaction in polyethylene may occur only on anomalous structural units, such as branching points, etc., which reduce the cross-linking efficiency. Tertiary alkyl radicals originating from these branching points undergo more fragmentation and disproportionation than secondary ones. Polypropylene compared to poly ethylene may be, therefore, cross-linked only by a 20-fold higher concentration of peroxide.[15] Polyisobutylene, in the presence of oxyradicals, undergoes degradation.

The time of cross-linking usually corresponds to four values of the halflife of peroxide decomposition. The subsequent decomposition of the residual amount (ca. 5%) of peroxide is lengthy and does not lead to an effective increase of cross-linking efficiency. At a low concentration of peroxide in the system, the low stationary concentration of formed radicals will bring about the enhancement of the course of side monomolecular reactions taking place to the detriment of the bimolecular combination of macroradicals. From this viewpoint, the faster decomposition of peroxide at a higher temperature and for a shorter time interval is more advantageous for cross-linking.

Although the cross-linking process of macroradical dimerization is necessary either in the use of ionization radiation and ultraviolet light or at organic peroxide initiation, each of the procedures has its own peculiarities and precedences. When using ultraviolet light and sensitizers, the cross-linking of macromolecules occurs only in superficial layers of material.[16]

Cross-linking by peroxides may be realized selectively in amorphous regions of macromolecular chains while radiation cross-linking creates the cross-links throughout the whole polymer sample. The parallel occurrence of destruction reactions is a drawback of radiation cross-linking at some applications as, e.g., at high strength polyethylene fibers and blends, where peroxide cross-linking is most preferred.[17]

III. REACTIONS OF RADICAL PAIRS ON TRANSITION ELEMENTS

As was already pointed out, the presence of transition metal elements in reaction systems may accelerate or inhibit many chemical or biochemical reactions.[18] Up to 80% of all useful organic reactions on the industrial scale are controlled by transition metal catalysts, and there would be no life without essential transition elements. Generally, the acceleration of radical reactions by transition metal ions may be due to their participation in the activation of reagents and in the generation of free radicals leading to the increase of free radical concentration in the system.[19,20] The latter was already dealt with in the generation of free radicals. The principles of the catalytic effect which are connected with the formation and unstability of coordination compounds were underlined in the preceding chapter. We present now some examples of catalytic reactions, indicating that there are mechanisms which may proceed via radical species, which are somehow masked and different from that of really free radicals.

The possibility of formation of coordination compounds and their specific effect ensue from the structure and properties of transition elements and their compounds. It consists in an uncomplete filling of d-orbitals of transition metals as well as in energetic levels of these orbitals comparable with those of many organic reactants. This supports the easy formation of a relatively unstable bond between the catalytically efficient transition metal and a given reactant.

Very numerous reactions involve, for example, insertion, which is based upon the inter-action of ligand R bound to a transition element with a weak covalent σ-bond and of a reactant coordinated to the same element (Equation 7).

$$
\begin{array}{ccc}
\text{R} & \text{C} & \\
| & \| & || \\
L_nM \dots C & \rightarrow & L_nM\text{–CC–R} \\
& & ||
\end{array}
\tag{7}
$$

Ligand R is usually an alkyl or H atom, while a reactant may be an unsaturated hydrocarbon, CO, etc. Insertion probably proceeds via a four-centered cyclic activated complex in which the homolytic cleavage of an old and the formation of a new bond are concerted.[21] Active intermediates are not thus free radicals but at most very short-lived radical pairs.

Coordination of the alkene and successive insertion leads to the growth of alkyl radicals. Reproduction of insertion reactions is determined by relative rates of insertion and β-elim-ination reaction (Equation 8). The tendency to this

$$
L_nMCH_2CH_2R \rightarrow L_nMH + CH_2{=}CHR
\tag{8}
$$

reaction depends on ligands L, transition metal M, and its oxidation state. The competition of insertion and β-elimination sets the formation of dimers, oligomers, or polymers.[22]

At present, we know many catalytic systems which are able to initiate the polymerization of alkenes, dienes, and alkines. They are called Ziegler-Natta catalysts. They involve the combination of alkyls, hydrides, and halogenides of elements of I to III Group (organometallic compounds) and complexes of transition elements from IV to VIII Group including Sc, Ir, and U. The appearance of active centers on a catalyst is due to a mutual interaction of organometallic compound (procatalyst) with a complex of a transition element.[23] Most used industrially are heterogeneous catalytic systems of titanium compounds. Polymerization is induced by alkyl radicals covalently bound to a transition element, which, however, do not abandon the ligand field of a catalyst. This type of fixed radical is still capable of reaction with a monomer molecule attached to the transition element by a coordination bond. After insertion of a monomer into the unstable bond transition element-alkyl radical, the active site of a catalyst is regenerated and a new coordination bond catalyst-monomer appears. For sustained catalytic activity, the energetic levels of d-orbitals of transition elements should be situated between π-bonding and π*-antibonding orbitals of a monomer.

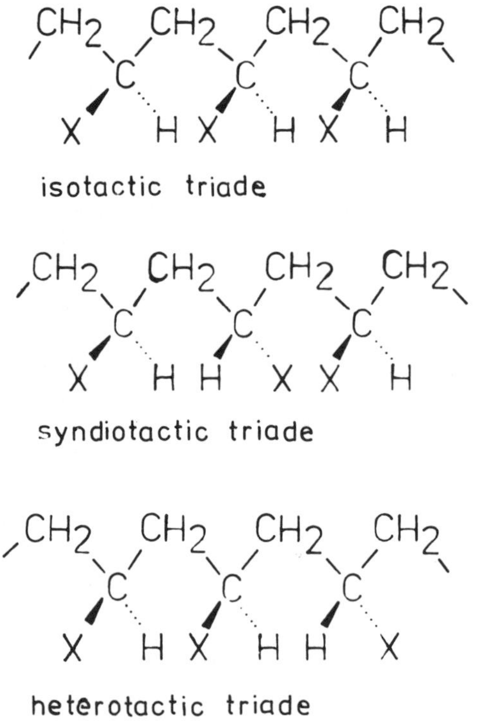

isotactic triade

syndiotactic triade

heterotactic triade

FIGURE 1.   Stereoisomer configuration of vinyl polymer. Carbon atoms of macromolecule backbone are in the plane; substituents are above the plane (thick arrow) or below the plane (dashed line).

At the polymerization of α-alkenes and 1,3-dienes on Ziegler-Natta catalysts, stereoisomeric polymers are obtained. The stereospecific effect of catalysts is due to symmetric properties of orbitals which condition regular binding of a monomer to a catalyst and defined geometry of an active catalytic center.[24,25] Varying with the structure of the active site a respective tacticity of the growing alkyl radical arises (Figure 1). If the structure of the active site facilitates irregular insertion of monomer units into the bond transition metal-alkyl, an atactic polymer is formed. Besides insertion reactions of alkenes and alkines, many other reactions[21] exist which may occur similarly by a mechanism of unstable contact radical pairs. The existence of these types of radical pairs was demonstrated at the hydrogenation of unsaturated substrates by mononuclear transition metal hydrides.[26]

At the hydrogenation of α-metyl styrene (Equation 9)

$$C_6H_5C(CH_3)=CH_2 + 2\ HMn(CO)_5 \rightarrow C_6H_5CH(CH_3)_2 + Mn(CO)_{10} \tag{9}$$

the effect of chemically induced dynamic nuclear polarization (CIDNP) was observed. In this reaction, an $L_nM \cdot \cdot R$ radical pair is formed by H-atom transfer (Equation 10), which leads to the

$$L_nMH + >C=C< \rightarrow L_nM \cdot + \cdot\overset{|\ \ |}{\underset{|\ \ |}{C-CH}} \tag{10}$$

occurrence of CIDNP through singlet-triplet mixing and spin selective reactions. Such examples are more numerous today. Of particular interest is the observation of CIDNP in the hydrogenation of alkines by a binuclear rhodium hydride. Here, the radical pair responsible for CIDNP must be a metal-centered biradical.[27]

Central to the problem of understanding the reactivity of the catalytic center is the requirement of characterizing the properties of the ligand-to-metal bond. These bonds are intriguing in that the ligand is supplying both of the required electrons. The bonding orbitals of C–C and C–H bonds in alkenes have insufficient energy to perform effective bond interaction with d-orbitals of the transition element. The decisive role in the coordination of alkenes to transition metal ions is, therefore, played by donor interaction of bonding π-orbitals of alkene with vacant d-orbitals or sometimes with s- and p-orbitals of the transition element. The coordination bond is complemented by an acceptor interaction of antibonding π* orbitals of alkene with occupied d- or p-orbitals of the transition element. Donor interactions π-M form σ-bond metal-alkene, while acceptor interactions M-π* lead to π bond metal-alkene (Chapter 4, Figure 9). Both kinds of interactions reduce the strength of the C–C bond due to the diminution of electronic populations on π-bonding orbitals and to an increase of electron populations of π*-antibonding orbitals of alkene, which increase the bond length C=C compared with free noncoordinated alkene. The increase of bond length corresponds to alkene activation. The coordinated alkene thus acquires properties of an electronically excited state.

Although a discussion of the reactions of contact radical pairs partially digresses from the main subject of the book, they cannot be forgotten. Some apparently negligible amount of radicals from contact pairs may after all become free, which might have an important consequence on the biochemical reactions of the natural aging of cells.

REFERENCES

1. **Mirkind, L. A.**, Anode reactions of dimerization addition and substitution of organic compounds (Russian), *Usp. Khim.*, 44, 2088, 1975.
2. **Feldhues, M. and Schafer, H. J.**, Selective mixed coupling of carboxylic acids. Electrolysis, thermolysis and photolysis of unsymmetrical diacyl peroxides with alkenyl, halo, keton, carbonyl groups and chiral-carbon. Comparison with the mixed Kolbe electrolysis, *Tetrahedron*, 41, 4195, 1985.
3. **Korshak, V. V.**, Die Nichtgleichgewichts-Polykondensation, *J. Prakt. Chemie*, 313, 422, 1971.
4. **Hocker, H. and Lottermannz, G.**, Polycombinations reactions propagated by electron transfer. A new type of polymerization reaction, *J. Polym. Sci. Symp.*, 54, 361, 1976.
5. **Hiratsuka, H., Akovali, G., Shen, M., and Bell, A. T.**, Plasma polymerization of some simple saturated hydrocarbons, *J. Appl. Polym. Sci.*, 22, 917, 1978.
6. **Baird, W. G., Joonase, P., Rose, A. B., Jr., and Helman, W. Ph.**, Bibliographies on radiation crosslinking of polymers, *Radiat. Phys. Chem.*, 19, 339, 1982.
7. **Dole, M.**, History of the irradiation crosslinking of polyethylene, *J. Macromol. Sci. Chem.*, A15, 1403, 1981.
8. **Mitsui, H., Hosoi, F., and Kagyia, T.**, Accelerating effect of acetylene on the γ-radiation induced crosslinking of polyethylene, *Polym. J.*, 6, 20, 1974.
9. **Leen, D. W. and Braun, D.**, Strahlenvernetzung von Polyethylene in Gegenwart von Polymerisierbaren Monomeren, *Angew. Makromol. Chem.*, 68, 199, 1978.
10. **Silvermann, J., Zoepfl, F. J., Randal, J. C., and Markovich, V.**, The mechanism of radiation induced linking phenomena in polyethylene, *Radiat. Phys. Chem.*, 22, 583, 1983.
11. **Chapiro, A.**, *Radiation Chemistry of Polymeric Systems*, Interscience, New York, 1962.
12. **Charlesby, A.**, Crosslinking and degradation of polymers, *Radiat. Phys. Chem.*, 18, 59, 1981.
13. **Van Dine, G. W. and Shaw, R. C.**, Crosslinking of polyethylene by polyesters, *Polym. Prepr.*, 12, 713, 1971.

14. **Van Drumpt, J. D. and Osterwijk, H. H. J.,** Kinetics and mechanism of the thermal reaction between tert.butyl perbenzoate and n-alkanes. A model system for the crosslinking of polyethylene, *J. Polym. Sci. Polym. Chem. Ed.,*. 14, 1485, 1976.

15. **Chodak, I. and Lazar, M.,** Effect of the type of radical initiator on crosslinking of polypropylene, *Angew. Makromol. Chem.,* 106, 153, 1982.

16. **Pukschanski, M. D., Zyuzina, L. I., Khaikin, S. I., Sirota, A. G., Kachan, A. A., and Goldenberg, A. L.,** The pecularities of photochemical crosslinking of polyethylene in the presence of tetrachloroethylene (Russian), *Vysokomol. Soed.,* A14, 2096, 1972.

17. **Matsuo, M. and Sawatari, Ch.,** Crosslinking of ultrahigh molecular weight polyethylene fibres produced by gelation crystallization from solution under elongation process, *Macromolecules,* 19, 2028, 1986.

18. **Halpern, J.,** Free radical mechanism in organometallic and bioorganometallic chemistry, *Pure Appl. Chem.,* 58, 575, 1986.

19. **Roesky, H. W.,** Catalysis and coordination compounds involving electron-rich main group elements, *Chem. Soc. Rev.,* 15, 309, 1986.

20. **Mortreux, A. and Petit, F.,** A new route to coordination catalysis by electrogeneration of organometal transition active species. A review, *Appl. Catal.,* 24, 1, 1986.

21. **Halpern, J.,** Free radical mechanism in coordination chemistry, *Pure Appl. Chem.,* 51, 2171, 1979.

22. **Beleckaia, I. P., Artemkina, G. A., and Reutov, O. A.,** Vzaimodejstvije metaloorganiceskich proizvodnych s organiceskimi galogenidami, *Usp. Khim.,* 45, 661, 1976.

23. **Henrici-Olive, G. and Olive, S.,** *Catalyzed Hydrogenation of Carbon Monoxide,* Springer-Verlag, Berlin, 1984.

24. **Bogdanovic, B.,** Selectivity control in nickel-catalyzed olefin oligomerization, *Adv. Organometal. Chem.,* 17, 105, 1979.

25. **Boor, J.,** *Ziegler-Natta Catalysts and Polymerizations,* Academic Press, New York, 1979.

26. **Zakharov, I. I. and Zakharov, V. A.,** Influence of the coordination state of organic compounds of transition metals and their catalytic properties in polymerization of olefins, *React. Kinet. Catal. Lett.,* 1, 61, 1983.

27. **Cassoux, P., Grasnier, F., and Labarre, J.,** $TiMeCl_3C_2H_4$ system as a model for quantitative evaluation of the Cossee mechanism, *J. Organomet. Chem.,* 165, 303, 1979.

28. **Sweany, R. L. and Halpern, J.,** Hydrogenation of α-methylstyrene by hydridopentacarbonylmanganese. I. Evidence for a free-radical mechanism, *J. Am. Chem. Soc.,* 99, 8335, 1977.

29. **Hommeltoft, S. I., Berry, D. H., and Eisenberg, A.,** A metal-centered radical-pair mechanism for alkyne hydrogenation with a binuclear rhodium hydride complex. CIDNP without organic radicals, *J. Am. Chem. Soc.,* 108, 5345, 1986.

Chapter 6

NONBRANCHED CHAIN REACTIONS

I. INTRODUCTION

Regeneration of radicals and cyclic repetition of some elementary reactions is the main feature of chain reactions. The number of cycles of elementary reactions per one generated radical is called the kinetic length of the chain reaction. It may be expressed as a ratio of the rate of propagation and a radical decay. Its maximum value can be $10.^7$ Reaction products of chain reactions are formed in the propagation step while those formed in termination reactions represent the negligible part.

The important aid for the interpretation of the course and mechanism of chain radical reactions is the idea of steady state (Bodenstein), which suggests the constant concentration of respective radicals in the system given by the equality of the rates of free radical formation and decay. To demonstrate what such simplification means, the simple reaction scheme of the transformation of reactant A to product B via the intermediate radical R· is analyzed in more detail (Scheme 1). In order to ensure the approximately constant

$$A \xrightarrow{k_1} R \cdot \xrightarrow{k_2} B$$

SCHEME 1

concentration of radicals, the rate constant k_2 of the above scheme has to be considerably higher than k_1. For most reagent molecules and radicals, differing in reactivity considerably, the assumption $k_2 \gg k_1$ is logical. In the opposite case $k_2 < k_1$, the radical concentration in time exhibits a maximum. The time changes of concentration of reactant [A] may be expressed by the differential Equation 1, which

$$-d[A]/dt = k_1[A] \tag{1}$$

corresponds to the reaction of the first order. If the initial concentration of reactant is $[A]_0$, its concentration [A] in time t is described by the relation in Equation 2. The rate of

$$[A] = [A]_0 \exp(-k_1 t) \tag{2}$$

formation of product B may be expressed as Equation 3, while the

$$d[B]/dt = k_2[R \cdot] \tag{3}$$

changes of radicals R· are given by the rates of their formation and decay (Equation 4). The concentration of radicals R· in

$$d[R \cdot]/dt = k_1[A] - k_2[R \cdot] \tag{4}$$

time t found from the above equations is expressed in Equation 5. If we

$$[R \cdot] = k_1[A]_0 (k_2 - k_1)^{-1} (\exp(-k_1 t) - \exp(-k_2 t)) \tag{5}$$

assume the existence of a stationary state (Equation 6), from the very

$$d[R\cdot]/dt = 0 = k_1[A] - k_2[R\cdot] \qquad (6)$$

beginning of the reaction, the concentration of free radicals R· can be described simply by Equation 7. From this textbook example we

$$[R\cdot] = k_1/k_2[A] \qquad (7)$$

may see that the resulting kinetic equations may be considerably more simple,[1] differential equations being replaced by algebraic equations in the stationary state approach. The more complicated the mechanism of a radical reaction, the more relative simplification is brought about by the steady-state idea which thus assumes a very important place at the analysis of the mechanism of radical chain reactions.

Apart from the great success of this approach, we should bear in mind its limitations, too. The difference in rates of formation and decay of free radicals is obviously very small when compared with the absolute value of the reaction rate, but it is not zero. Sometimes, as in the case of explosion or of other branched-chain reactions, the rate of radical appearance may considerably exceed that of decay, and the steady state simplification cannot be used at all. For the exact kinetic estimations and analysis, the differential equations should then be solved in their original form. This is, however, possible only with the use of a computer.

The above relatively abstract pattern of nonbranched chain reactions will be documented on some examples. The significance of chain reactions does not consist only in the study of mechanisms. Worldwide efforts of scientific and research teams were oriented also to the practice. From nonbranched chain reactions, the practical application found, e.g., the polymerization of ethylene and some of its derivatives. The study of thermal degradation anticipates the usage of polymer wastes to the back production of low molecular chemicals. Of great industrial importance is the thermal cracking of hydrocarbons, which is used for the production of ethylene, propylene, higher alkenes, gasoline, and other motor fuels. Also the widely used synthesis of halogen derivatives of hydrocarbons belongs to the group of chain reactions. Radical reactions of alkanes are probably the only efficient way to transform them to some more complex compounds. The distinct inert response to ionic reagents, which were employed in the early days of chemistry to induce the chemical transformations of alkanes, has even caused them to be called paraffins, which reflected their low affinity to reactants. Chain radical reactions also found their application in the synthesis of new compounds.

II. POLYMERIZATION

The term polymer designates compounds having molecules composed of hundreds to millions of atoms mutually linked by chemical bonds. Radical polymerization belongs to one of the built-up polyreactions forming a polymer chain from the monomer molecules. Almost all vinyl $CH_2{=}CHX$ and vinylidene $CH_2{=}CYX$ derivatives, ethylene, alkenes, alkines, ketenes, dienes, and other compounds with multiple bonds, can be used as monomers. Symmetrical 1,2-disubstituted derivatives of ethylene polymerize less willingly than 1,1-derivatives. For steric reasons, trisubstituted and tetrasubstituted ethylenes do not polymerize at all, except for fluorinated and perfluorinated ethylenes, where less bulky fluorine atoms do not impede the growth of the polymer chain. Monomers, which do not polymerize can, however, copolymerize with other monomers.

At each chain reaction, the polymerization has three fundamental stages, namely, the initiation, i.e., the generation of free radicals, propagation reactions of radicals, and termination, respectively. All elementary reactions of these three stages take place simultaneously in a homogeneous polymerization system. The decisive elementary reaction of a propagation stage is a repeating addition of the growing polymerization radical to a monomer. The radical growth may be sometimes combined with the isomerization of the growing radical. These reactions are accompanied by the release of the prevailing part of reaction heat, and the entropy of the system changes, too. The polymerization heat liberated at the addition of a macroradical to the C=C bond of a monomer is equal to the difference of dissociation energies of a single σ-bond and π-bond carbon-carbon, which is 90 kJ/mol, approximately. This value is within the experimentally observed values of polymerization heat (from 35 to 135 kJ/mol). The observed scattering of measured polymerization heat is due to the substituent effect on the strength of respective bonds in both the monomer and polymer chain.

The polymerization, as an aggregation process, is generally accompanied by a decrease of the entropy of the system, which is brought about by the restriction of the possibilities of the translational motion of monomer units after their incorporation into the polymer chain. At the polymerization of liquid vinyl and vinylidene monomers, entropy decreases by 100 to 170 kJ mol^{-1} K^{-1}. The considerably higher rate of addition over fragmentation of macroradicals is the self-evident condition of the polymerization course. Besides the rate constant of the addition reaction, the rate of polymerization is determined also by a monomer concentration as well as by the concentration of radicals depending on both the rate of initiation and of termination. The isomerization reactions of growing radicals influence the resulting structure of the macromolecule. As for the molecular weight of the formed polymer, the important role is played by transfer and fragmentation reactions, which also belong to propagation reactions.

Macromolecules already formed may also affect the further polymerization of a monomer. They either dissolve in it and thus increase the viscosity of the system or, if they are insoluble, form a new phase in which polymerization may also occur. In both cases, the initial conditions of the polymerization change, which has its impact on the polymerization rate (which usually increases) and on the molecular weight of the polymer in the later stages of the process. Provided that a macroradical is formed from a macromolecule by a subsequent transfer reaction, it may also enter the addition reactions with a monomer and the originally linear macromolecule becomes branched. The process of branching of macromolecule structure is usually undesirable in a polymer preparation and should be suppressed to a minimum by the proper reaction conditions.

Under heterogeneous conditions, the elementary reactions do not occur simultaneously. In emulsion polymerization, e.g., the initiation takes place in the continuous phase of the whole system, while in micelles of emulgator only addition reactions occur. The average size of the formed macromolecule is then determined by the number of radicals permeating the micelle as well as by the chemical properties of the polymerizing monomer. The rate of polymerization depends on the amount of growing particles in the system. If we succeed in interrupting the supply of radicals into the micelles, the polymerization in individual separated particles induced by one radical will proceed until all monomer molecules are consumed and may go on even after the supplementary addition of a new monomer of different chemical structure.[2] Under the above conditions of so-called living radical polymerization, the sizes of a macromolecule will be controlled by transfer and fragmentation reactions. The decay of radicals in isolated particles may be rather slow and will be brought about predominantly by the diffusion of lower molecular radicals through the interphase or by interpenetration of emulgated particles of the polymerizing monomer. The average lifetime of the living radical is considerably higher than 1 sec, which is the mean time interval of the overall sequence of elementary reactions leading to one new macromolecule in homogeneous polymerization.

The mutual interrelation of subsequent and competition processes in the chain reaction of polymerization depends not only on the homogeneity of the reaction medium, but also on various alternatives of respective elementary reactions.

A. Initiation

The initiation of the polymerization is connected with the generation of reactive radicals in the system. As was already pointed out, this may be ensured by thermal initiators, compounds having chemical bonds of dissociation energy from 120 to 170 kJ/mol in a molecule. This limitation is given by the necessity to achieve a certain rate of radical production in the temperature range from 20 to 200°C. The dissociation energies of the above values have covalent bonds oxygen-oxygen, oxygen-nitrogen, nitrogen-carbon, sulfur-sulfur, and some others. The most frequent initiators of free radical polymerization are peroxidic compounds and α, α'-azobis(isobutyronitrile); from peroxides it is chiefly dibenzoylperoxide.

The chemical initiation of polymerization consists in the decomposition of the initiator (Equation 8) and in addition of formed

$$\underset{(CH_3)_2C-N=N-C(CH_3)_2}{\overset{\overset{\displaystyle CN}{|} \qquad \overset{\displaystyle CN}{|}}{}} \xrightarrow{\text{50 to 70°C}} \underset{(CH_3)_3C\cdot}{\overset{\overset{\displaystyle CN}{|}}{2}} + N_2 \tag{8}$$

radicals to a monomer as, e.g., styrene (Equation 9).

$$(CH_3)_2(CN)C\cdot + CH_2=CHC_6H_5 \rightarrow (CH_3)_2(CN)CCH_2\overset{\cdot}{C}HC_6H_5 \tag{9}$$

The primary initiating radicals are thus incorporated into the structure of the growing radical.[3,4]

The initiation reaction is not, however, always so unambiguous. Addition on a less accessible carbon of the double bond of the monomer may occur. The primary radical of the initiator can, moreover, react with the reactive site of the side group. The transformation of the formed radical has several reaction pathways, the addition reaction being the most likely.

At the initiation of polymerization of vinyl monomers by dibenzoylperoxide, (from 75 to 96%) the expected addition of primary radicals on CH_2 group prevails.[5] Addition on substituted carbon of ethylene is relatively important at the polymerization of vinylacetate, where about 25% of benzoyloxy radicals are added. From the reactions of substituents, the addition of benzoyloxy radicals to the phenyl group of styrene is of importance too, since about 10% of primary benzoyloxy radicals react in such a way. The abstraction of hydrogen from a molecule by a primary radical is worth noticing at the polymerization of methyl methacrylate by cumyloxy radicals.[6] Approximately one half of cumyloxy radicals add to the double bond on the monomer, while one fifth abstract hydrogen. The hydrogen of the α-methyl group of methyl methacrylate, which is abstracted about 10 times faster than that of the ester methyl group, is the most reactive site in the transfer reaction. One quarter of potentially formed cumyloxy radicals fragmentate to acetophenone and methyl radicals.

This is also the case of primary benzoyloxy radicals which decompose to phenyl radicals and carbon dioxide. For more reactive monomers, the ratio of primary radicals is higher while less reactive monomers support the fragmentation. This is also the reason why at the initiation of polymerization of the sequence of monomers, such as styrene, less reactive methyl methacrylate, and acrylonitrile, 4, 28, and 70% of primary benzoyloxy radicals fragmentate.

The ratio of various initiation reactions depends not only on the kind of initiator and monomer but also on experimental conditions of polymerization. It should be remembered that some fraction of formed radicals is also consumed in termination reactions. The term of efficiency of the initiation reaction was introduced, which expresses the ratio of radicals entering the addition reaction with a monomer related to the amount of decomposed initiator. The efficiency increases with the increasing reactivity of primary radicals and with the concentration of the reacting monomer. For current initiators and monomers the efficiency of initiation lies within the interval from 0.3 to 0.7.

The concentration of thermal initiators necessary for polymerization of current monomers is 10^{-3} to 10^{-5} mols per 1 mol of monomer. About 10% of the added initiator only is usually used up for the complete course of polymerization. Some excess of the initiator is required in order to achieve a sufficiently high rate of radical production. At the stepwise dosage of initiator to the polymerization system, the amount of initiator may be somewhat reduced.

Redox systems, where nonbonding electrons from ions or atoms are transferred to an unstable bond of, e.g., peroxidic compound and facilitate thus its decomposition, are very frequent initiators. As an example, we may present the decomposition of hydrogen peroxide catalyzed by Fe^{2+} ions or a reaction of dibenzoyl peroxide with dimethyl aniline referred to earlier, and the decomposition of peroxodisulfate by Fe^{2+} ions (Equation 10). Similar reactions occur with various compounds.[7]

$$^-O_3S-OO-SO_3^- + Fe^{2+} \rightarrow {}^-OSO_2O\cdot + SO_4^{2-} + Fe^{3+} \qquad (10)$$

The generation of radicals by redox initiation is largely used on an industrial scale, particularly, in the polymerization of monomers carried out in a water medium at 0 to 30°C.

Polymerization reactions may be initiated in many other different ways. Provided that ions are generated in parallel with radicals in the reaction system, the mechanism of polymerization depends on reaction conditions. The irradiation of styrene by γ-rays produces, e.g., 40 times more radicals than ions. Despite this fact, the polymerization of styrene, carefully purified from a trace amount of water and other cationic scavengers, is of an ionic nature. This is due to considerably larger rate constants for both free cations and free anions than free radical propagation.[8] It follows from numerous papers that the prevailing mechanism of radiation polymerization is cationic in a dry system and radical in a wet system.

The particular method of initiation may be demonstrated on mechano or thermally activated inorganic compounds. The reactive $-Si\cdot$ and $-SiO\cdot$ radicals capable of initiation of polymerization of some monomers may be, e.g., generated by the mechanical milling of silica.[9] The milling of silica to a powder of specific surface of about 10 m²/g gives about 10^{14} to 10^{15} radicals per 1 m². Even higher concentrations (10^{17} to 10^{18} spins/m²) may be achieved by heating natural calcium carbonate up to 700 to 1000 K. The formed active centers can initiate copolymerization of styrene and methyl methacrylate which indicates the radical character of the reaction.[10] The surface radicals generated on inorganic materials may be used for a formation of a thin layer of surface coatings which could ameliorate the compatibility of the inorganic filler with the polymer.[11]

B. Initiators of Living Polymerization

Until now, the addition of formed radicals to a monomer in an initiation reaction occurred irreversibly, and growing polymer chains, when terminating, became inactive. Another type of initiation may be performed with radicals of low reactivity which initiate and terminate kinetic chains with comparable probability. The transiently formed end group from a radical may be cleaved back here and a polymerization has a living radical character. This may well be documented on photopolymerization of styrene initiated with tetraethylthiuramdisulfide (TETD).[12]

By photolysis of TETD, (Equation 11)

$$[(C_2H_5)_2NCS_2^-]_2 \rightarrow 2 \ (C_2H_5)_2N(C{=}S)S\cdot \tag{11}$$

primary radicals are formed which slowly initiate polymerization and, simultaneously, also participate in termination. The end groups thus formed, however, undergo photolysis, too. Besides primary radicals of low reactivity, this photolysis produces reactive alkyl radicals of growing polystyrene chains which continue to add further molecules of styrene. The process may be temporarily stopped by termination with thiuram radicals and regenerated again by photolysis of the product. The molecular weight of the forming polymer increases and its polymolecularity becomes wider in the course of polymerization. When consuming a monomer, polymerization may be restored by an addition of a new monomer. Provided that this monomer is chemically different, polymerization leads to block copolymers.

The similar initiation reaction may be performed also with some thermal initiators, such as, e.g., tetraphenylethane derivatives (Scheme 2) which are capable of initiating

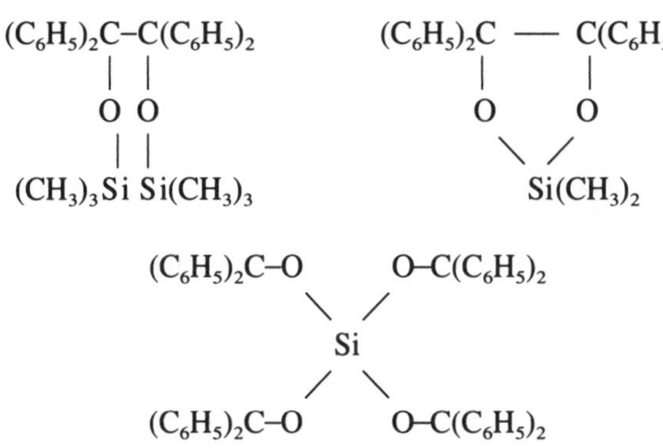

$$(C_6H_5)_2C{-}C(C_6H_5)_2$$
$$| \qquad |$$
$$O \quad O$$
$$| \qquad |$$
$$(CH_3)_3Si \ Si(CH_3)_3$$

$$(C_6H_5)_2C \ {-\!-\!-} \ C(C_6H_5)_2$$
$$| \qquad\qquad |$$
$$O \qquad\quad O$$
$$\diagdown \quad \diagup$$
$$Si(CH_3)_2$$

$$(C_6H_5)_2C{-}O \qquad O{-}C(C_6H_5)_2$$
$$\diagdown \qquad\quad \diagup$$
$$Si$$
$$\diagup \qquad\quad \diagdown$$
$$(C_6H_5)_2C{-}O \qquad O{-}C(C_6H_5)_2$$

SCHEME 2

polymerization of styrene at 100°C.[13] At the initial amount of initiator about 2.5% of w., the conversion of the monomer to a polymer decreases after 1 hr with an increase of the initiator functionality (13, 11, and 8%). On the other hand, the average molecular weight shows an opposite tendency (19000, 91000, 250000 g/mol). The decrease of the polymerization rate may be due to the tendency of active centers situated in mutual close proximity to the monomer. The increase of molecular weight of the formed polymer is obvious from the mechanism of the initiation reaction, but it does not correspond qualitatively to the increase of functionality. The relatively larger increase of molecular weight of polystyrene when compared with the multiples of functionality (1:2:4) may be brought about by interpenetration and entanglement of the potentially formed cyclic macromolecules. This explanation was, however, neither verified by independent experiment nor was the formation of cyclic macromolecules by the above mechanism ascertained. The open question remains, how many molecules enter the living radical pair before its termination?

Apart from this uncertainty, the initiation of the process of living radical polymerization is of interest since the monomer enters the initiation center in the repeating reaction cycles. The radical center of a growing radical does not escape the primary radical as it does in the case of classical free radical polymerization, but growing and primary radicals form a radical pair which terminates, in fact, via addition of monomer units. Addition of a monomer to a growing radical does not change the structure of the active moiety of the radical pair; only the length of the polymer chain increases. A certain analogy with the initiation of polymerization catalyzed by transition metal complexes exists here.

C. Propagation

The multiple repetition of a radical addition to a double bond of a monomer gives macroradicals which, after termination or transfer reaction, are transformed to saturated macromolecules. Before the end of its growth, the macroradical can undergo various isomerizations taking place in the time lag between respective addition steps.

Addition of a radical to a monomer of lower symmetry than ethylene gives "head to tail" conformations of the polymer chain (Equation 12) which are thermodynamically more favorable than

$$\sim CH_2-\overset{\cdot}{C}H + CH_2{=}CH \rightarrow \sim CH_2-CHCH_2-\overset{\cdot}{C}H \qquad (12)$$
$$\underset{X}{|}\phantom{+ CH_2{=}}\underset{X}{|}\underset{X}{|}\underset{X}{|}$$

"head to head" isomers (Equation 13). The formation of a certain

$$\sim CH_2\overset{\cdot}{C}H + CH_2{=}CH \rightarrow \sim CH_2-CH-CH-\overset{\cdot}{C}H_2 \qquad (13)$$
$$\underset{X}{|}\phantom{+ CH_2{=}}\underset{X}{|}\underset{X}{|}\underset{X}{|}$$

ratio of less advantageous conformers "head to head" or "tail to tail" cannot be avoided, however, in the polymerizing system.[14]

Decreasing the polymerization temperature and increasing the bulkiness of substituent X lead to an increase of regularity of the formed structure of a polymer chain. If X is the polyatomic substituent, the ratio of "head to head" conformers is about 1%. Some fluorocarbon polymers such as polyvinylidene fluoride and polyvinyl fluoride have an even higher content of these defect structures (26 to 32% and 8 to 12%, respectively). This relatively large ratio of irregular links is mainly due to small sizes of fluorine atoms.

The addition reaction of a monomer in a "head to tail" arrangement gives two possible mirror configurations (Scheme 3).

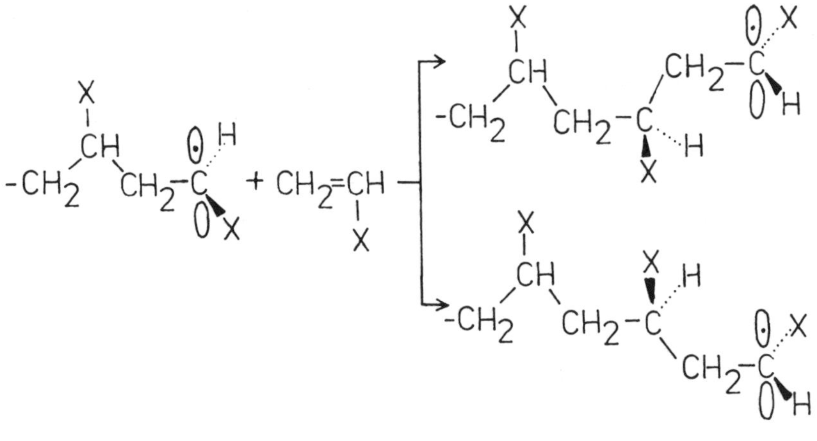

SCHEME 3

The subsequent addition may occur either randomly or with certain regularity and a different degree of macromolecule symmetry is thus achieved (Chapter 5, Figure 1). Regularity of the growth reaction may be influenced by the structure of both the macroradical and monomer or by their steric and polar factors as well as by experimental conditions of polymerization. To avoid confusion, it should be pointed out, however, that at the addition of a radical to a monomer, rather irregular chains are formed compared with other polymerization mechanisms used for the regulation of polymer tacticity. In a macromolecule prepared by a radical polymerization, a prevailing number of syndiotactic triades appear along with heterotactic ones, the chain being atactic.

Stereoisomer arrangement of the polymer chain is not affected by primary radicals which add to monomer irreversibly. By the repeated addition, the end group originating from the initiator rapidly gets far away from the active center of the growing macroradical. A different situation could arise at the addition of a monomer to the radical pair of a living macroradical. These problems were elaborated particularly at polymerizations catalyzed by transition metal complexes.

Some role at the formation of stereoregular polymers could be played by an "a priori" prepared regular arrrangement of monomer molecules. This can be achieved by polymerization carried out on a proper matrix or in channels of crystalline compounds.[15] The distance between pre-oriented monomer units can be, however, quickly disrupted by polymerization, and their initial arrangement thus impaired so that the effect on the formation of the stereoregular structure should be low. If this is not so, and a regular arrangement of the monomer units is maintained during the addition reaction, the effect on stereoisomerism may be quite distinct.

The addition reaction of dienes includes more possibilities. Besides "head to tail" and "head to head" linkages, the polymer chain may involve other isomer structures, too. Derivatives of butadiene polymerize, e.g., by 1,4-addition (Equation 14) along

$$R \cdot \ + \ CH_2=CXCH=CH_2 \ \rightarrow \ RCH_2CX=C\dot{H}CH_2 \tag{14}$$

with 1,2 and 3,4-additions (Equation 15). 1,4-Addition has the

$$R \cdot \ + \ CH_2=CXCH=CH_2- \begin{cases} \longrightarrow RCH_2\dot{C}XCH=CH_2 \\ \\ \longrightarrow RCH_2\dot{C}HCX=CH_2 \end{cases} \tag{15}$$

lowest activation energy and predominates over the other two alternatives at low temperature polymerization. 1,4-Addition may occur as a *cis* or *trans* process (Scheme 4). Addition

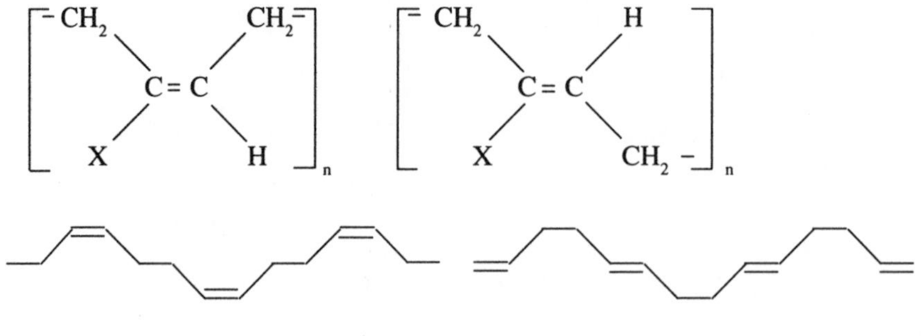

cis   *trans*

SCHEME 4

reactions 1,2 and 3,4 link structural units of different stereoisomerism of growing macroradicals similarly as for vinyl and asymmetrical vinylidene monomers.

Even in the case of a monomer having one reactive double bond C=C, the forming polymer chain reflects different alternatives of the propagation reaction. Besides the reactive C=C bond, there are less reactive multiple bonds in some monomers, which may also enter the addition reaction. This is, e.g., the case of monomers with aromatic groups (styrene, vinylcarbazol), with nitrile groups (acrylonitrile), or with carbonyl group (ester acrylates). A competitive addition reaction occurring on these groups is, however, less probable but leads to the formation of anomalous (defect) structural units. It is more probable at higher polymerization temperatures or when either excited radicals or monomers are present in the reaction system.

Provided that the polymerization batch contains several monofunctional monomers, the possibilities of addition reactions increase so that only relative rates of respective additions should be noticed at first, regardless of the isomerism of structural units. The process itself is called copolymerization. For an illustration, take the case of two monomers M_A and M_B which may enter four kinds of growing addition reactions (Equations 16 to 19).

$$P_A\cdot + M_A \rightarrow P_A\cdot \qquad k_{AA} \tag{16}$$

$$P_A\cdot + M_B \rightarrow P_B\cdot \qquad k_{AB} \tag{17}$$

$$P_B\cdot + M_B \rightarrow P_B\cdot \qquad k_{BB} \tag{18}$$

$$P_B\cdot + M_A \rightarrow P_A\cdot \qquad k_{BA} \tag{19}$$

The relative ratio of the respective reactions is determined by the concentrations and corresponding rate constants of the reaction of radicals and monomers. The ratios of rate constants k_{AA}/k_{AB} and k_{BB}/k_{BA} are called copolymerization parameters (r_A or r_B). If the parameter $r_A > 1$ and $r_B < 1$, the forming copolymer contains more structural units M_A, while the batch is enriched by the monomer M_B. The gradual change of the composition of the copolymerization batch changes stepwise the copolymer composition. The ideal case of copolymerization occurs for $r_A = r_B = 1$, where the participation of respective addition reactions is controlled by the concentration of monomers in the copolymerization mixture. Provided that copolymerization parameters are close to zero, the regular alternations of structural units in the copolymer chain predominate; at high values of copolymerization parameters, copolymers having large blocks of homopolymer structural units are formed. The above model of four addition reactions represents a good approximation for the most free radical copolymerizations. For the kinetic description, however, this type of model is unsufficient and 8 addition reactions for four types of radicals have been proposed.[16]

An increased number of kinetic equations leads to good accordance between experimental and theoretically assumed values. The four types of radicals determining the course of the copolymerization reaction were introduced upon the basis of an idea that the reactivity of the radical is not determined by the end unit only, but also by the unit next to the end. The extent of the effect of this penultimate structural unit on radical reactivity is, however, questionable. The more important effect seems to be the unequal orientation of monomer units in different end radicals.

The irregularities occurring in addition reactions of two monomers may be due to their electron donor or electron acceptor interactions. Donor-acceptor interactions of monomers may bring about other supplementary changes of radical reactivity as, e.g., the low effect of inhibitors and transfer agents on rate constants of crossover growing reactions, etc.[17]

The change in the reactivity of monomers in addition reactions may be caused by all possible intermolecular interactions in the polymerization system. Acrylonitril is a particular case among vinyl monomers regarding the solvent dependence of the addition rate constant in homogeneous polymerization. The highest rate of its addition reaction was observed for water. The important peculiarity of the water molecule resides in its ability to form two hydrogen bonds simultaneously, i.e., one with the ultimate $-C\equiv N$ group of a growing polymer radical and one with a monomer.[18]

This arrangement not only would increase the effective local concentration of a monomer in the neighborhood of the radical, but also might activate the monomer. A similar template effect, i.e., the mutual approach and activation of reactants is in some cases performed even by the formed polymer itself.[19]

Addition reactions of most of the vinyl and vinylidene monomers have an activation energy of about 20 kJ/mol and frequency factor from 10^6 to 10^7 $dm^3 \cdot mol^{-1} \cdot sec^{-1}$. The low activation energy of the growth reaction allows the polymerization course to proceed at a moderately low temperature. Relatively low frequency factors compared with those of monomer and radical collisions ($\sim 10^{10}$ $dm^3 \cdot mol^{-1} \cdot sec^{-1}$) indicate the inevitability of proper orientation of monomers with respect to a radical center taking place before the addition reaction.

D. Nonaddition Propagation Reactions

Isomerization of macroradicals as an integral part of the propagation step comes into the play particularly at the ring opening polymerizations. Polymerization of 2-methyl-1,3-dioxolane may be shown as an example. After the addition of a radical to a cyclic ketene acetal (Equation 20), the ring is cleaved

$$R\cdot + CH_2=C \overset{\displaystyle O-CH_2}{\underset{\displaystyle O-CH_2}{\Big\langle}} \quad \rightarrow RCH_2\overset{\displaystyle\cdot}{C} \overset{\displaystyle O-CH_2}{\underset{\displaystyle O-CH_2}{\Big\langle}} \tag{20}$$

in β-position to a radical center (Equation 21)

$$RCH_2\overset{\displaystyle\cdot}{C} \overset{\displaystyle O-CH_2}{\underset{\displaystyle O-CH_2}{\Big\langle}} \quad \rightarrow RCH_2COOCH_2CH_2\cdot \tag{21}$$

and the already formed alkyl radical adds again to the methylene group of the monomer cyclic acetal. The alternative pathway of addition and isomerization reactions goes on until the macroradical terminates.

The fast cleavage of the ring is supported by the formation of a stable $C=O$ bond in the isomerized alkyl radical. If isomerization is not sufficiently fast, the addition proceeds via incorporation of original rings into the polymer chain. The degree of isomerization of macroradical depends on the monomer used and on the reaction temperature. Taking into account the higher activation energy of isomerization, the increased temperature hastens the rate of isomerization more than the rate of addition.

Among the monomers that have been shown to undergo such ring opening polymerization, cyclic ketene acetals, cyclic ketene aminals, cyclic vinyl ethers, vinylcyclopropane, etc. can be included.[20]

Isomerization of macroradicals occurs also during the polymerization of noncyclic monomers, but to a lesser degree (Equation 22).

$$R(CH_2)_3\overset{\displaystyle\cdot}{C}H_2 \rightarrow R\overset{\displaystyle\cdot}{C}H(CH_2)_2CH_3 \tag{22}$$

It leads to the appearance of short side chains attached to the main chain (Equation 23).

$$\overset{\cdot}{RCH}(CH_2)_2CH_3 + n\ CH_2{=}CH_2 \rightarrow R(C_3H_7)CH(CH_2)_{2n-1}\overset{\cdot}{CH_2} \tag{23}$$

Fragmentation reactions occur during polymerization to the same extent as at reactions of low molecular radicals. Depolymerization, the reverse reaction to addition, is very frequent fragmentation. Fragmentation giving a macromolecule and low molecular radical is also probable, but will be usually overlapped with transfer to a solvent or to a monomer. Decreasing the concentration of the monomer in the system supports the course of transfer and fragmentation reactions when compared with competitive addition reactions. The former reactions decrease the sizes of formed macromolecules.

The transfer reactions shortening the length of the forming polymer chains are undesirable processes at the polymer synthesis. Since they have usually higher activation energy than the growth addition reaction (the difference is from 20 to 60 kJ/mol), a temperature as low as possible is recommended for the polymer preparation. On the other hand, the transfer reactions serve for a control of the chain length of polymer molecules and for the preparation of a polymer of certain molecular weight. The compounds which are purposefully added to the polymerization system for regulation of the molecular weight are called regulators. Organic sulfur compounds such as, e.g., disulfides or thiols belong to the most frequent ones. Their effect consists in reactions with growing radicals $P\cdot$ (Equations 24 and 25). The formed radicals $RS\cdot$ are capable of further

$$P\cdot + RSSR \rightarrow PSR + RS\cdot \tag{24}$$

$$P\cdot + RSH \rightarrow PH + RS\cdot \tag{25}$$

initiation of the new chains. The compounds, having the transfer rate constant approximately the same as that of the addition reaction are the most suitable. At the same value of both rate constants, the regulator and monomer are consumed equally and a polymer of the same molecular weight is formed throughout the overall course of polymerization.

The transfer reaction may occur, of course, not only with added regulators, but also with fundamental components of the polymerization system, namely, with the monomer, initiator, and the formed polymer. The transfer to the most monomers is a rather slow reaction compared to addition. From the viewpoint of the formed macromolecule, it is, however, the key process which determines the number of structural units in the polymer chain. The rate of the transfer reaction is determined by the alkene substituent, namely, by its capability to split out the mobile hydrogen or halogen. The abstraction of hydrogen from carbons of a double bond is considerably less probable. Provided that the substituent is the methyl or $-CH_2R$ group (R is phenyl, alkyl, or halogen), the transfer reaction may achieve such a degree that it suppresses totally the macromolecule formation. The transfer reaction is supported here by the formation of allyl radicals, which enter terminations predominantly and are not capable of reinitiating a new addition reaction. This type of transfer to a monomer is called degradation transfer. In vinylidene monomers, the methyl or substituted methyl group does not have such significant impact on degradation transfer since the substitution has a more profound effect on the addition reaction than on the degradation transfer.

At the transfer to initiator, besides the decrease of molecular weight, the initiator is inefficiently removed from the polymerization system. This formally corresponds to the induced decomposition of peroxide, quoted earlier.

The transfer reaction to a newly formed polymer affects the structure of formed macromolecules. The participation of the chain-centered macroradical in addition reactions brings about the formation of long branches in a macromolecule. Since the transfer reaction to a polymer comes into play only at a higher concentration of macromolecules in the system, the branching of polymer chains (which changes the polymer properties) takes place particularly at higher conversions of monomer to polymer.

E. Termination

Many mutual collisions of polymer radicals end as combination (Equation 26) or as disproportionation (Equation 27).

$$2 \text{ RCH}_2\dot{\text{C}}\text{HX} \rightarrow \text{RCH}_2\text{-CH(X)-CH(X)-CH}_2\text{R} \tag{26}$$

$$2 \text{ RCH}_2\dot{\text{C}}\text{HX} \rightarrow \text{RCH=CHX} + \text{XCH}_2\text{CH}_2\text{R} \tag{27}$$

For most monomers, both termination processes occur in parallel. Vinylidene derivatives undergo more disproportionation, while vinyl derivatives combine predominantly. The preponderance of disproportionation is caused by the higher amount of hydrogen atoms surrounding the radical site as well as by the sterical hindrance of the termination.

Disproportionation gives macromolecules of one half the degree of polymerization than combination. The combination reaction leads to the appearance of one supplementary structural unit "head to head" in a polymer chain, while at the disproportionation one half of formed macromolecules contain one double bond.

The reaction of a macroradical with primary radicals of the initiator is an unwanted termination reaction. It takes place particularly in the case of nonreactive monomers. Its importance increases with the increasing conversion of a monomer to a polymer where primary radicals have a lowered possibility of reacting with the monomer. Another reason is the increasing viscosity of the system which impedes the mutual collisions of two macroradicals more than those of macroradicals with low molecular primary radicals. For less reactive monomers, however, even at low conversions, the relatively high ratio of initiator residues may be incorporated into the polymer chain due to termination with primary initiator radicals.[21]

The termination of growing macroradicals is governed by different mechanisms. The rate-determining step in polymer/monomer mixtures of low to moderate viscosity is the translational diffusion of growing polymer chains, which is influenced *inter alia* by chain entanglements. In this case, the termination rate coefficient will depend on the length of the growing chains. A number of models exist which give the analytical dependence of termination rate constant k_t on the chain length of two reacting radicals. These models can be simplified to the relationship (Equation 28) where $k_{t(n,m)}$ is the specific rate constant of termination between two radicals of size n and m and k_{t0} and a are constants.[22]

$$k_{t(n,m)} = k_{t0}(n,m)^{-a} \tag{28}$$

The chain length dependence of the rate constant of termination leads to marked deviations from the predictions of the so-called classical polymerization kinetics with chain-length independent termination.[23]

Residual termination, which is determined by propagation processes, would be independent of the molecular weight of the propagating species. Even when chain segments are completely frozen, there will be a residual termination. This arises from the volume swept out by the propagating free radicals, which implies that there would be a nonzero rate of termination even in a glassy system, as long as the propagation rate coefficient remained nonzero.[24]

When polymer radicals are so small (as, e.g., for polystyrene of an average chain length less than 300) that they do not entangle with the surrounding polymer molecules, the termination rate does not depend on chain length, but on the critical free volume, which either permits or impedes segmental motion of the polymer chain.[25]

The same or very similar reason for the termination rate constant lowering appears at the polymerization of monomers in some solvents, as e.g., in H_3PO_4, which restricts the mobility of macroradicals. Under such conditions, therefore, the polymerization of acrylates may proceed by the mechanism of living macroradicals.[17] The increase in participation of termination reactions with regard to addition reactions is due to the decrease in reactivity of propagating radicals in transfer or in other type of addition reactions.

The compounds which form nonreactive radicals in reaction with the propagating species and thus completely stop the polymerization are inhibitors, while those which only slow down the process are retarders. They function so, until they are consumed in the initiation reaction. The effect of an inhibitor may be quantitatively characterized by the inhibition time during which the polymerization does not proceed noticeably at a certain concentration of the initiator. Among the most frequent retarders or inhibitors, compounds such as quinones, substituted phenols, arylamines, and aromatic nitro compounds may be quoted. Quinones and aromatic nitro compounds give more stable radicals in addition reactions with propagating species, while phenols and arylamines do the same in transfer reactions. The addition of radicals to quinones may take place either on the carbonyl group or, more frequently, on the double bond C=C. The formed semistable radicals decay in self reactions or with reactive radicals by the disproportionation mechanism. Instead of inhibitors from which nonreactive radicals are formed during polymerization, stable radicals may be used, too. Diphenylpicryl hydrazyl is most suitable for such a purpose.

Oxygen molecules which have a triplet ground state, belong to inhibitors of polymerization, too. They react easily with the growing polymerization radicals and give peroxy radicals which may retard or even stop the polymerization. The effect of oxygen is not, however, always so unambiguous. Peroxides or hydroperoxides forming in subsequent reactions of peroxy radicals may decompose at an elevated temperature to reactive radicals and the rate of polymerization may sometimes even increase.[26,27] The insertion of peroxidic oxygen into the main chain of the macromolecule is harmful from the viewpoint of the quality of the resulting polymer, which has lowered thermal and weathering stability.

F. The Quantification of the Polymerization Process

The requirements for the reliable control of the polymerization process in big industrial reactors and the effort to comprehend it more deeply have stimulated more detailed kinetic measurements. To a first approximation, it may be assumed that a monomer is consumed practically quantitatively in the repeating addition reaction of macroradicals. The rate of polymerization v_p may be expressed by the equation (Equation 29)

$$v_p = -d[M]/dt = k_p[M][P\cdot] \qquad (29)$$

where k_p is the rate constant of the addition reaction and $[M]$ is the concentration of the monomer. The concentration of macroradicals $[P\cdot]$ may well be replaced by the expression involving the rate of initiation v_i and the rate constant of termination k_t, derived from the steady state condition (Equation 30).

$$[P\cdot] = (v_i/k_t)^{0.5} \qquad (30)$$

In the steady state, the rate of radical decay is equal to the rate of initiation $v_i = k_t[P\cdot]^2$. The character of initiation thus determines the overall rate of polymerization. In the case of chemical initiation we have Equation 31

$$v_p = k_p[M](2k_df[I]/k_t)^{0.5} \qquad (31)$$

where k_d denotes the rate constant of initiator [I] decomposition and f is the efficiency of initiation; other symbols were quoted earlier. From the above equation, it is seen that the overall rate of polymerization is directly proportional to the concentration of the monomer and to the square root of the initiator concentration. Even though the relation between v_p and the square root of the initiator concentration is valid in many cases, there are also a number of exceptions. When the concentration of primary radicals increases, they start to compete in termination with macroradicals. For an illustration, if we consider the limiting case of termination by primary radicals R· only, then the condition of steady state is as follows from Equation 32

$$k_i[R·][M] = k_t[R·][P·] \tag{32}$$

and the equation for the rate of polymerization changes to Equation 33. In such a case, the rate of polymerization does not

$$v_p = k_p \, k_i/k_t[M]^2 \tag{33}$$

depend on the initiator concentration, but the dependence on the monomer concentration is more pronounced.

Another case can be observed when the growing radicals decay in the reaction with present retarder X. Then, in the stationary state, Equations 34 and 35 are valid, and the rate of

$$v_i = k_t[P·][X] \tag{34}$$

$$v_p = v_i(k_p/k_t)([M]/[X]) \tag{35}$$

polymerization is directly proportional to the initiator concentration and inversely proportional to the concentration of the retarder.

In the respective experimental measurements, the above limiting dependences are observed only seldomly. The value or the tendency of the change of the exponent above the concentration of the initiator indicates the probable mechanism of termination and initiation. For primary radicals of low reactivity, the probability of the termination by initiator radicals is higher, which gives exponent values lower than 0.5. Provided that the decay of macroradicals is of the first order related to their concentration, the observed exponent is higher than 0.5 The latter case may be observed at retarded polymerization or under similar conditions of radical deactivation, i.e., at the occlusion of macroradicals, at the degradation transfer to monomer, etc.

The effect of temperature on the polymerization course may be appreciated from the overall activation energy (E_c) of the polymerization. The relationship between activation energies of elementary reactions and the overall activation energy may be obtained simply from the logarithm of the complex coefficient of respective rate constants in the expression for the overall rate of polymerization, the individual rate constants being expressed by the Arrhenius equation.

For "normal" chemical initiation we thus obtain (Equation 36)

$$E_c = E_p + 0.5(E_d - E_t) \tag{36}$$

while for initiation with radicals of low reactivity (Equation 37)

$$E_c = E_p + E_i - E_t \tag{37}$$

and for monomolecular termination (Equation 38).

$$E_c = E_p + E_d - E_t \qquad (38)$$

If we take activation energy of addition as $E_p = 30$ kJ/mol, we may see that the main contribution to the overall activation energy comes from the activation energy E_d of initiator decomposition, which for peroxidic initiators has the average value of 125 kJ/mol. For polymerization involving bimolecular termination of macroradicals and peroxidic initiation, E_c is thus 90 kJ/mol, while for redox initiation E_c is only about 50 kJ/mol. For initiation with radicals of low reactivity and primary radical termimation, E_c is about 60 kJ/mol; for monomolecular termination, E_c attains 150 kJ/mol. We recall that the increase of temperature by 10 K in the temperature range of 300 to 400 K accelerates the reaction two times approximately, if E_c is about 80 kJ/mol.

The above considerations on the effect of temperature on the polymerization rate are valid for such monomers that form macroradicals which depolymerize negligibly at the given temperature. The equal rate of both the polymerization and depolymerization is attained at ceiling temperature, T_c (Equation 39).

$$T_c = (E_p - E_{dp})([M]Rln A_p/A_{dp})^{-1} \qquad (39)$$

This temperature may be found from the identity between the rate of addition and depolymerization, where the rate constants are defined by the Arrhenius equation. The equilibrium concentration of a monomer follows from the ratio of the rate constant of depolymerization and the addition of a macroradical to a monomer.

At the polymerization of noncyclic monomers the volume contraction occurs; one should not be, therefore, surprised that the pressure usually accelerates the polymerization. The pressure effect on the polymerization rate is mediated through various parameters such as the viscosity or density of the medium or parameters of the configuration of activated complexes in elementary reactions. If the volume of activated complexes is lower than that of reactants, the increasing pressure accelerates the reaction.

The expression for the overall activation volume of polymerization and those of individual partial reactions (Equation 40) is

$$\Delta V_c^{+} = \Delta V_p^{+} + 0.5\Delta V_i^{+} - 0.5\Delta V_t^{+} \qquad (40)$$

analogical to that for activation energies. For vinyl monomers $\Delta V_c^{+} = -17$ cm^3·mol^{-1} at 30°C; ΔV_p^{+} is also of negative value, being about 10% higher absolutely than ΔV_c^{+}. The activation volumes of initiation and termination are approximately equal and positive which means that increasing pressure decreases the respective rate constants. The increase of pressure from 0.1 MPa to 100 MPa increases the polymerization rate twice, approximately. A pressure higher than 1000 MPa may, however, sometimes considerably decelerate the polymerization since the increased pressure may bring about the solidification of the monomer.

Simultaneously with the knowledge concerning kinetics, it is often important to know the relations determining the changes of molecular weight of a polymer varying with experimental conditions. The average number of monomer units (n) incorporated into the polymer chain is expressed by the ratio of the rate of the addition reaction and by the sum of the rate of termination and transfer reactions (Equation 41) as well as by

$$n = (1 + \lambda)k_p[P\cdot][M](k_t[P\cdot]^2 + k_{tr}[P\cdot][X])^{-1} \qquad (41)$$

the ratio of combination and disproportionation (Equation 42)

$$\lambda = k_c(k_c + k_d)^{-1} \qquad (42)$$

If the concentration of macroradicals is expressed from Equation 29 and the only transfer agent X is monomer M, initiator I, solvent S, and formed polymer P, and if the ratio of constants k_{trx}/k_p is replaced by C_M, C_I, C_S and C_P terms, respectively, the equation for n has its final form as Equation 43.

$$n^{-1} = (1 + \lambda)^{-1}((k_t v_p)/(k_p[M]^2 + C_M + C_S[S]/[M] + C_I[I]/[M] + C_P[P]/[M] \quad (43)$$

At the estimation of the polymerization degree we take into account the tabulated data[28] for respective constants which at 50°C lie in the interval: $C_I = 10^{-4}$ to 10^{-1}, $C_S = 10^{-6}$ to 1, C_M and $C_P = 10^{-5}$ to 10^{-3}, and $k_t/k_p^2 = 10^2$ to 10^3 mol dm^{-3} sec.$^{-1}$ Increasing temperature increases predominantly the transfer constants C and decreases the value of n.

The change of temperature does not influence the forming polymer, however, only via corresponding transfer constants. This will become more obvious when we ignore the transfer reactions (Equation 44). In systems where activation energy

$$n = (1 + \lambda)v_p/v_t = (1 + \lambda)v_p/v_i \quad (44)$$

of initiation is higher than that of polymerization, increasing temperature decreases n. In the opposite case of photo or radiation initiation, increasing temperature increases the number n of mers in the polymer chain.

The distribution of macromolecules with a different amount of structural units also corresponds to the mechanism of polymerization. Recombination leads to the more narrow distribution curve than disproportionation. The change of polymerization conditions, such as the change of temperature, consumption of initiator, the increase of viscosity, and the participation of the formed monomer in transfer reactions, contribute to the broader distribution. Another reason for broader distribution consists in the simultaneous course of polymerization at several places of a heterogeneous system (polymerization in the monomer solution on the surface or inside of forming polymer particles).

Besides these very simply formulated relationships, a wealth of knowledge is disclosed in numerous papers which specify the respective polymerizing systems in much more detail.[29]

The apparently voluminous literature dealing with polymerization might create the impression that all problems of radical polymerization have been solved to date. That impression would not, however, be correct. To illustrate it, we demonstrate here the interpretation of the low rate of copolymerization of some monomers, which homopolymerize rather fast. The retardation effect of a reactive monomer may be qualitatively understood in such a way that it easily enters the growth reaction with the macroradical of its comonomer, while the formed macroradical is nonactive in subsequent growth reaction. An addition reaction of a nonreactive macroradical with a reactive monomer is less probable because of its lower concentration in the system. The quantitative estimation of such an assumption, however, leads to the rate constant of cross termination of unequal radicals by two orders higher than corresponds to the geometrical mean value of the termination rate constants for identical radicals. This contradicts data on reactions of low molecular radicals as well as direct mesurements of the termination rate constants for the system of two monomers.[30]

As was already pointed out, there is an effort to explain it by the penultimate model, in which the reactivity of radicals depends on both the end and the next to the end structural units. The higher number of macroradicals involved in copolymerization may be understood, however, also on the basis of various addition steps of monomers present in the system. This explanation of the different reactivity of macroradicals seems to be more plausible.

III. TELOMERIZATION

The term telomerization, originating from the Greek word *telos* (end), expresses the method of preparation of α, ω-difunctional hydrocarbons. The end groups X and Y are initially present in telogen XY which reacts by chain reaction with monomer M (called also taxogen) (Equation 45). The mechanism of

$$XY + nM \rightarrow X(M)_n Y \tag{45}$$

telomerization depends on the mode of initiation and on the type of propagation reactions. The latter may involve alternating addition-transfer and addition-fragmentation or isomerization reactions. Besides the thermal decomposition of peroxides and other traditional methods of free radical generation, the initiation may be performed by compounds of metals of transition valency. The proper choice of catalyst is often a decisive factor in the telomerization course with the respective taxogen, otherwise a polymer instead of a telomer is obtained.

The incorporation of monomer units into the chemical bond of telogen XY occurs after the addition of the radical derived from telogen to taxogen and after a subsequent polymerization process (Equation 46) terminated by competitive transfer

$$\cdot CCl_3 + n\ CH_2{=}CH_2 \rightarrow CCl_3(CH_2CH_2)_{n-1}CH_2CH_2 \cdot \tag{46}$$

reaction of the growing radical with the telogen molecule (Equation 47).

$$CCl_3(CH_2)_{2n-1}CH_2 \cdot + CCl_4 \rightarrow CCl_3(CH_2)_mCl + \cdot CCl_3 \tag{47}$$

The regenerated trichloromethyl radical of CCl_4 telogen starts a new cycle of telomerization.

The ratio of rate constants k_{47}/k_{46} depends on number n of monomer units in the growing radical.[31] In the above case of ethylene and CCl_4, the ratio k_{47}/k_{46} increases by 2 orders when going from n = 1 to n = 5. The change of rate constant is connected with a polar effect in transfer and growth reactions. The ratio of the rate constants depends significantly also on the nature of the telogen and the monomer.[32]

Very reactive telogens are: molecular chlorine and bromine, tetraiodomethane and tetrabromoethane, bromotrichloromethane, HBr, HCl, $SOCl_2$, and triarylhydrides of four-valent metals. They yield telomers which have very low values of n, sometimes only 1. Telogens of medium reactivity have the ratio of rate constant of transfer and growth reactions close to 1, which is the optimum case of regulation of the amount of monomer units in the telomer. CCl_4, $CHCl_3$, HCl, alkyliodides, thiols, and aldehydes are the most frequent representatives of this group; while monochloralkanes, monobromalkanes, alcohols, and carboxylic acids belong to telogens of low reactivity. The amount of already examined telogens exceeds 200 compounds. These compounds in telomerization reactions split out hydrogen from carbon, sulfur, silicon, phosphorus, tin, or more frequently halogen from carbon, nitrogen, sulfur, and phosphorus. Only in exceptional cases of scission of S-S or Sn-Sn bonds, the whole group of atoms is abstracted.

The effect of the structure of the polymerizing taxogen on telomerization reflects itself in the ratio of rate constant of growth (k_p) and termination (k_t) reactions. Telogens combined with certain taxogens are more reactive in telomerization if the monomer has a lower value of k_p/k_t ratio. About 50 monomers have been attempted in telomerization reactions to date.

The transfer reaction of a radical with telogen may be accelerated by metal ions of transition valency.[33] Under such conditions the process of telomerization (Equations 48 to 51)

$$\text{telogen} + M^{n+} \rightarrow R \cdot + M^{(n+1)+} + Y^- \tag{48}$$

$$M^{(n+1)+} + Y^- \rightarrow MY^{n+} \tag{49}$$

$$R\cdot + >C=C< \rightarrow R-\overset{|}{\underset{|}{C}}-\overset{|}{\underset{|}{C}}\cdot \tag{50}$$

$$R-\overset{|}{\underset{|}{C}}-\overset{|}{\underset{|}{C}}\cdot + MY^{n+} \rightarrow R-\overset{|}{\underset{|}{C}}-\overset{|}{\underset{|}{C}}-Y + M^{n+} \tag{51}$$

involves several kinds of elementary reactions where Y is Cl, Br, I, CN, SCN, N_3, S_2O_3 or SC(S)OR and M is a transition element.

The catalytic effect of transition metal ions may be documented on telomerization of styrene by tetrachlormethane. In the usual way of free radical initiation, the telomerization proceeds with difficulties since styrene undergoes polymerization predominantly and a chlorine atom from tetrachlormethane is abstracted only to a small extent. In the present of $FeCl_3$ in the polymerization system, however, the growing radical of styrene reacts with it at least by 6 orders faster than with tetrachlormethane (Equation 52) and $FeCl_3$ regenerates

$$CCl_3(CH_2CHC_6H_5)_nCH_2\overset{\cdot}{C}HC_6H_5 + FeCl_3 \rightarrow$$

$$CCl_3(CH_2CHC_6H_5)_nCH_2CHClC_6H_5 + FeCl_2 \tag{52}$$

by reaction with the telogen (Equation 53).

$$FeCl_2 + CCl_4 \rightarrow FeCl_3 + \cdot CCl_3 \tag{53}$$

A new trichloromethyl radical is formed initiating the formation of a molecule of telomer. A similar mechanism is valid for many other catalysts of telomerization.[34]

Another type of telomerization proceeds by alternation of addition and fragmentation reactions.[35] This is the case of ethylene and tetrachlorethylene (Equations 54 and 55).

$$Cl(CH_2CH_2)_{n-1}CH_2\overset{\cdot}{C}H_2 + CCl_2=CCl_2 \rightarrow Cl(CH_2)_{2n}CCl_2\overset{\cdot}{C}Cl_2 \tag{54}$$

$$Cl(CH_2)_{2n}CCl_2\overset{\cdot}{C}Cl_2 \rightarrow Cl(CH_2)_{2n}CCl=CCl_2 + Cl\cdot \tag{55}$$

By its addition to ethylene, a chlorine atom regenerates alkyl radicals in the system. The unstability of the bound halogen depends on its character as well as on the structure of a radical. For the easy course of telomerization, only splitting out the halogen does not suffice. The released atom has to be reactive enough to enter the addition reaction and thus to go on in the chain reaction. Taking this fact into account, chlor- and bromalkanes are suitable for telomerization, but not fluoralkanes or iodoalkanes.

Telomerization involving fragmentation of halogenalkyl radicals is not, thus, different from the former case only in elementary reaction steps but also in the formation of the $C=C$ double bond in one end group.

The end group of a telomer is determined by both the telogen and catalyst used. Telomers having identical groups at both ends of molecules are obtained when halogens or some other compounds are used.

The formation of a telomer having identical functional groups at both ends of the molecule is obvious when halogens are used. Less obvious is the appearance of identical end groups at telomerization by thionylchloride. Symmetrical binding of only halogen atoms at both ends of telomers is due to fragmentation of telogen radicals. A new cycle of telomerization reactions does not start from transiently formed radicals of telogen ($\cdot$SOC1) but from split out chlorine. Telogens as $SOCl_2$ are therefore functioning as halogen carriers.

The role of catalyst at the formation of end groups may be seen on the telomerization of ethylene with chloroform. Provided that peroxide is used as the initiating agent, one end group is CCl_3, the second CH_3. At the same kind of telomerization performed in the presence of $FeCl_3$, one end group is $CHCl_2$, the second CH_2Cl. They are formed by an analogous sequence of reactions as in Equations 52 and 53.

The end trichlormethyl groups of telomers may be hydrolyzed to a carboxy group while chlorine in the chlormethyl group is replaced by an amino group. Aminoacids prepared in such a way could serve for the preparation of polymers. It is more advantageous, however, to incorporate the required group directly into the telomer. This is, e.g., the case of telomerization of acrylonitrile with carboxyalkyl radicals (Equation 56) performed

$$HOOC(CH_2)_4\overset{\cdot}{C}H_2 + n\ CH_2{=}CHCN \rightarrow HOOC(CH_2)_5(CH_2CHCN)_{n-1}CH_2\overset{\cdot}{C}HCN \quad (56)$$

in the presence of complexes[36] of transition metals (Ti(III), CR(II), V(II), Fe(II), Cu(I), etc.). The telomer radical is reduced by the transition metal complex and neutralized with protons to telomer molecules (Equation 57). Carboxyalkyl radicals can be

$$\overset{\cdot}{R}HCN + Cr(II) + H^+ \rightarrow RCH_2CN + Cr(III) \quad (57)$$

prepared by reduction of the telogen with the complex of the transition metal in a lower oxidation state (Equation 58). The hydroxycyclohexyloxy

$$(58)$$

radical isomerizes (Equation 59) to a ω-carboxypentyl radical.

$$\rightarrow HOOC(CH_2)_4CH_2\cdot$$

$$(59)$$

Provided that the ratio of concentration of hydrogen peroxide and cyclohexanone attains 1:1, the redox decomposition of peroxidic adduct in the medium of acrylonitrile yields ω-cyanoalkyl radicals and telomer that contains two end cyano groups. The polymerization degree of the telomer is determined by the concentration and by the kind of transition metal complex.

Aromatic groups with different substituents may be attached to one end of a telomer molecule by catalytic decomposition of diazonium salts[33] (Equation 60 to 63).

$$ArN_2^+ + Cu(I) \rightarrow ArN{=}N\cdot + Cu(II) \quad (60)$$

$$ArN{=}N\cdot \rightarrow Ar\cdot + N_2 \quad (61)$$

$$Ar\cdot \ + \ n >C=C< \ \rightarrow \ Ar\left(\begin{array}{c}|\ \ |\\ C\text{--}C\\ |\ \ |\end{array}\right)_{n-1} \begin{array}{c}|\ \ |\\ C\text{--}C\\ |\ \ |\end{array}$$

(62)

$$Ar\left(\begin{array}{c}|\ \ |\\ C\text{--}C\\ |\ \ |\end{array}\right)_{n-1} \begin{array}{c}|\ \ |\\ C\text{--}C\cdot\\ |\ \ |\end{array} + \ Cu(II)Cl \rightarrow Ar\left(\begin{array}{c}|\ \ |\\ C\text{--}C\\ |\ \ |\end{array}\right)_{n} Cl \ + \ Cu(I)$$

(63)

An alternative reaction is the addition of *N*-chloramines to alkenes (Equations 64 to 66).

$$R_2N\text{--}Cl \ + \ Fe(II) \rightarrow R_2N\cdot \ + \ [Fe(III)Cl]^{2+}$$

(64)

$$R_2N\cdot \ + \ >C=C< \rightarrow R_2N\text{--}\overset{\displaystyle |}{C}\text{--}\overset{\displaystyle |}{C}\cdot$$

(65)

$$R_2N\text{--}\overset{\displaystyle |}{C}\text{--}\overset{\displaystyle |}{C}\cdot \ + \ [Fe(III)Cl]^{2+} \rightarrow R_2N\text{--}\overset{\displaystyle |}{C}\text{--}\overset{\displaystyle |}{C}\text{--}Cl \ + \ Fe(II)$$

(66)

Depending on the character of R and the reaction medium, stereoisomer products are formed. At the addition on cyclohexene in an acid medium there are, e.g., formed both the *cis* and *trans* products while in basic medium, nonprotonated *N*-chloramines are added so that only a *cis* isomer is formed. Such stereoselectivity may be explained by the coordination of the nonprotonated amino group to Fe(III) complex (Equation 67) which

$$R_2NCl \ + \ Fe(II) \rightarrow [R_2N\cdot \ ....Fe(III)Cl]^{2+}$$

(67)

facilitates the chlorine atom transfer. The protonated amino group does not give such a complex, and nonstereospecific addition, therefore, may be observed in such a case. As may be seen from this incomplete survey of telomerizations realized until now, the possibilities of selection of different end groups in telomer molecules are indeed numerous and experimentally are relatively well explored.

The structure of the intrinsic part of a telomer is determined by the taxogen used or by the composition of the mixture of chemically different monomers (cotelomerization). The example of cotelomerization of the mixture of acrylic acid and butadiene with methylcarboxyethyl radicals performed in the presence of Fe(II) and Cu(II) complexes demonstrates the competition of the most important elementary reactions.[37] The nucleophile methylcarboxy alkyl radical adds to acrylic acid preponderantly (Equation 68), while its addition to butadiene is

$$CH_3OCO(CH_2)_4\overset{\displaystyle .}{C}H_2 \ + \ CH_2{=}CHCOOH \rightarrow CH_3OCO(CH_2)_6\overset{\displaystyle .}{C}HCOOH$$

(68)

practically negligible. The newly formed radical is more electrophile and a subsequent addition step, therefore, occurs with butadiene (Equation 69). Further copolymerization is impeded by a fast

$$CH_3OCO(CH_2)_6\overset{\displaystyle .}{C}HCOOH \ + \ CH_2{=}CHCH{-}CH_2 \rightarrow$$

$$CH_3OCO(CH_2)_6CH(COOH)CH_2\overset{\displaystyle .}{C}HCH{=}CH_2$$

(69)

oxidation of alkyl radicals with Cu(II) which is, however, rather selective. The acrylate radical is not oxidized at all; the alkyl radical derived from telogen (peroxidic adduct of cyclohexane, methanol, and H_2O_2) reacts only to a limited extent, while the allyl radical is oxidized quantitatively. For the generation of the methylcarboxy alkyl radical (Equation 70, 71) the complex

$$CH_3O \quad OOH \qquad\qquad CH_3O \quad O^{\cdot}$$

$$\text{(cyclohexane ring)} \quad + \text{ Fe(II) } ---> \quad \text{(cyclohexane ring)} \quad + \text{ Fe(III) } + \ ^{-}OH \tag{70}$$

$$CH_3O \quad O^{\cdot}$$

$$\text{(cyclohexane ring)} \qquad \rightarrow CH_3OCO(CH_2)_4CH_2^{\cdot} \tag{71}$$

of the peroxidic adduct with Fe(II) is used. The return of the complex to the necessary oxidation state is ensured by reaction 72.

$$Fe(III) + Cu(I) \rightarrow Fe(II) + Cu(II) \tag{72}$$

At homotelomerization the structure of the inner part of the telomer is influenced by the stereoselectivity of an addition reaction.[38] In the case of esters of methacrylic acid and bromotrichloromethane, syndiotactic and isotactic dimers are formed in the ratio 3 to 10:1, depending on the reaction temperature and on the kind of ester group of the taxogen.

The telomer molecules formed during telomerization have different molecular weights; a mixture is formed and not a chemical individuum. The distribution of molecular weights depends on the character of taxogen, telogen and initiator, on the presence of a catalyst, mutual ratio of reactants, temperature, and many other less important factors.

The main drawback of telomerization as a synthetic method is the necessity for subsequent separation of the prepared mixture of telomer molecules. If products of telomerization may be used as a mixture of telomers as, e.g., in the case of fluorocarbon oils, which are prepared by telomerization of trifluorchlorethylene with CCl_4, the immediate industrial application is not surprising. To increase the resistance of oils against the effect of thermal heating and of corrosive compounds, the end groups of telomers are fluorinated.

Low molecular perfluorocarbons with one reactive end group can be prepared by telomerization of tetrafluoroethylene in the presence of trifluoroiodomethane.[39]

The complications in preparation of α, ω-difunctional compounds explain why from so many theoretical studies only a few found application in practice. Apart from these difficulties, however, one may expect that from the large number of procedures of telomer preparation attempted to date, some of them will surely be useful. The increased demand for telomers used as plasticizers, surfactants, macromonomers, specialty oils, and waxes and the steady progress in separation techniques as well as in the knowledge of selective methods of telomerization are strong indications of such an idea. Moreover, since many telomers have unstable substituent groups on both ends, they can be considered as the starting chemicals for further reactions.

IV. HIGH TEMPERATURE REACTIONS OF HYDROCARBONS

The research of the pyrolysis of paraffins played an important part in the recognition of kinetic laws of gas phase reactions. Its practical consequence, the increased yield of gasoline obtained at thermal decomposition (cracking) of hydrocarbons from the crude oil as well as the production of ethylene and other fundamental chemicals of petrochemistry, motivated the considerable development of this field of chemistry. The term cracking, of course, does not cover all important reactions taking place at the pyrolysis of hydrocarbons. Besides the low molecular hydrocarbons, branched alkanes and polynuclear aromatic compounds are simultaneously formed. The process is accompanied by a gradual dehydrogenation of hydrocarbons giving elementary carbon as a soot or coke. The latter products are not formed only at the pyrolysis of crude oil or coal, but at the decomposition of low molecular hydrocarbons, too. This indicates that polymerization reactions may also occur at pyrolysis.[40]

Noncatalyzed cracking requires a temperature from 500 to 600°C; the catalytic process is performed in the presence of heterogeneous acid catalysts ($SiO_2 - Al_2O_3$, zeolites) at about 450°C, both processes being endothermic. Exothermic hydrogenation cracking (hydrocracking) which proceeds in the presence of hydrogen at the pressure 10 to 20 MPa and at temperatures from 280 to 400°C, uses nickel, tungsten, and other catalysts. Since no carbon residue is formed at the hydrocracking, it is, therefore, suitable for the production of gasoline from higher crude oil fractions. Cracking starts by the carbon-carbon bond scission. For longer molecules, the probability of the scission is higher inside the chain than at its end. Taking into account the relatively low temperature of the catalyzed process, the fragmentation of alkyl radicals to methane and ethylene is usually slower. The presence of hydrogen in the system suppresses the dehydrogenation of hydrocarbons as well as the formation of aromates, alkenes, and carbon residues. Acid sites of the catalysts support the formation of cations and cation radicals which undergo isomerization to a much higher extent than neutral radicals.

In spite of the fact that pyrolysis of coking coal has been used in practice for a considerably longer time than hydrocarbon cracking, its chemistry is less understood. The general pattern of coal pyrolysis seems to be quite obvious, but the quantitative description of elementary reactions and their mutual interrelations are far from being clarified.[41] One may, however, expect that the continuous effort to use it industrially will lead to the investigation of respective chemical procedures in a deeper way. The pyrolysis of bituminous coal gives coke and coking gas composed of hydrogen and methane, predominantly.

When compared with pyrolysis of crude oil, due to the considerably lower content of alkanes in the parent material, the gaseous products of coal pyrolysis contain markedly less ethylene. (The coal is preponderantly composed of polynuclear methylaromatic hydrocarbons linked mutually by short aliphatic chains.)

A. Pyrolysis of Simple Alkanes, Alkenes, and Aromates

Methane as the first member of the homological series of hydrocarbons is extraordinarily thermally stable. The relatively simple molecule, its symmetry, and low number of alternatives of radical reactions, cause its decomposition to start at temperatures higher than 1000°C. Provided that methyl radicals are generated from methane below 1000°C, the transfer reaction to methane regenerates methane back. A relay-like course of reactions may be observed with tetradeuteromethane additions.[42] In the first stage of the reaction, monodeuteromethane and trideuteromethane are formed preponderantly.

Ethane is decomposed at 480°C to ethylene and hydrogen. The higher alkanes are cracked to the mixture of hydrocarbons with a lower number of carbons and partially dehydrogenate. The higher the number of carbon atoms in the hydrocarbon molecule, the more likely is cleavage of the carbon-carbon bond.

The mechanism of the pyrolysis of alkanes may be demonstrated on propane.[43] Since the dissociation energy of the C-C bond is in the region 325 to 350 kJ/mol while that for the C-H bond is between 380 to 430 kJ/mol, for all alkanes except methane, the primary cleavage occurs at the C-C bond. In the case of propane, there is thus formed methyl and ethyl radicals. Methyl radicals can abstract hydrogen from propane and give the stable product of pyrolysis, methane. For ethyl radicals, the transfer to propane ($k = 4 \cdot 10^7 \exp(-40,000/RT)$ $dm^3 mol^{-1} sec^{-1}$) competes with their fragmentation ($k = 8 \; 10^{12} \exp(-159 \; 000/RT)$ sec^{-1}) (Equation 73). At temperatures higher than 1000°C and

$$\cdot CH_2CH_3 \rightarrow CH_2=CH_2 + H \cdot \qquad (73)$$

pressure of propane below 0.1 MPa, fragmentation prevails. Hydrogen atoms react with propane or with other hydrocarbons very fast ($k = 10^{11} \exp(-30 \; 000/RT)$ $dm^3 mol^{-1}$ sec^{-1}) and give propyl radicals and molecular hydrogen. Recombination of hydrogen atoms is less probable. The course of the chain reaction continues with β-fragmentation of propyl radicals (Equations 74 to 76) which are regenerated by transfer

$$\cdot CH_2CH_2CH_3 \rightarrow CH_2=CH_2 + \cdot CH_3 \qquad (74)$$
$$\longrightarrow CH_2=CHCH_3 + H \cdot \qquad (75)$$

$$CH_3\dot{C}HCH_3 \rightarrow CH_3CH=CH_2 + H \cdot \qquad (76)$$

reactions of hydrogen atoms and methyl radicals to propane.

The rate of chain reaction of alkane decomposition is slowed down by the presence of alkenes either supplied to the reaction system or formed directly in the reaction. Their retardation effect consists in the transfer reaction of reactive radicals to alkene and in the formation of less reactive allyl radicals. The retardation of cracking by inhibitors of free radical rections, such as, e.g., NO is one of the indications of a chain mechanism of pyrolysis of alkanes. Another convincing argument in favor of the chain mechanism is the effect of initiators of free radical reaction on the pyrolysis course. In their presence, alkanes are decomposed at considerably lower temperatures (by 150 to 200°C). Initiators generate free radicals much more easily than alkane itself and free radicals attack the substrate and facilitate its decomposition. Organometallic compounds, peroxides, ethyleneoxide, etc. may be used as cracking initiators. Also ionization radiation has an initiating effect; regarding zero activation energy of initiation reaction, the lowest activation energy of pyrolysis from all the above ways of initiation may be thus achieved. Initiated pyrolysis differs from a purely thermal one by the composition of formed products. The lower temperature of the former brings about the lower rate of formation of branched alkanes and alkenes.

In the course of pyrolysis, besides dehydrogenation and C-C bond scission, also isomerization, addition and dehydrocyclization reactions proceed. This is the reason why branched and aromatic hydrocarbons appear in these reaction products simultaneously with products of lower molecular weight and with dehydrogenated hydrocarbons. Since isomerization reactions were dealt with in the respective part of elementary reactions of radicals, we concentrate here mainly on the formation and disappearance of aromatic cycles as well as on acetylene formation.

The formation of aromatic compounds from alkene may be demonstrated on pyrolysis of 1,3-butadiene.[44] Below 700 K, the main product is 4-vinylcyclohexene, which is formed in cycloaddition reaction (Equation 77). At higher temperatures,

$$2 \; CH_2=CH-CH=CH_2 \rightarrow \begin{array}{c} CH=CH \\ / \quad \backslash \\ CH_2 \qquad CH_2 \\ \backslash \quad / \\ CH_2-CH-CH=CH_2 \end{array} \qquad (77)$$

butadiene is split out to vinyl radicals which add to butadiene and allylakyl radicals are formed (Equation 78). By their

$$\cdot C_2H_3 + C_4H_6 \rightarrow CH_2{=}CHCH_2CH{\cdots}CH{\cdots}CH_2 \tag{78}$$

cyclization and subsequent dehydrogenation we obtain cyclohexadiene. The further dehydrogenation of this cyclic diene leading to the more stable product, benzene, may proceed in several reaction pathways. Transfer reactions of vinyl radicals with butadiene produce ethylene and butadienyl radicals. The further step of the propagation reaction consists in fragmentation of butadienyl radicals (Equation 79) where vinyl radicals and

$$\cdot C_4H_5 \rightarrow \cdot C_2H_3 + C_2H_2 \tag{79}$$

acetylene are formed and the above steps of propagation reactions may repeat again. Acetylene is the precursor of the soot.

Although it is not always so, the thermolysis of alkyl-aromatic compounds usually starts also by C-C bond cleavage. The exception is the toluene pyrolysis where benzyl radicals are formed by homolysis of C-H bond, which is the main initiation reaction. On the other hand, in the case of ethyl benzene, benzyl radicals are formed, too, but the reaction in which their formation proceeds is the cleavage of C-C bond.[45] Above 1000 K, the benzyl radical fragmentates to a considerable extent to acetylene and propargyl radicals. This reaction (Equation 80), which

$$C_6H_5CH_2\cdot \rightarrow \cdot C_3H_3 + 2\ C_2H_2 \tag{80}$$

has an activation energy of about 350 kJ/mol, is an example of an opening of the aromatic ring in a radical chain reaction. Fragmentation of phenyl radicals to acetylene and corresponding four carbon radicals has similar activation energy. After dehydrogenation, radicals give hydrocarbons with low content of hydrogen which may polymerize to soot.

In the case of toluene, hydrogen atoms regenerate benzyl radicals and chain pyrolysis may go on. Dimerization of propargyl radicals and the subsequent cyclization of the formed dimer yields benzene. This reaction step, however, decreases the kinetic length of the chain reaction.

In such a way we may enumerate all possible alternatives of reaction steps but not put forward the general description of pyrolysis. This is due to the strong dependence of the reaction course on given experimental conditions, such as temperature, pressure, the rate of heating, the manner of reaction product removal, etc. The recognition of the effect of the above factors on individual reactions is a prerequisite that the conditions may be found where pyrolysis can become a relatively selective method of synthesis of new compounds.

B. Pyrolytical Production of Chemicals

The thermal decompositon of hydrocarbons oriented to the synthesis of ethylene belongs to the fundamental procedures of petrochemistry. Ethane and primary gasoline, the mixture of alkanes having from 4 to 12 carbon atoms, are the starting raw materials. At present, also higher oil fractions and hydrocracking products are pyrolysed to ethylene, too.

Ethylene is received by passage of respective hydrocarbons through a pipeline heated to 750 to 900°C. The residence time of hydrocarbon at a given temperature is 0.1 to 0.5 sec. To prevent the side reactions of ethylene the reaction products are then quickly cooled down to 350°C. Fragmentation of the primary alkyl radical giving ethylene competes with bimolecular transfer reactions leading to secondary and tertiary alkyl radicals. Since β-fragmentation is a monomolecular reaction, the decrease of the concentration (pressure) of hydrocarbons in the system supports the formation of ethylene when compared with other reaction products. This may be dangerous since due to the low pressure in the reactor, the surrounding air might be sucked up inside and the system will explode. The decrease of hydrocarbon concentration is, therefore, usually achieved by the supply of water vapors to the hydrocarbon mixture which react with carbon (Equation 81) and decrease its deposits on the reactor walls.

$$C + H_2O \rightarrow CO + H_2 \tag{81}$$

The yield of ethylene depends on the composition of starting hydrocarbons. Aromatic hydrocarbons diminish it and support the formation of acetylene and carbon residues. The ratio of ethylene may be increased by hydrogenation of hydrocarbons preceding pyrolysis.

From cyclic alkanes, ethylene and conjugated dienes are formed. An alkenyl radical is formed, e.g., by isomerization of the cyclohexyl radical which gives at first one molecule of ethylene (Equation 82) and the formed butenyl radical is

$$CH_2=CH(CH_2)_3\dot{C}H_2 \rightarrow CH_2=CHCH_2\dot{C}H_2 + CH_2=CH_2 \tag{82}$$

cleaved to 1,3-butadiene and hydrogen atoms (Equation 83).

$$CH_2=CHCH_2\dot{C}H_2 \rightarrow CH_2=CHCH=CH_2 + H\cdot \tag{83}$$

The reactions explain why at the pyrolysis of cycloalkanes higher amounts of dienes are formed than in the case of alkanes.

The mutual interrelations of reactions of radicals of different types at the simultaneous pyrolysis of several hydrocarbons in the mixture seems to be quite evident. The actual mechanism is not, however, very obvious. From the viewpoint of the synthetic strategy, interest is focused particularly on the coupling of several chain reactions which should decrease the number of reaction products. Even though such an optimistic idea seems to be somewhat exaggerated, we cannot exclude it totally.[46] Experimental results of pyrolysis of toluene alone and of its mixtures with hydrogen are convincing enough to believe that this kind of regulation of chain reactions may be very effective.

At the pyrolysis of toluene (840°C) benzene (45% mol), dibenzyl (20% mol), and styrene and methylstyrene (13% mol) are formed. Provided that the surplus of hydrogen is present in the system, benzene is formed in 98% yield. This result may be explained by the parallel production of phenyl, benzyl, and methyl radicals from toluene. The transfer reaction of benzyl radicals with hydrogen leads to the regeneration of toluene while phenyl radicals give benzene.

There are several such examples. The results of analysis of pyrolytic products confirm that the coupling of the chain reaction proves to be a good principle of their effective utilization and regulation. Until now, research in this field took into account incomplete quantitative data mainly and in some way also chemical intuition since the rate constants of respective reactions necessary for the exact approach were missing.

Appart from the problems concerned with the optimization of the yield of given compounds at pyrolytic chain reactions, the pyrolysis is widely used for the industrial production of chemicals. In addition to ethylene, styrene, propylene, dienes, cyclic hydrocarbons, etc. are also produced in such a way. By the proper choice of conditions of the respective operations, the composition of products may well be adapted. The prepared compounds should then be isolated from the hydrocarbon mixture by some separation technique.

High temperature reactions are used for the synthesis also on laboratory scale.[47] Many practical procedures are known, using isomerization and fragmentation reaction of compounds, isolated in advance, where the synthetic aspect prevails over that of understanding the reaction mechanism.

C. Thermal Destruction of Polymers

Heat treatment of polymers is one of the possible ways to disrupt the backbone of a macromolecule or how to change the number and configuration of atoms in side substituents. The slowest process in the purely thermal destruction of a polymer is the initiation, which requires relatively high energy. The macromolecule is split out either randomly when all its bonds are equivalent or at some weak site.[48]

The defect center in a macromolecule can be an anomalous arrangement of monomer units or end groups. Their existence in the polymer is mostly assumed since their direct determination is limited by the accuracy of available analytical methods.

The most serious argument in favor of the occurrence of defect structures in polymers is the lower thermal stability of macromolecules when compared with their low molecular models. The thermal stability of a polymer may be significantly influenced by the manner of its synthesis or by the additional intervention into its structure. Peroxidic oxygen inserting into the main chain in the polymerization reaction of a monomer in insufficiently inert atmosphere or side hydroperoxidic groups of oxidized polymer are the most frequent sources of the appearance of free radicals at the thermal destruction of polymers. The primary radicals formed in the polymer initiate the subsequent reactions, fragmentation and transfer reactions being the most important of them. Since the various reaction pathways can proceed simultaneously, the reaction products usually represent a very complex mixture. Fragmentation reactions increase the overall number of molecules in the system; the decrease of pressure, therefore, accelerates the destruction. The faster removal of polymerizable compounds decreases the probability of the opposite process. The increase of the rate of polymer transformation may be achieved by the acceleration of the initiation reaction, too.

The convincing indication of the mechanism of destruction follows from the shape of the plot of the molecular weight of a polymer on the degree of its degradation. When the length of all macromolecules decreases rapidly, the degradation of the backbone bonds in macromolecule is a random process. If the degradation occurs within the respective macromolecule with the formation of the parent monomer, the process is called depolymerization.

The important feature of random degradation of polyalkenes is, that after the formation of 1 to 2% of low molecular products, the average length of macromolecules drops to a value several orders lower. Provided that the distribution of polymer differs from the most probable one, polymolecularity changes, too. The low molecular reaction products contain different compounds including a monomer, which depends on the extent of competitive inter- and intramolecular transfer reactions.

The effect of a polyalkene structure on the temperature of the start of thermal degradation may be expressed by the more general statement, namely, that the branching of the polymer chain decreases the thermal stability of a polymer. At the same time, the branching supports depolymerization; under the same conditions of the thermal decomposition of the respective polymer more isobutylene than propylene and more propylene than ethylene is formed.

The thermal stability of the polymer is decreased also by the isolated double bonds in the macromolecule backbone. The strength of C-C bonds linking the structural units of 1,4-polydienes is lower than that in saturated hydrocarbons. The dissociation energy of these bonds is decreased by twofold value of the stabilization energy of allyl radicals which are primarily formed, here. Degradation will occur, therefore, at lower temperature and with higher probability of monomer formation from polydiene macroradical.

The amount of the formed monomer depends on conditions of degradation as well as on the thermal history of the polymer sample. At the pyrolysis of natural rubber, up to 65% of the monomer (isoprene) is formed provided that rubber is warmed up quickly to 800°C and low pressure is kept in the system (about 0.1 kPa), while at a temperature of 400°C, only 2% of the monomer may be found in reaction products. The high rate of heating is necessary just to provide a fast overcrossing of the temperature region 250 to 500°C, where polyisoprene undergoes cyclization to six-membered structural units. The cyclization increases the probability of other fragmentation reactions to the detriment of pure depolymerization. This is due to the disappearance of the respective double bond, which supports the depolymerization by weakening of a cleaved end.

The rate of thermal decomposition and the extent of monomer formation depend also on the structure of the side substituent of the main chain (Scheme 5).

Substituent X	H	CH_3	$COOCH_3$	C_6H_5
$[-CH_2CHX-]_n$	0.03	0.17	0.70	42
$[-CH_2C(CH_3)X-]_n$	0.17	18	91	97

The tendency to depolymerization of different polymers in percents of the formed monomer after the complete decomposition of the polymer.

SCHEME 5

The more stable the macroradical, the higher the amount of monomer formed at the degradation. The effect of the substituent structure on the yield of monomer becomes obvious as the different reactivity of the polymer in a transfer reaction.

The importance of the effect of transfer reactions on the composition of products may be demonstrated on the thermal decomposition of polyethylene and polytetrafluoroethylene. Although polytetrafluoroethylene is decomposed at a higher temperature than polyethylene, tetrafluoroethylene is formed here as an almost exclusive reaction product. The thermal stability of polytetrafluoroethylene follows from the strengthening of the C-C bond in the macromolecule backbone while the high yield of monomer from the strength of the C-F bond. If we replace one fluorine atom by chlorine in a structural unit of polytetrafluoroethylene, which readily enters the transfer reaction, we obtain a polymer which gives only 60% of monomer at its degradation.

The above substitution simultaneously will increase the rate of decomposition of the respective polymer (polytrifluorochloroethylene) by 5 orders at 350°C. This is due to the easier homolysis of the C-Cl bond and, consequently, to the higher rate of initiation at thermal degradation.

Dehydrochlorination of polyvinylchloride and degradation of polyvinylacetate, polyvinylfluoride, polyvinylidenchloride, and poly-*tert*-butylmethacrylate belong to the group of polymer destructions starting by the formation of a double bond in the macromolecule backbone and the elimination of a low molecular compound from the side group of the polymer.

The interest has been focused on the mechanism of thermal dehydrochlorination of polyvinylchloride particularly, since it is one of the polymers produced on a large scale, but least stable thermally. At 120°C, HCl is released from it and on the main chain, polyene structures are formed. The largest polyene sequences formed at the thermal dehydrochlorination of polyvinylchloride include 25 to 30 double bonds, the average length being from 3 to 15. The low thermal stability of polyvinylchloride is the result of a cooperative effect of various factors which increase the rate of initiation.[49,50] Among them, some metal cations, the polymerization initiator residues, strongly basic compounds, double bonds, peroxidic bonds incorporated into the polymer during its synthesis and storage, and some structural anomalies such as tertiary carbon-chlorine bonds, carbonyls in β-position to the C=C bond, are the most important. Propagation of the chain reaction involves several alternatives. It may be, e.g., the chlorine atom attack of the polymer chain (Equation 84), where HCl is formed. The fragmentation reaction

$$Cl\cdot + -CH_2CHCl- \rightarrow -\dot{C}HCHCl- + HCl \qquad (84)$$

of the alkyl macroradical releases a new chlorine atom into the propagation cycle (Equation 85) and a double bond is formed on the

$$-CH_2CHCl\dot{C}HCHCl- \rightarrow -CH_2CH=CHCHCl- + Cl\cdot \qquad (85)$$

polymer backbone. The double bond facilitates the subsequent elimination of neighboring chlorine atoms and allyl radicals are thus formed either in initiation or in propagation reaction. Such a mechanism leads to the accumulation of double bonds in the polymer, accelerating the further course of dehydrochlorination and, consequently, the polyene structures become the precursor of the benzene formation.

The radical mechanism of thermal decomposition of polyvinylchloride is not commonly accepted since it is not able to explain all observed facts, such as, e.g., the catalytic effect of HCl molecules on the rate of further reaction, etc. Another alternative is the molecular reaction (Equation 86)

$$
\begin{array}{c}
\overset{\displaystyle Cl\;\;-H}{\vdots\quad\vdots} \\
H\quad Cl \quad \rightarrow -CH=CH- + 2\ HCl \\
|\quad | \\
-CH-CH-
\end{array}
\qquad (86)
$$

involving the cyclic activated complex.

The thermolysis of cellulose is an example of combination of the reaction pathway composed of depolymerization and side groups elimination. It starts at 180°C. Up to 280°C, however, only water is eliminated; dehydrocellulose and carbon residue are formed.[51] At 450 to 500°C, cellulose is decomposed to 1,6-anhydro-β-D-glucopyranose (levoglucosan) and hydroxyacetaldehyde which are the main reaction products in the mixture of many other products of relatively less yield.[52]

The initial thermal decomposition of cellulose proceeds by disruption of the cellulose polymer chain at the glycosidic bond between glucose monomeric units (Equation 87).

$$(87)$$

The direct depolymerization of macroradicals such as is known at vinyl and vinylidene polymers is not possible since β-fragmentation leads to the cleavage of the cycle. In the case of primary alkoxy radicals, it is reaction 88 and secondary alkyl

$$\text{(88)}$$

radicals eliminate hydroxyacetaldehyde (Equation 89).

$$+ \ HOCH_2CH=O \qquad \text{(89)}$$

The formation of levoglucosan may be explained by the sequence of 1,5-isomerization of alkoxy macroradical (Equation 90) and of

$$\text{(90)}$$

intramolecular cyclization accompanied by fragmentation which regenerate by one structural unit shorter original alkoxy radical (Equation 91).

$$+ \ \cdot OR \qquad \text{(91)}$$

An analogical scheme may explain the formation of various other products. Their experimental proof, however, requires more independent experiments. It is also not excluded that hydrolytic reactions of released water play some role here.

As the method of the backward synthesis of monomers, pyrolysis of polymers has not found large practical utilization. Although such an idea seemed to be attractive, it was applied only to polymethyl methacrylate and polytetrafluorethylene where practically pure monomer is formed. In the case of other polymers besides monomer, there arises usually the complex mixture of sometimes unwanted products.

The interest devoted to the thermal destruction of polymers was stimulated mainly by the necessity to recognize the general relationships between properties and the structure of polymers oriented particularly to the synthesis of thermally resistant polymers. For the proper choice of structural units of such a polymer, the knowledge of the thermal stability of low molecular compounds is used as well as the principle of strengthening the backbone by the synthesis of ladder-like polymers. Its meaning consists in the fact that for disruption of the macromolecule, at least two bonds of one ladder unit should be broken down in parallel. In the case of aromatic polymers this is less probable than the random disruption of two bonds in different cycles.

V. HALOGENATION OF HYDROCARBONS

The propagation in radical halogenation of hydrocarbons by molecular halogen involves two types of transfer reactions (Equations 92 and 93), namely, that of halogen atom X with hydrocarbon RH

$$X \cdot + RH \rightarrow R \cdot + HX \tag{92}$$

and of formed alkyl radical R· with the molecule of halogen

$$R \cdot + X_2 \rightarrow X \cdot + RX \tag{93}$$

where halogenated hydrocarbon RX and hydrogen halide are formed.

The propagation steps may be more numerous when another halogenation reagent is used instead of molecular halogen or if some of the formed radicals undergo isomerization or fragmentation. The selectivity of the halogenation reaction depends on the nature of the radical implementing transfer to hydrocarbon.

The facility of halogenation of saturated hydrocarbons may well be deduced from the approximate data of calculated reaction heat ΔH of the respective propagation reactions of halogen atoms X with hydrocarbon and alkyl radicals with halogen molecules X_2.

Halogen:	F·	Cl·	Br·	I·	F_2	Cl_2	Br_2	I_2
ΔH, kJ/mol:	−150	−20	+50	+110	−280	+5	−95	−75

Fluorination is so exothermic that it occurs as combustion. The controlled course of fluorination of organic compounds requires the considerable dilution of reactants. On the other hand, iodination occurs only with difficulties and, moreover, the alkyl iodide once formed is easily reduced by HI back to hydrocarbon and molecular iodine (Equation 94)

$$RI + HI \rightarrow RH + I_2 \tag{94}$$

At the halogenation of unsaturated hydrocarbons, there proceed two main propagation reactions, one being the addition (Equation 95) and the second the transfer. The primary reaction

$$X \cdot + {>}C{=}C{<} \rightarrow X{-}\overset{|}{\underset{|}{C}}{-}\overset{|}{\underset{|}{C}} \cdot \tag{95}$$

product is then a dihalogenide of the organic compound. The addition of the fluorine atom is the most exothermic (160 kJ/mol), additions of chlorine and bromine (−60 and −10 kJ/mol) are less exothermic, while that of iodine atom is endothermic (+30 kJ/mol). Since the reaction heat of the transfer reaction of alkyl radicals to the molecule of halogen will be approximately the same for both, saturated and unsaturated hydrocarbons, the halogenation of the latter will be usually faster because of the higher rate of the addition reaction when compared with the H-atom abstraction from hydrocarbon.

Such a generalization is valid only for unsaturated hydrocarbons with isolated noncon-jugated double bonds. In the case of aromatic hydrocarbons, the radical addition of halogens is more complicated and may be even slower than the transfer because of the necessity to disrupt the energetically favorable conjugation of double bonds.

Halogenation of hydrocarbons can be initiated either thermally or by light-induced dis-sociation of halogen molecules. Also additives generating free radicals in the system may be used. Of particular importance is the interaction of halogen molecules with hydrocarbons.

A. Fluorination

The abstraction of hydrogen by atomic fluorine as well as the scission of molecular fluorine by alkyl radicals is an enormously fast process. Its activation energy is approximately zero and frequency factor has a value of about $5 \cdot 10^9$ mol dm^{-3} sec^{-1}. The kinetic length of the chain fluorination attains a value above 10^7 and may be even increased by an additional branching.

Although the branching in fluorination of methane derivatives has been supposed long ago, all conclusions in favor of it have been drawn from indirect evidence such as the existence of limiting phenomena, ignition delays, and analysis of final products. The possible branching species was experimentally found at the fluorination of difluoromethane.[53] In this reaction (Equation 96), the excited fluoroform is formed which is

$$\cdot CHF_2 + F_2 \rightarrow CHF_3^* + F \cdot \tag{96}$$

decomposed to a branching intermediate (Equation 97) and in the

$$CHF_3^* \rightarrow |CF_2 + HF \tag{97}$$

subsequent step (Equation 98) it increases the number of free radicals in the system.

$$F_2 + |CHF_2 \rightarrow \cdot CF_3 + F \cdot \tag{98}$$

The internal cause of a vigorous course of fluorination consists in the high strength of forming H-F and C-F bonds, in the low dissociation energy of the F-F bond, as well as in the high kinetic length of the chain reaction. Fluorination can thus hardly be controlled unless the reactants are not sufficiently diluted by inert gases.

Initiation of fluorination consists in the bimolecular interaction of molecular fluorine and hydrocabon RH (Equation 99).

$$F_2 + RH \rightarrow R \cdot + F \cdot + HF \quad (\Delta H = 20 \text{ kJ/mol}) \tag{99}$$

This formation of radicals is more probable than the direct dissociation of fluorine molecules ($\Delta H = 155$ kJ/mol), which is more endothermic. The above bimolecular initiation is preferred only in the case of fluorination. Provided that alkene is present in the system, the bimolecular initiation is even more facile. The reaction of fluorine with ethylene giving fluorine atoms and fluoroethyl radicals is already slightly exothermic ($\Delta H = -12$ kJ/mol).

The regulated course of fluorination of hydrocarbons may be seen at the surface reaction of polyethylene, polyvinyl fluoride, and polyvinylidenefluoride with diluted fluorine.[54] Hy-drogen of the polymer chain is substituted by fluorine even at a pressure 200 Pa (1.5 torr) of the helium (95%) and fluroine (5%) mixture. The fluorination may be accelerated by the plasma discharge which generates the polymer radicals and fluorine atoms. Even though the photodissociation of fluorine molecules is not a primary factor in these UV reactions, this does not imply that atomic fluorine is insignificant in the plasma environment.

B. Chlorination

1. General

The chlorination of hydrocarbons belongs to the fast and at the same time easily controllable reactions which led to the accumulation of numerous experimental data.[55,56] Substitution chlorination of the first several aliphatic hydrocarbons requires a temperature of about 200°C when we are working with molecular chlorine and in the dark. At this temperature, a small part of the chlorine molecules dissociate to atoms which initiate the chain reaction with a kinetic length of chain about 10^7. The chlorination temperature may be considerably lowered when free radical initiators are used or the system is irradiated with γ-rays or ultraviolet light of wavelength 487 nm. Alkenes, which with molecular chlorine give free radicals in a reaction (Equation 100) which reminds one of

$$R-CH=CH_2 + Cl_2 \rightarrow R-CH(Cl)\dot{C}H_2 + Cl\cdot \tag{100}$$

backward disproportionation, may serve as an example of initiators of alkane chlorination. Oxygen retards chlorination significantly because of the fast transformation of alkyl radicals to less reactive peroxyl radicals.

The activation energy of the transfer reaction of chlorine atoms to hydrocarbon substrates is in the interval from 1 to 15 kJ/mol. It depends on the dissociation energy of the attacked C-H bond, increasing with the increasing strength of cleaved bond. The frequency factor of this reaction is of the order of 10^{10} dm^3 mol^{-1} sec^{-1}. Also, the reaction of alkyl radicals with molecular chlorine, which gives a chlorinated product and chlorine atoms, is particularly fast. Its activation energy lies in the range of 2 to 20 kJ/mol, increasing with the degree of chlorine atom substitution in the respective alkyl radical,[57] frequency factor being about 10^8 dm^3 mol^{-1} sec^{-1}. High rates of both propagation steps are also the reason of the high kinetic length of the chain reaction.

From the Arrhenius parameters of the above two propagation reaction it may be deduced that the rate constant of hydrogen abstraction by the chlorine atom is higher than that for the alkyl radical and chlorine molecule. In the stationary state, the rate of both reactions will be equal (Equation 101).

$$k_a[Cl\cdot][C-H] = k_b[R\cdot][Cl_2] \tag{101}$$

Since $k_a > k_b$ and the concentration of C-H bonds, the potential sites of hydrogen abstraction is higher than the concentration of chlorine molecules; the amount of alkyl radicals in the system is higher than that of chlorine atoms.

The increase of the activation energy for halogen alkyl radicals with chlorine molecules corresponds to the deceleration of the propagation reaction. At the same time, the rate of reaction of the chlorine atom with the molecule of partially halogenated hydrocarbon will be lower. The decrease of both rate constants will slow the chlorination down.

The effect of the vicinal substituents on the reactivity of the functional group may be expressed quantitatively.[58] Assuming that equivalent structural units A in the polyfunctional molecule $(A)_n$ are transformed to B units, the overall rate may be described in terms of three rate constants. It is the rate constant k_o for A→B transformation in the sequence AAA, k_1 for triads AAB or BAA, and finally the rate constant k_2 for the triad BAB. For photochemically initiated chlorination of polyethylene in chlorbenzene at 50°C, the best agreement between the experimental and theoretical curve has been found for the ratio of rate constants $k_o:k_1:k_2 = 1:0.35:0.08$. A similar ratio was obtained also for reactions of long linear paraffins and cycloparaffins. As was shown in the chlorination of α-deuterated polyvinyl chloride, where the deuterium content in the polymer remained constant with an increasing amount of bound chlorine only one chlorine atom substitutes the methylene hydrogen in the first reaction stage.[59] The polar effect appearing in the propagation steps of the chain chlorination is obvious also from the fact that at 90% mol of reacted CH_2 groups in polyvinyl chloride, only 2.6% mol of 1,1-dichlormethylene units are formed.

The above difference in the rate constants of the propagation reaction of chlorine atom attack on hydrocarbon displays the selectivity of the chlorination. As was already pointed out, chlorine atoms are very reactive and rather unselective radicals. A certain selectivity is brought about by their electrophile character which prefers such reaction sites which have relatively high electron density. In saturated hydrocarbons, however, almost all C-H bonds are substituted by chlorine by a rate not differing more than by 1 order of magnitude, except for methane and cyclopropane which are chlorinated considerably more slowly. Also, hydrogens in aromatic compounds, in ethylene, and acetylene are less reactive. In alkyl aromatic hydrocarbons, hydrogens of alkyl groups are substituted preferably. The phenyl group increases slightly the reactivity of neighboring C-H bonds of a respective alkyl, and simultaneously the selectivity of the reaction on a given C-H bond increases. This is supported by the formation of a complex between the chlorine atom and phenyl group of the alkyl aromatic hydrocarbon. The regulated course of chlorination to a certain halogenated product requires the increased selectivity of the transfer reaction to hydrocarbon. Numerous studies, however, have revealed that neither temperature nor the aggregation state (gaseous or liquid) of the chlorination had significant effect on the selectivity of hydrogen atom abstraction. At the photoinitiated reaction the selectivity of the chlorination may be thus increased only by the suitable chlorination agent or by the solvent used.

At the chlorination of alkenes, ionic reaction competes with a radical addition particularly in such a case as when carbons of double bonds are substituted by three or four alkyls.[60] Radical addition is a dominating process for less substituted alkenes and in the absence of oxygen. The rate of radical chlorination of alkenes depends only slightly on the kind of their substituents. Also, between the reactivity of the double or triple carbon-carbon bond of the substrate, there is no large difference. The chlorination of aromatic compounds occurs more slowly.

Low selectivity of the chlorination reaction appears not only at the abstraction of hydrogen from hydrocarbons and at the addition to multiple bonds, but also at the mutual competition of addition and transfer reactions. The control of the reaction course in the direction of either addition or substitution may be achieved by the temperature change. This may be demonstrated on the course of chlorination of propylene in the gaseous phase where below 200°C a product of addition 1,2-dichlorpropane is formed, while above 400°C, allylchloride is formed by a substitution reaction.

2. Chlorination Agents

The solvents capable of forming complexes with chlorine atoms increase the selectivity of both the substitution and addition chlorination. The complex between chlorine atom and the solvent which has a lifetime of microseconds is formed by the partial charge transfer to chlorine atoms. The increased concentration of the complexing solvent shifts the equilibrium between chlorine atoms and solvent molecules to the right and the selectivity of the chlorination increases. The ratio of the reactivity of tertiary and primary hydrogen at the photochlorination of 2,3-dimethyl butane at 250°C increases thus tenfold in benzene and 50-fold in CS_2, the initial value being 4. The relative reactivity of primary and secondary hydrogens at the photochlorination of heptane in the complexing solvents does not increase so markedly. In benzene solution, the ratio of secondary/primary hydrogen reactivity increases from 3 to 9, while in CS_2 to 30. The high selectivity may be, however, achieved by infinite dilution. The increased selectivity of chlorination may be due to the heat liberated at the complexation of the chlorine atom by a solvent which decreases the exothermicity of the chlorine atom transfer. The reaction is then slowed down and becomes more sensitive to the structure of chlorinated hydrocarbon.

In addition to chlorine, either free or bound into a relatively unstable complex with a polarizable solvent, there exist other compounds which may be used for the radical chlorination of hydrocarbons. On the laboratory scale it is, e.g., sulfuryl chloride (b.p. 69°C) which needs a simultaneous application of peroxidic initiator or photoinitiator. The only ultraviolet light which is weakly absorbed by sulfuryl chloride is not very suitable.

The chlorination of hydrocarbons by sulfuryl chloride may be depicted by the summary Equation 102

$$RH + SO_2Cl_2 \rightarrow RCl + HCl + SO_2 \tag{102}$$

which shows that all side reaction products are gaseous under ambient conditions; their gradual liberation deprives the system of oxygen and supresses thus its inhibition effect. The reaction is somewhat more selective than chlorination by molecular chlorine. Its propagation stage involves five elementary steps (Equations 103 to 106). The selectivity of the reaction is

$$R \cdot + SO_2Cl_2 \rightarrow RCl + \cdot SO_2Cl \tag{103}$$

$$RH + \cdot SO_2Cl \rightarrow R \cdot + HCl + SO_2 \tag{104}$$

$$\cdot SO_2Cl \leftrightarrow SO_2 + Cl \cdot \tag{105}$$

$$RH + Cl \cdot \rightarrow R \cdot + HCl \tag{106}$$

increased by the transfer reaction of sulfuryl radicals (Equation 104). Chlorination of alkenes by sulfuryl chloride is accompanied by the formation of β-chlorsulfones appearing as the result of chlorosulfuryl radical addition to a double bond.

Another chlorination agent which is stable in the dark is *tert*-butyl hypochlorite. Its decomposition is initiated by ultraviolet light. The propagation cycle involves the abstraction of hydrogen from hydrocarbon by *tert*-butoxy radical (Equation 107) and the reaction of alkyl radicals with hypochlorite

$$(CH_3)_3CO \cdot + RH \rightarrow R \cdot + (CH_3)_3COH \tag{107}$$

(Equation 108). It is of interest that the chlorine atom does not

$$R \cdot + (CH_3)_3COCl \rightarrow RCl + (CH_3)_3CO \cdot \tag{108}$$

participate in the propagation stage as an attacking radical. The abstraction of hydrogen by *tert*-butoxy radical is the slower process having about 25 kJ/mol higher activation energy than the similar reaction of the chlorine atom. The transfer reaction of alkyl radicals to hypochlorite as compared with molecular chlorine is thermochemically more favorable since the dissociation energy of the O-Cl bond is by 40 to 60 kJ/mol lower than that for the Cl-Cl bond. Both propagation reactions are sufficiently fast, and the high kinetic length of chain chlorination for saturated hydrocarbons ($\sim 10^4$) may be achieved.

The lower reactivity of the *tert*-butoxy radical compared with chlorine atoms increases the selectivity of chlorination. Since the propagating *tert*-butoxy radical abstracts alkyl hydrogens before it enters the addition reaction to a double bond, *tert*-butyl hypochlorite is suitable for allyl chlorination of alkenes. The ratio of selectivity for primary, secondary, and tertiary hydrogen in aliphatic hydrocarbons is 1:7:45 at 40°C, which is approximately the same as in the case of chlorination with chlorine complexed by a proper solvent. At this time, however, the process does not require the large dilution.

The relatively high selectivity is performed by trichloromethyl sulfuryl chloride. The ratio of reactivity of secondary and primary hydrogens is from 20 to 30 while that for tertiary and primary even 110 to 130. The selectivity of the reaction is given by the participation of trichloromethyl radicals in the propagation transfer reaction (Equation 109). The reaction of

$$\cdot CCl_3 + RH \rightarrow R\cdot + CHCl_3 \tag{109}$$

alkyl radicals with trichloromethyl sulfuryl chloride (Equation 110)

$$R\cdot + CCl_3SO_2Cl \rightarrow RCl + CCl_3SO_2\cdot \tag{110}$$

is slow which decreases the kinetic chain length. Cumyl radicals, e.g., react so slowly that chlorination does not have the character of a chain reaction. Even much more reactive cyclohexyl radicals do not react sufficiently fast when compared to termination reactions, and the kinetic chain length is only about 7. This requires a large amount of the initiator and the reaction products are contaminated by the residual products of its decomposition.

Trichloromethyl radicals appear in the reaction system as products of trichloromethyl sulfuryl radical fragmentation. This reaction competes with the transfer reaction to hydrocarbon and, depending on reaction conditions, sets the degree of selectivity of chlorination.

For chlorination, the reaction between halocarbons (particularly CCl_4) and hydrocarbon RH in the presence of peroxides or a range of low valent metal complexes M can also be used. Such chlorination proceeds by a chain route (Equations 111 to 113) in which the metal complex (particularly $Re_2 (CO)_{10}$), acts solely as an initiator.[61]

$$M + CCl_4 \rightarrow MCl + CCl_3 \tag{111}$$

$$\cdot CCl_3 + RH \rightarrow CHCl_3 + R\cdot \tag{112}$$

$$R\cdot + CCl_4 \rightarrow RCl + \cdot CCl_3 \tag{113}$$

From the other examined chlorination compounds, iodobenzenedichloride, which forms when mixing iodobenzene and chlorine, is worth noticing. At photoinitiation, alkanes are chlorinated with this agent much more seletively (secondary/primary = 20, tertiary/primary = 350). Iodobenzenedichloride chlorinates alkenes to a high yield of dichlorides which may have a stereochemical configuration different from products of photochlorination by molecular chlorine.

N-Chloroalkylamines are particularly important chlorinating compounds.[62-64] Chlorination by N-chloroalkyl amines is performed in diluted sulfuric acid where both the substrate and N-chloroalkylamine are protonated. In the propagation stage of chlorination, the key role is played by protonated amminium radical $R_2NH^{+}\cdot$ or less reactive ammonium radical $R_3N^{+}\cdot$ which abstract hydrogen atoms from hydrocarbons (Equation 114). Alkyl radicals then react with N-chloroalkylamine

$$R_2NH^{+}\cdot + RH \rightarrow R_2\overset{+}{N}H_2 + R\cdot \tag{114}$$

and nitrogen cation radicals are regenerated (Equation 115). The

$$R\cdot + R_2NClH^{+} \rightarrow RCl + R_2NH^{+}\cdot \tag{115}$$

chain chlorination has a relatively high kinetic chain length, but lower than in reaction with molecular chlorine.

For initiation of the chain reaction, photolytic decomposition of N-chloroalkyl amines is used which gives a chlorine atom and N-cation radicals. At the redox initiation by complexes of transition metal ions, the selectivity of chlorination depends on the type of the transition metal ion as well as on the purity of the reaction system. The results indicate that for a respective mode of catalyzed decomposition of N-chloroalkylamine, chlorine atoms are formed predominantly as reaction intermediates.

N-Cation radicals derived from N-alkylamines are about 5 times more selective in the reaction with secondary hydrogens than chlorine atoms. Selectivity depends very much on the sizes of alkyl substituents in N-chloroalkylamine. Provided that the bulkiness of substituents close to the reaction site is large, as it is, e.g., in di-*tert*-butyl N-chloramine, then primary hydrogen of isopentane becomes more reactive than the tertiary one in the transfer reaction of cation radicals. This type of selectivity of N-chloroalkylamines reminds one of the selectivity of the enzymatic catalysts.

The strongly polar medium of sulfuric acid plays a certain role, too, which is not, of course, isolated from other factors. Provided that sulfuric acid is used as the medium and chlorine is the chlorinating agent, the resulting effect of medium polarity on selectivity may be neglected. It becomes much more evident when ammonium or amminium cation radicals are chain carriers. In the more polar solvents, the polar effect is then more pronounced, presumably since the charge separation inherent in the transition state is more readily supported in the acidic solvent.

C. Bromination
1. General

Bromine atoms are less reactive than chlorine atoms. Radical bromination thus requires a higher temperature and depends to a larger extent on the hydrocarbon structure. The relatively high selectivity of bromination found considerable use in organic synthesis and relatively much knowledge is available about these halogenation reactions.[65] With regard to the lower reactivity of bromine atoms, the complexing solvents influence the selectivity of radical bromination less than in the case of chlorination. Besides substitution and addition reactions of bromine, the bromination may proceed also as a particular type of chain radical reaction of HBr addition to unsaturated hydrocarbons.

The substitution bromination starts by abstraction of hydrogen by bromine atoms, which is an endothermic reaction in most cases having a relatively high activation energy (30 to 80 kJ/mol) and a frequency factor $\sim 2 \cdot 10^{10}$ dm^3 mol^{-1} sec^{-1}. The relatively steep dependence of the activation energy of the transfer reaction of the bromine atom on the dissociation energy of a broken bond manifests itself in a large selectivity of the reaction site on hydrocarbon. While the selectivity of chlorine atom transfer to primary and tertiary hydrogen lies in the extent of one order only, that of the bromine atom differs by 4 to 6 orders of magnitude depending on the kind of substituents on the tertiary atom. The high selectivity of hydrogen abstraction from paraffins does not appear only in the sequence of primary, secondary, and *tert*-hydrogen, but we may also distinguish the difference in reactivity for the same kind of hydrogen in different hydrocarbons. At the bromination of heptane, position 2 is preferred for substitution by bromine ahead of position 3, even though the secondary C-H bonds are considered to be energetically equivalent.

The rules concerning the relative reactivity of a certain type of C-H bond have some other exceptions. Apart from the large difference in reactivity of secondary and tertiary hydrogen atoms in paraffins which represents about two orders, in some bicyclic saturated hydrocarbons as, for example, in norbornanes, the secondary hydrogens react first. The low reactivity of tertiary hydrogens in norborane (Equation 116) follows from the sterically

$$\triangle\!\!\!\!\square\ + Br_2\ \longrightarrow\ \text{(structure) Br} + \text{(structure) Br} + \text{(structure) Br}$$

| 75 % | 25 % | 0 % | (116) |

hindered and rigid structure of bicyclic hydrocarbons at tertiary carbon atom. The resulting selectivity could be influenced also by a different restriction of the reaction of respective alkyl radicals with more bulky bromine molecules. This, however, should presume the transfer of hydrogen to tertiary cycloalkyl radicals. The reactivity of secondary exo-hydrogens in norbornane with respect to bromine is approximately the same as for cyclohexane.

In bromination the polar effect is asserted which may be documented on the reactions of fluoro- and chlorobutanes. Electrophile bromine and also chlorine atoms prefer the reaction sites which are more distant from the electronegative substituent. At the bromination of monobromderivatives of hydrocarbons another effect appears determining the site of substitution. Bromination of bromobutanes and bromocycloalkanes leads to a high yield of vicinal dibromides where bromine substituents are at neighboring carbon atoms. The regulating effect of the already bound bromine substituent on the further reaction of vicinal hydrogen may be probably connected with the ability of the first bormine atom to stabilize the forming radical. The weakening of the β-bond is distinct at the bromination of bromalkanes only since this reaction is sensitive enough to changes of the C-H bond strength. The chlorination does not show a similar effect since the higher reactivity of chlorine atoms and polar effect prevail, here. The stabilization effect of the bromine substituent on the β-halogen alkyl radical follows from larger spin polarizability of bromine when compared to chlorine and consequently from possibility of interactions with unpaired β-electron. The specific course of bromination may be explained by the formation of intermediate bridged halogen-alkyl radicals. The bonding in symmetrically bridged radicals arises from the three-electron interaction between the singly occupied molecular orbital of the radical and the alkene π-orbital. It is of interest that the theoretical approach favors the classical open structure while the bridged structure is reasonable only in the case of protonated β-halogenethyl radicals.[6]

The second propagation reaction in bromination, the transfer of alkyl radical to molecular bromine, is a very fast exothermic reaction. This may be seen on the bromination of optically active 1-brom (or 1-chlor)-2-methyl butane which leads to optically active 1,2-dibromo (or 1-chloro-2-bromo) 2-methyl butane. The optical activity of the reaction product starts to decrease when concentration of bromine in the reaction system decreases below 0.05 mol dm^{-3}. The exchange of hydrogen by chlorine on optically active carbon is a slow process which is not able to preserve the configuration of atoms in nonattacked molecule moiety. It, therefore, results in the optically nonactive mixture of dichloro enantiomers. The transiently formed radical has time enough to manifold inversion of pyramidal configuration and an equal amount of both isomers of chlorinated hydrocarbon may thus arise. The above results indicate unequivocally that alkyl radicals react with bromine molecules faster than with chlorine. This does not allow settling of the equilibrium (Equation 117), which is the reason for the

$$\underset{C\cdot}{\overset{CH_3}{\underset{|}{C_2H_5 - CH_2Cl}}} \rightleftharpoons \underset{CH_3}{\overset{\dot{C}}{\underset{|}{C_2H_5 - CH_2Cl}}} \tag{117}$$

formation of a racemate at the chlorination of optically active butane.

Faster reaction of alkyl radicals with the molecule of bromine when compared with chlorine may also interpret the apparently controversial results of halogenation of cyclohexane and toluene. As expected, bromine reacts 60 times faster with toluene than with cyclohexane. On the other hand, chlorine reacts 11 times faster with cyclohexane compared with toluene. Even though the direct comparison may be misleading, after the recalculation of reactivity of a respective molecule related to the number of hydrogen atoms prone to abstraction, we get the opposite tendency in the reactivity of both hydrocarbons with halogens, namely.

	Cyclohexane	Toluene
Bromine	1	240
Chlorine	3	1

The overall rate of bromination is thus governed by the rate of transfer reaction of bromine to hydrocarbon while the chlorination is controlled by the reaction of alkyl radicals with halogen. The rate of transfer reaction of bromine atoms determines also the kinetic length of the chain reaction. The more difficult is the course of this transfer, the more bromine atoms dimerize back to molecules and, consequently, the kinetic length of the chain will be lower.

The high reactivity of alkyl radicals to bromine molecules and the slow course of the bromine atoms transfer to hydrocarbons is a reason for the high concentration of bromine atoms in the system. The air oxygen thus interferes with the propagation cycle of bromination only to a small extent.

Besides the above two fundamental reactions, other elementary reactions also come into consideration in bromination of hydrocarbons. Fragmentation of β-bromoalkyl radicals occurs at chlorination of 1-bromobutene where about 3% of 1,2-dichlorobutane is formed. This unexpected product is explained by the elimination of bromine from β-bromalkyl radicals and by the addition of chlorine to a formed alkene. If fragmentation of the β-bromoalkyl radical takes place in bromination, the reaction product corresponds to 1,2-dibromide.

Isomerization of 2-bromoisobutyl alkyl radicals (Equation 118)

$$(CH_3)_2CBrCH_2 \cdot \rightarrow (CH_3)_2\overset{\cdot}{C}CH_2Br \tag{118}$$

is assumed to occur in chlorination of *tert*-butyl bromide where 1-brom-2-chlor isobutane appears among reaction products.

At the addition course of bromination, the cycle of the chain reaction involves the exothermic addition of the bromine atom to a double bond of alkene and an even more exothermic reaction of the bromoalkyl radical with the molecule of bromine. Instead of bromine, HBr may enter the propagation reaction, too. In such a case, alkyl radicals abstract hydrogen from HBr and the bromine atom is regenerated back. The chain radical reaction with HBr prevails. Its peculiarity consists in exothermicity of both propagation steps. At the chain reaction of HF, the addition reaction of fluorine is highly exothermic while that of the alkyl radical with HF is strongly endothermic. As for HCl, the situation is qualitatively the same, but quantitatively it is more favorable than in the case of HF since the endothermicity of the second propagation step is considerably lower ($\Delta H \sim 20$ kJ/mol).

On the other hand, the addition of iodine atom is endothermic, and the transfer reaction of hydrogen exothermic. A radical hydrobromination at the initiation effect of γ-rays is used for the production of ethylbromide from ethylene and HBr.[67] Hydrobromination of ethylene has a kinetic length of chains of about 10^5. The relative reactivity of different alkenes varies to the extent of 4 orders. The rate of the reaction of halogenated alkenes is lower while that of alkyl alkenes is higher when compared with ethylene. The hydrochlorination reaction is of lower kinetic chain length than hydrobromination.

Taking into account the overall exothermicity of the reaction of hydrogenhalides with alkenes, one may expect that with a radical chain reaction there will occur nonradical and radical nonchain addition reactions, too. In the noninitiated reaction of gaseous 2-methyl propene and gaseous hydrogen chloride the surface catalysis is required for the formation of chloro-2-methyl propane. The reaction, which occurs at the walls, takes place most probably between strongly adsorbed hydrogen chloride and weakly adsorbed 2-methyl propene.[68] The mechanism of the reaction appears to be of the nonchain and ionic nature.

Besides radical addition of bromine and HBr ionic addition reactions are also possible. While radical and ionic addition of bromine give identical products, the addition of HBr to asymmetric dienes leads to different products. Ionic addition of HBr to allyl bromide yields, e.g., 1,2-dibrompropane, while radical reaction gives 1,3-dibrompropane.

According to Markovnikov's rule, the protonation proceeds towards the formation of the most stable carbonium ion. The bromide anion, therefore, adds to the more substituted carbon. By the radical addition, the most stable alkyl radical is formed, and bromine is bound to the least substituted carbon of double bond.

The predominance of the radical course of alkenes hydrobromination may be achieved by the presence of free radical initiators, by light or other kind of radiation. Ionic hydrobromation occurs particularly in polar solvents, at high concentrations of HBr and with alkenes which may be easily protonated.

2. Brominating Compounds

The most important brominating agent is N-bromosuccinimide which is used for the substitution of allylic hydrogen by bromine. The specific course of substitution is due to the higher reversibility of the addition when compared with substitution reaction (Scheme 6).

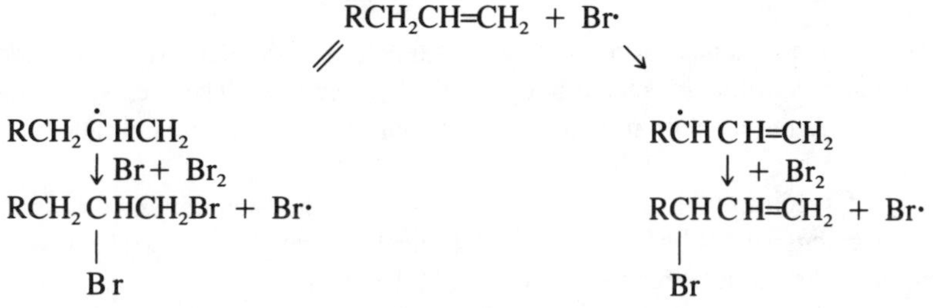

$$RCH_2CH{=}CH_2 \; + \; Br\cdot$$

$$RCH_2\overset{\cdot}{C}HCH_2 \qquad\qquad\qquad R\overset{\cdot}{C}H\,CH{=}CH_2$$
$$\downarrow Br + \; Br_2 \qquad\qquad\qquad\qquad \downarrow \; + \; Br_2$$
$$RCH_2\,C\,HCH_2Br \; + \; Br\cdot \qquad\qquad RCH\,CH{=}CH_2 \; + \; Br\cdot$$
$$\mid \qquad\qquad\qquad\qquad\qquad\qquad \mid$$
$$Br \qquad\qquad\qquad\qquad\qquad\qquad Br$$

SCHEME 6

Provided that the concentration of bromine molecules is low, bromoalkyl radicals succeed in decomposing back to bromine atoms and alkene. An irreversible substitution reaction then prevails. The low level of molecular bromine is maintained by the reaction of N-bromosuccinimide with HBr (Equation 119) which is

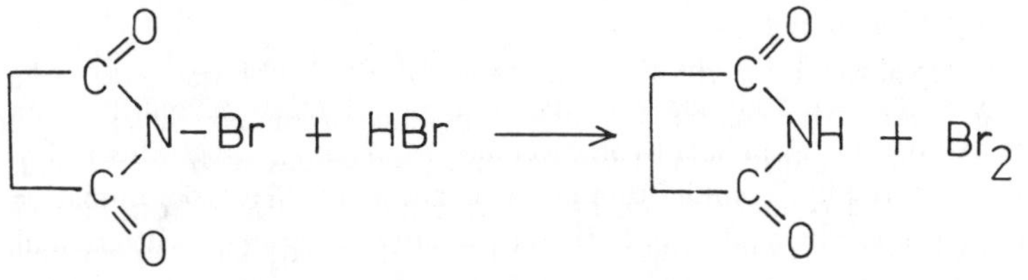

$$\left[\begin{array}{c}-C{\overset{O}{\diagup}} \\ \\ -C{\underset{O}{\diagdown}}\end{array} N{-}Br\right] \;+\; HBr \;\longrightarrow\; \left[\begin{array}{c}-C{\overset{O}{\diagup}} \\ \\ -C{\underset{O}{\diagdown}}\end{array} NH\right] \;+\; Br_2 \qquad (119)$$

one of the side products of substitution bromination.

A similar course may be observed with bromine alone provided that it is kept in the system at some low level. In allyl bromination with *N*-bromosuccinimide, the radical chain reaction is propagated by bromine atoms and alkyl radicals. The elimination of bromine atoms by alkyl radicals from *N*-bromosuccinimide is restricted because of the low solubility of *N*-bromosuccinimide in CCl_4 which is mostly used as a reaction medium.

When good solvents for *N*-bromosuccinimide and a small amount of some alkene without allylic hydrogen, which scavenge bromine, is added to the reaction system, the pattern of chain bromination of hydrocarbons is fundamentally changed. In the propagation of alkyl radicals, the tranfser reaction to *N*-bromosuccinimide prevails and imidyl radicals, which enter the propagation reaction, appear in the system.[69] The substitution reaction of bromine atoms is replaced by that of succinimidyl radicals. Neither HBr nor molecular bromine are present at that time in the system. Alkyl radicals attack thus *N*-bromosuccinimide and not bromine. The replacement of chain propagating radicals results in the change of selectivity. While at the photoreaction of butane with bromine, 2-brombutane is formed as the only monobromide; at the bromination with *N*-bromosuccinimide 75% of 2-brombutane and 25% of 1-brombutane are formed. A similar selectivity may be achieved at reactions of *N*-chlorsuccinimide and *N*-iodosuccinimide as well which indicates the exclusive participation of succinimidyl radicals in the propagation reaction stage. The succinimidyl radical belongs to reactive electrophile radicals which, besides abstraction of hydrogen from hydrocarbons, undergoes other reactions. Its dominating side reaction is ring-opening (Equation 120), which may become the major reaction

$$(120)$$

channel — in some instances accounting for as much as 96%. The degree of side reaction depends on the conditions of succinimidyl radical generation. There existed large controversies concerning the excitation of this radical, but until now no unambiguous conclusion can be made.[70]

The study of succinimidyl radicals is complicated by the low solubility of *N*-bromosuccinimide and extensive loss of succinimidyl radicals by ring opening. To a large extent, these complications do not occur with glutarimidyl radicals.[71]

Hypobromites belong to another group of brominating agents, the most used being *tert*-butyl hypobromite. Since *tert*-butoxy radical abstracts hydrogen from alkenes preferably, it may be used for substitution bromination of unsaturated hydrocarbons. The reactivity of hypobromite with alkyl radicals (Equation 121) is approximately the same as with molecular

$$R\cdot + C_4H_9OBr \rightarrow RBr + C_4H_9O\cdot \qquad (121)$$

bromine which is evidenced by the preservation of optical activity when optically active derivatives are brominated.

For substrates of low reactivity the mixture of chlorine and bromine as brominating agent may be used. Nonreactive compounds such as methane or methyl silanes may be brominated in such a way. Photochlorination of low reactivity substrates with BrCl occurs mainly with $Cl\cdot$ selectivity.[72] The primary attack by chlorine atoms follows from considerably higher reactivity of Cl than Br atoms with respect to the C-H bond. The rate constant of hydrogen abstraction by chlorine atoms for primary hydrogens is by 9 orders, for secondary hydrogens by 7 orders, and, finally, for tertiary hydrogens by 5 orders higher than that for bromine atoms. On the other hand, alkyl radicals react more rapidly with molecules of bromine than with chlorine and, accordingly, more bromoalkane is formed than chloroalkane. At an equal concentration of molecular bromine and chlorine, about 10^5-fold larger amount of bromine than chlorine atoms is present in the reaction system at 20°C. In the presence of sufficiently reactive substrates, the selectivity in the attack of tertiary or benzylic hydrogens by bromine atoms may, therefore, appear.

Polymeric brominating compounds are easily dosed and separated from reaction mixture. Less is, however, known about their reactivity and mechanism of their effect. As an example of a good donor of bromine, particularly for addition reactions, the complex of bromine with crosslinked polyvinylpyridine may be quoted.

D. Iodination

The reactivity of iodine atoms is rather low, the transfer reaction to hydrocarbons being performed only with difficulties. The reaction is endothermic and occurs at high temperatures only (often above 600°C), when alkyl iodides are not capable of surviving and decompose to alkenes and HI.

The problems with substitution iodination consists also in the fact that iodine atoms attack the formed alkyliodide (Equation 122) and decrease its concentration in the reaction

$$I\cdot + I\text{--}CH(CH_3)C_2H_5 \Leftrightarrow \cdot CH(CH_3)C_2H_5 + I_2 \tag{122}$$

medium. The reaction is reversible, which may be documented by the transformation of optically active alkyliodide to optically inactive product (racemate). When iodine isotope labels are used, the radioactive racemate is formed. Addition iodination is rather rarely attempted, too. From all halogens, only the addition of iodine atoms is endothermic (~ 30 kJ/mol). The transient addition of iodine to alkenes may be mostly assumed from the course of *cis-trans* isomerization of alkenes. The same is valid for addition hydroiodination.

Halogenation reactions of hydrocarbons are widely used reactions of organic chemistry. Chlorination and bromination may be performed in ways which best correspond to the depth of knowledge and industrial applications. The use of monochloroalkanes as the original raw materials for the synthesis of important detergents, alkylbenzenesulfonates, acquired, e.g., large practical meaning. Hydrocarbons chlorinated to the higher degree are used as solvents in organic chemistry as well as intermediate products of dyes, insecticides, herbicides, many medications, and important plastics products. We don't exaggerate, therefore, when we say that halogenation has influenced the development of civilization considerably, but regrettably not only in the positive sense. Halogenated hydrocarbons as a new element of our environment impairs the ecological equilibrium of living organisms as well as the chemistry of the atmosphere.

E. Chain Electron Transfer Reactions

The propagation step of chain reactions may involve even an electron transfer. Such a mechanism of chain reactions seems to be valid in the synthesis of some organic compounds, reactions of inorganic materials, as well as in some biochemical processes.[73,74] The initiation may occur as the transfer of a single electron to the substrate giving an odd-electron intermediate. The transfer of electrons is realized within the charge transfer complex or exciplex of reactants or electrochemically.[75,76]

These reactions may be illustrated as follows (Equation 123)

$$RX + e^- \Leftrightarrow R\dot{X}^- \tag{123}$$

The anion radical of reactant RX is successively cleaved to anion X^- and radical $R\cdot$, which is then coupled with nucleophile Nu^-. Addition of a radical to nucleophile reactant Nu^- (Equation 124) is the key step of the propagation reaction leading

$$R\cdot + Nu^- \rightarrow R\dot{N}u^- \tag{124}$$

to the anion radical of the final substituted product.

As a mediator of an electron transfer there may act reactant RX, itself; since through the reaction 125, $R\dot{N}u^-$ is

$$R\dot{N}u^- + RX \Leftrightarrow RNu + R\dot{X}^- \tag{125}$$

reoxidized into the reaction product RNu and the anion radical $R\dot{X}^-$ is regenerated back and may start a new propagation cycle.

The chain reaction is inhibited by stronger acceptors of electrons than the starting compound RX and by compounds which react with radicals $R\cdot$ faster than nucleophile Nu. The above mechanism may alternatively interpret the substitution of halogens by cyanide ions in arylhalogenides (Equation 126),[76,77] the reaction of alkyl radicals with

$$ArX + CN^- \rightarrow ArCN + X^- \tag{126}$$

carbanions,[78] the decomposition of peroxides,[79] decarboxylation of organic acids,[80] the reactions of Grignard reagents with phenyl-substituted methyl chlorides, one-electron transfer reactions of piperidine nitroxides with some biologically active substances,[81] and many other reactions.

The mechanism of chain electron transfer reactions is not always as uniform as depicted. At the photosensitized decomposition of diazonium salts conducted in the presence of oxalate anions,[82] e.g., unstable aryldiazo radicals (Equation 127) are formed which quickly eliminate nitrogen

$$ArN_2^+ + e^- \rightarrow ArN_2\cdot \tag{127}$$

and give aryl radicals. The latter abstract hydrogen from oxalate anions (Equation 128) from which carbondioxide radicals

$$Ar\cdot + HCOO^- \rightarrow ArH + CO_2^{\ominus}\cdot \tag{128}$$

arise as new electron donors and chain decomposition may go on. The kinetic length of the chain reaction (or quantum yield of photolysis) amounts to 300. Such quantitative values are relatively seldom reported in the literature, even though they might be of use also in other electron transfer chain reactions.

Besides the schemes referred to earlier, other sequences of elementary steps in electron transfer chain reactions may be put forward, too. However, the general pattern of the process, namely, an electron transfer followed by reactions of the odd-electron intermediates to give by coupling the second-electron transfer, fragmentation or disproportionation of even-electron products, remains unchanged. The ratio of individual elementary events involved depends on the structure and concentration of reactants, the rate of ion radical production and on the medium quality. Many investigations are now emphasizing the importance of solvent-assisted electron transfer, because when a charged or neutral species is in solution,[83] there are always some interactions between the species and its solvation shell. This shell simultaneously enhances or reduces donor-acceptor properties of reactants which affect the rate constants of electron transfer reactions. The reaction mechanism could be better understood by the detection of the ion radical and radical intermediates which are involved in the electron transfer chain reactions. Introduction of suitable detection methods may thus show that the reactions considered as purely radical or ionic ones are in fact ion radical processes.

REFERENCES

1. **Portern, M. D. and Skinner, G. B.**, The steady state approximations in free radical calculations, *J. Chem. Educ.*, 53, 366, 1976.
2. **Horie, K. and Mikulasova, D.**, Living radical polymerization of styrene and its block copolymerization with methyl methacrylate, *Makromol. Chem.*, 175, 2091, 1974.
3. **Bevington, J. C., Huckerby, T. N., and Varma, S. C.**, Reactivities of monomers towards the 2-cyano-2-propyl radical at 100°C, *Eur. Polym. J.*, 22, 427, 1986.
4. **Rubio, L. H. G., Ro, N., and Patel, R. D.**, UV analysis of benzoyl peroxide-initiated styrene polymerizations and copolymerizations, *Macromolecules*, 17, 1998, 1984.
5. **Moad, G., Rizzardo, E., and Solomon, D. H.**, Reactions of benzoyloxy radicals with some common vinyl monomers, *Macromol. Chem. Rapid Commun.*, 3, 533, 1982.
6. **Rizzardo, E., Serelis, A. K., and Solomon, D. H.**, Initiation in radical polymerizations. Reaction of cumyloxy radicals with methyl methacrylate and styrene, *Aust. J. Chem.*, 2013, 1982.
7. **Nayak, P. L. and Lenka, S.**, Redox polymerization initiated by metal ions, *J. Macromol. Sci. Rev.*, C19, 83, 1980.
8. **Hayashi, K.**, Polymerization, *Radiat. Phys. Chem.*, 18, 183, 1981.
9. **Melnikov, V. P., Gulyaeva, L. S., and Markevich, A. M.**, The determination of the rate constant of propagation for polymerization of tetrafluoroethylene initiated by surface active radicals (Russian), *Khim. Phys.*, 4, 1528, 1985.
10. **Paus, K. F. and Klyuchnikova, N. V.**, The formation of surface polyradicals at thermolysis of natural calcium carbonate (Russian), *Dok. Akad. Nauk U.S.S.R.*, 289, 135, 1986.
11. **Ivanchev, S. S. and Dmitrenko, A. V.**, The way of free radical polymerization filling for preparation of composite materials (Russian), *Usp. Khim.*, 51, 1178, 1982.
12. **Otsu, T., Yoshids, M., and Kuriyama, A.**, Living radical polymerizations in homogeneous solutions by using organic sulfides as photoinitiators, *Pol. Bull.*, 7, 45, 1982.
13. **Crivello, J. V., Lee, J. L., and Conion, D. A.**, Cyclic silyl pinacol ethers. A new class of multifunctional free radical initiators, *Polymer. Bull.*, 16, 95, 1986.
14. **Kammerer, H. and Pachta, J.**, Nebenreaktionen bei der Radikalischen Intramolekularen Cycloaddition von 2,6-Bis(2-Methyloxy-5-methylbenzyl-4-methyl phenyl methacrylate, *Colloid. Polymer. Sci.*, 255, 656, 1977.
15. **Takemoto, K. and Miyata, M.**, Polymerizaton of vinyl and diene monomers in canal complexes, *J. Macromol. Sci., Rev. Macromol. Chem.*, C18, 83, 1980.
16. **Fukuda, T., Ma, Y. D., and Inagaki, H.**, Free radical copolymerization. 3. Determination of rate constants of propagation and termination for the styrene/methyl methacrylate system. A critical test of terminal model kinetics, *Macrolecules*, 18, 17, 1985.
17. **Topchiev, D. R.**, Untersuchungen Sowjetischer Forscher auf dem Gebiet der Radikalischen Polymerization, *Acta Polym.*, 36, 585, 1985.
18. **Henrici-Olive, G. and Olive, S.**, The template effect of water in acrylonitrile polymerization, *Polym. Bull.*, 1, 47, 1978.
19. **Chapiro, A.**, Autoacceleration in free radical polymerization caused by oriented monomer-polymer association complexes, *Pure Appl. Chem.*, 53, 643, 1981.
20. **Bailey, W. J.**, Free radical ring opening polymerization, *Macromol. Chem. Suppl.*, 13, 171, 1985.
21. **Okieimen, E. F.**, Studies in vinyl chloride polymerization. Determination of primary radical termination by initiator fragment method, *Eur. Pol. J.*, 19, 255, 1983.
22. **Mahabadi, M. Kh.**, Effects of chain length dependence of termination rate constant on the kinetics of free radical polymerization. 1. Evaluation of an analytical expression relating the apparent rate constant of termination to the number average degree of polymerization, *Macromolecules*, 18, 1319, 1985.
23. **Olaj, O. F., Zifferer, G., Gleiner, G., and Stickler, M.**, Termination processes in free radical polymerization. 7. The treatment of chain length dependent termination in presence of chain transfer, *Eur. Pol. J.*, 22, 585, 1986.

24. **Ballard, M. J., Napper, D. H., Gilbert, R. G., and Sangster, D. F.,** Termination rate coefficient in methyl methacrylate polymerizations, *J. Polym. Sci. A, Polym. Chem.,* 24, 1027, 1986.

25. **Ito, K.,** Kinetics of radical polymerization of styrene at high concentration of diisopropyl dicarbonate, *Eur. Pol. J.,* 22, 253, 1986.

26. **Mogilevich, M. M.,** Oxidation polymerization of vinyl monomers (Russian), *Usp. Khim.,* 48, 362, 1979.

27. **Tatsukami, Y., Takahashi, T., and Yoshioka, H.,** Reaction mechanism of oxygen-initiated ethylene polymerization at high pressure, *Makromol. Chem.,* 181, 1107, 1980.

28. **Brandrup, H. E. and Immergut, H. E., Eds.,** *Polymer Handbook,* Interscience, New York, 1975.

29. **McGrath, J. E.,** Chain reaction polymerization, *J. Chem. Educ.,* 58, 844, 1981.

30. **Ma, Y. D., Fukuda, T., and Inagaki, H.,** Free radical copolymerization. 4. Rate constants of propagation and termination for the *p*-Chlorostyrene/methyl acrylate system, *Macromolecules,* 18, 26, 1985.

31. **Starks, Ch. M.,** *Free Radical Telomerization,* Academic Press, New York, 1974.

32. **Greenley, R. Z.,** Q and e values of telogens, *J. Macromol. Sci. Chem.,* A11, 933, 1977.

33. **Minisci, F.,** Free radical additions to olefins in the presence of redox systems, *Acc. Chem. Res.,* 8, 165, 1975.

34. **Boutevin, B., Maubert, C., Mebkhout, A., and Pietrosanta, Y.,** Telomerization kinetics by redox catalysis, *J. Polym. Sci. Polym. Chem. Educ.,* 19, 499, 1981.

35. **Velichko, F. K., Vasileva, T. T., Petrova, R. G., and Terentjev, A. B.,** Radical telomerization with halogen containing compounds. Isomerization of short-living radicals (Russian), *Zh. Vses. Khim. Obschest.,* 24, 181, 1979.

36. **Citterio, A., Minisci, F., and Serravalla, M.,** Polar effects in free radical reactions. Reductive alkylation of olefins with electron withdrawing groups, *J. Chem. Res.,* (S), 198, 1981.

37. **Cittero, A., Arnoldi, A., and Minisci, F.,** Nucleophilic character of alkyl radicals. 18. Absolute rate constants, *J. Org. Chem.,* 44, 2674, 1979.

38. **Kimura, T., Nakanishi, I., and Hamashima, M.,** Study on radical telomerization of esters of methacrylic acid by using bromotrichloromethane and characteristic of the resulting telomers. III. Aryl methacrylates, *Polym. J.,* 18, 689, 1986.

39. **Ashton, D. S., Tedder, J. M., and Walton, J. C.,** Free radical addition to olefins. 12. Telomerization of tetrafluorethylene with dibromodifluoromethane and trifluoroiodomethane, *J. Chem. Soc., Faraday Trans. I.,* 70, 299, 1974.

40. **Weitkamp, J.,** Hydrocracken, Cracken und Isomerisieren von Kohlen-wasserstoffen Erdol, Kohle, Erdgas, *Petrochem.,* 31, 13, 1978.

41. **Schafer, H. G.,** Uber den Mechanismus der Pyrolyse von Kohlen, *Chem. Zeit.,* 109, 401, 1985.

42. **Chen, C. J., Back, M. H., and Back, R. A.,** The thermal decomposition of methane. III. Methyl radical exchange in Ch_4-CD_4 mixtures, *Can. J. Chem.,* 55, 1624, 1977.

43. **Bradley, J. N.,** A general mechanism for the high temperature pyrolysis of alkanes. The pyrolysis of isobutane, *Proc. R. Soc. Lond.,* A337, 199, 1974.

44. **Kiefer, J. H., Wei, H. C., Kern, R. D., and Wu, C. H.,** The high temperature pyrolysis of 1,3-butadiene. Heat of formation and rate of dissociation of vinyl radical, *Int. J. Chem. Kinet.,* 17, 225, 1985.

45. **Mizerka, L. J. and Kiefer, J. H.,** The high temperature pyrolysis of ethylbenzene. Evidence for dissociation to benzyl and methyl radicals, *Int. J. Chem. Kinet.,* 18, 363, 1986.

46. **Emanuel, N. M.,** Problems of selectivity of chemical reactions (Russian), *Usp. Khim.,* 47, 1329, 1978.

47. **Karpf, M.,** Organic synthesis at high temperatures. Gas phase flow thermolysis, *Agnew. Chemie Int. Ed. Engl.,* 25, 414, 1986.

48. **Schnabel, W.,** *Polymer Degradation, Principles and Practical Application,* Hanser International, Munchen, 1981.

49. **Ayrey, G., Head, B. C., and Poller, R. C.,** The thermal dehydrochlorination and stabilization of PVC, *Macromol. Rev.,* 8, 1, 1974.

50. **Hjetberg, T. and Sorvik, E. M.,** Formation of anomalous structures in poly(vinylchloride) and their influence on the thermal stability, *ACS Symp. Series,* 280, 259, 1985.

51. **Golova, D. P.,** Chemical transformation of cellulose at the processing temperatures (Russian), *Usp. Khim.,* 44, 1454, 1975.

52. **Piskorz, J., Radlen, D., and Scott, D. S.,** On the mechanism of the rapid pyrolysis of cellulose, *J. Anal. Appl. Pyrol.,* 9, 121, 1986.

53. **Seeger, C., Rotzoll, G., Lubbert, A., and Schugerl, K.,** Direct detection of CF_2 and computed modelling of its appearance in the fluorination of CH_2F_2, *Int. J. Chem. Kinet.,* 14, 457, 1982.

54. **Corbin, G. A., Cohen, R. E., and Baddor, R. F.,** Surface fluorination of polymers in a glow discharge plasma: photochemistry, *Macromolecules,* 18, 98, 1985.

55. **Postma, M. L.,** Free radical chlorination of organic molecules, in *Methods in Free Radical Chemistry I,* Huyser, E. S., Ed., Marcel Decker, New York, 1969, 80.

56. **Timonen, R., Kalliorinno, K., and Koskikallio, J.,** Kinetics of reactions of methyl and ethyl radicals with chlorine in the gas phase studied by photochlorination of methane, *Acta Chim. Scand.,* A40, 459, 1986.

57. **Timonen, R. S., Russell, J. J., and Gutman, D.,** Kinetics of the reactions of halogenated methyl radicals with molecular chlorine, *Int. J. Chem. Kinet.,* 18, 1193, 1986.

58. **Plate, N. A.,** Problems of polymer modification and reactivity of functional groups in macromolecules, *Pure Appl. Chem.,* 46, 49, 1976.

59. **Kolinsky, M., Doskocilova, D., Schneider, B., Stork, J., Drahoradova, E., and Kuska, V.,** Structure of chlorinated polyvinyl chloride. Determination of the mechanism of chlorination from infrared and NMR spectra, *J. Polym. Sci.,* A1, 9, 791, 1971.

60. **Serguchev, Yu. A. and Konyuschenko, V. P.,** The criterion of the polar and molecular mechanism of chlorination of unsaturated compounds in their solutions (Russian), *Z. Obsch. Khim.,* 11, 1353, 1975.

61. **Davis, R., Durrant, J. L. A., and Rowland, C. C.,** Free radical halogenation of alkanes initiated by transition metal complexes, *J. Organomet. Chem.,* 316, 147, 1986.

62. **Minisci, F., Galli, R., Galli, A., and Bernardi, R.,** A new highly selective type of radical chlorination, *Tetrahedron Lett.,* 23, 2207, 1967.

63. **Deno, N. C.,** Free radical chlorination via nitrogen cation radicals, in *Methods in Free Radical Chemistry, Vol. 3,* Huyser, E. S., Ed., Marcel Decker, New York, 1979, 135.

64. **Tanner, D. D., Arhart, R., and Meintzer, Ch. P.,** Polar Radicals.17. On the mechanism of chlorination by *N*-chloramines — intermolecular and intramolecular abstraction, *Tetrahedron,* 41, 4261, 1985.

65. **Thaler, W. A.,** Free radical brominations, in *Methods in Free Radical Chemistry,* Vol. 2, Huyser, E. S., Ed., M. Decker, New York, 1967, 121.

66. **Clark, T. and Symons, M. C. R.,** Protonated β-halogen-ethyl radicals, *J. Chem. Soc., Chem. Commun.,* 96, 1986.

67. **Drave, H.,** Perspektiven der Angewandten Strahlenchemie, *Chem. Zeit.,* 109, 367, 1985.

68. **Costello, F., Dalton, D. R., and Poole, J. A.,** Hydrochlorination of alkenes. 2. Reaction of gaseous HCl and 2-methyl propene, *J. Phys. Chem.,* 90, 5352, 1986.

69. **Luning, U. and Skell, P. S.,** Imidyl radicals, *Tetrahedron,* 41, 4289, 1985.

70. **Skell, P. S., Luning, U., McBain, D. S., and Tanko, J. M.,** Ground and excited state succinimidyl radicals in chain reactions: a reexamination, *J. Am. Chem. Soc.,* 108, 121, 1986.

71. **Luning, U., Seshadr, S., and Skell, P. S.,** Glutarimidyl chemistry. Substitution reactions. Mechanism of Ziegler brominations, *J. Org. Chem.,* 51, 2071, 1986.

72. **Skell, P. S. and Bater, H. N., III, and Tanko, J. M.,** Reactions of BrCl with alkyl radicals, *Tetrahedron Lett.,* 27/43, 5181, 1986.

73. **Juliard, M. and Chanon, M.,** Photoelectron-transfer catalysis: its connnections with thermal and electrochemical analogues, *Chem. Rev.,* 83, 425, 1983.

74. **Chanon, M. and Tobe, M. L.,** ETC - a mechanistic concept for inorganic and organic chemistry, *Angew. Chem. Int. Ed. Engl.,* 211, 1, 1982.

75. **Kavarnos, G. J. and Turro, N. J.,** Photosensitization by reversible electron transfer: theories, experimental evidence, and examples, *Chem. Rev.,* 86, 401, 1986.

76. **Amatore, Ch., Combellas, C., Robveille, S., Saveant, J. M., and Thiebault, A.,** Electrochemically catalyzed aromatic nucleophilic substitution. Reactivity of cyanide ions toward aryl radicals in liquid ammonia, *J. Am. Chem. Soc.,* 108, 4754, 1986.

77. **Eberson, L., Jonsson, L., and Wistrand, L. G.,** The $S_{on}2$ mechanism a non-oxidative reaction that is initiated by electron transfer oxidation, *Tetrahedron Symp.,* 38, 1087, 1982.

78. **Russell, G. A. and Khanna, R. K.,** The reaction of carbanions with tert. butyl radicals, *Tetrahedron,* 41, 4133, 1985.

79. **Zupancic, J. J., Horn, K. A., and Schuster, G. B.,** Electron-transfer-initiated reactions of organic peroxides. Reaction of phthaloyl peroxide with olefins and other electron donors, *J. Am. Chem. Soc.,* 102, 5279, 1980.

80. **Brimage, D. R., Davidson, R. S., and Steiner, P. R.,** Use of heterocyclic compounds as photosensitizers for the decarboxylation of carboxylic acids, *J. Chem. Soc. Perkin Trans. I.,* 526, 1973.

81. **Liu, Y. Ch., Dang, H. S., and Liu, Z. L.,** Some recent studies on electron transfer reactions at Lanzhou University, *Rev. Chem. Intermediates,* 7, 111, 1986.

82. **Fomin, G. V., Mordvintsev, P. I., Mkhitarov, R. A., and Gordina, T. A.,** The chain radical decomposition of diazonium salts photosensitized by anthraquinone and benzoquinone derivatives. II. Reaction in formate, oxalate and EDTA solutions (Russian), *Zh. Fiz. Khim.,* 54, 240, 1980.

83. **Truong, T. B.,** Effect of solvent reorganization on the electron transfer reaction between donor-acceptor pairs in solution, *Pure Appl. Chem.,* 58, 1279, 1986.

Chapter 7

BRANCHED CHAIN REACTIONS

I. INTRODUCTION

Practically all radical oxidation reactions with the participation of oxygen, fluorine, nitrogen oxides, nitrates, or perchlorates have the character of branched chain reactions. The substrate may be hydrogen, carbon monoxide, hydrocarbons, alcohols, ketones, aldehydes, and other molecules. The principal feature of such processes is that under certain conditions the overall number of reactive free radicals in the system increases in an accelerating manner. Especially attractive behavior in this respect is demonstrated by systems which are apparently still and suddenly explode without an obvious reason.

According to the overall rate of the process or its change, the character of reactive intermediates, branching agent, and composition of products, we distinguish the branched chain reactions in the stage of slow oxidation and the combustion or explosion. This may well be illustrated by the classical approach to the characterization of ignitability of different fuels which consists in construction of so-called ignition diagrams. Provided that we have a closed vessel of some shape, kept at some temperature and at a known initial pressure inside, dosing fuel, we may find the critical parameters of ignition given in terms of the dependence of either pressure or composition (oxygen and the fuel amount) on temperature. The borderline dividing the zone of the slow and fast reaction is not usually simple and, varying with the complexity of fuel molecules, performs both the maximum and minimum, one or more bifurcations, lobes, etc. (Figure 1). Reactions in the respective region of parametric conditions of ignition diagrams of hydrocarbons have found applications in different fields of society, namely, in turbojet engines (region 1) and internal combustion engines (region 2), where conversion of chemical to mechanical energy is used; cool flames (region 3), where potentially useful chemicals such as O-heterocycles can be prepared; and finally slow oxidation (region 4), where the controlled conversion of substrate to useful chemicals (alcohols, peroxides, aldehydes, ketones) is largely explored. The dashed line on Figure 1 indicates some variants of an ignition diagram depending on the quality of the fuel. It is of interest that hydrocarbons have quite a different morphology of ignition diagram than, e.g., hydrogen. At a certain pressure three limiting temperatures of ignition (T_1, T_2, and T_3) exist, whereas the diagram of hydrogen ignition shows three limiting pressures of ignition (P_1, P_2 and P_3) at a certain temperature (Figure 2).

From the formal viewpoint it is important whether the ignition takes place spontaneously or whether we use some additional source of ignition such as spark, heat, etc.

According to the character of the propagation of the process, which may be steady or nonsteady, we receive either burning or explosion. Provided that the reaction starts to be spread by adiabatical compression of the nonreacted fuel, which is thus initiated in front of the explosion wave, the rate of the process is considerably higher than in the case of diffusionally controlled reactions and we speak about detonation. From this viewpoint, the steady flame may be considered as a standing explosion wave, the velocity of which is strongly influenced by the ratio of fuel and oxygen in the mixture.[1]

Schematically, the three fundamental phases of the chain oxidation may be illustrated accordingly:

Initiation	Ignition	— Spontaneous
		— Forced
Propagation	Propagation	— Steady
		— Nonsteady burning (explosion, detonation)
Termination	Extinction	

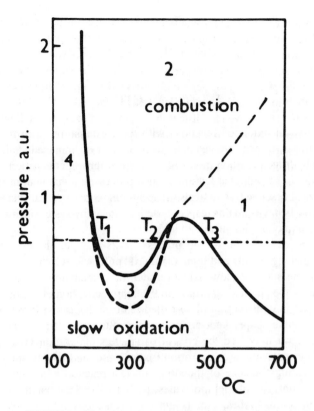

FIGURE 1.   The schematic ignition diagram of hydrocarbon-oxygen mixtures.

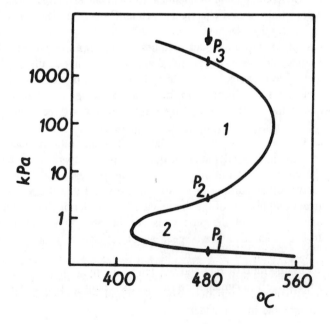

FIGURE 2.   Explosion limits of stoichiometric mixture of hydrogen and oxygen: 1 — slow reaction, 2 — explosion.

In the left column there are the expressions commonly used in nonbranched or slow chain reaction terminology, whereas the right column represents fast processes of burning (combustion) and explosion.

For the respective substrate, each of the stages of both the slow branched chain reaction or combustion may be positively or negatively affected by the additives called accelerators or propellants, if they increase the overall rate of the reaction, or inhibitors or retardants if they decrease the rate. The above classification, based upon the ignition diagram of branched chain reactions, depends on experimental conditions (temperature, pressure, concentration, presence of additives), and serves only as some orientation point, since overlap exists in respective reaction mechanisms. As may be seen from the general patterns of the ignition diagram, the slow processes usually occur at lower temperatures and at moderate or higher pressures, the composition of reaction products being more specific than in the case of noncontrolled burning or explosion, where carbon oxides, nitrogen oxides, water, and other simple molecules ultimately result.

II. OXIDATION OF HYDROCARBONS

Until now, the branched chain reaction of oxidation and combustion of hydrocarbons as a source of energy governed the activity and life of modern society significantly. With regard to the production of the number of chemicals by selective oxidation, the ''knock'' in the spark combustion engines, and other abnormal combustion phenomena, the oxidation of hydrocarbons in the temperature range 200 to 600°C has been widely studied.

The term degenerated chain branching should be introduced in hydrocarbon oxidation; it means the amplification of the number of reactive free radicals by the decomposition of relatively stable molecules such as alkyl hydroperoxides, dialkyl or diacyl peroxides, peracids, etc. These compounds which are formed as by-products of the oxidation reaction may be isolated from the reaction system. The branching induced by them is degenerated, because the concentration of such compounds being controlled by their decomposition cannot increase to infinity, as is theoretically possible in the case of nondegenerated branching such as, e.g., by hydrogen or oxygen atoms.

The initiation reaction of hydrocarbon oxidation includes every primary introduction of free radicals into the system. Schematically, it may be designated as in Equation 1

$$RH + O_2 \rightarrow R\cdot + HO_2 \tag{1}$$

and when the system is partially oxidized also as in Equations 2 or 3.

$$ROOH \rightarrow RO\cdot + \cdot OH \tag{2}$$

$$2\ ROOH \rightarrow RO\cdot + RO_2^{\cdot} + H_2O \tag{3}$$

Comparison of the nature of the primary products appearing up to 1% conversion at oxidation of butane, 2-methyl-pentane and other hydrocarbons has shown that the predominating chain propagation steps vary with the molecular weight and structure of hydrocarbon. The critical role here is obviously played by alkylperoxy radicals which theoretically have several pathways of their transformation. Principally, all elementary reactions discussed in the previous sections come into consideration.

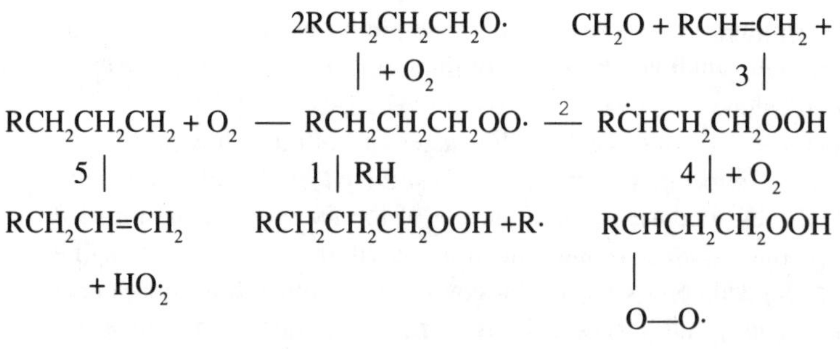

$$2RCH_2CH_2CH_2O\cdot \qquad CH_2O + RCH=CH_2 + \cdot OH$$

$$\big|\ +O_2 \qquad\qquad\qquad 3\big|$$

$$RCH_2CH_2CH_2 + O_2 \ -\!\!- \ RCH_2CH_2CH_2OO\cdot \ \xrightarrow{\ 2\ } \ R\dot{C}HCH_2CH_2OOH$$

$$5\big| \qquad\qquad 1\big|\ RH \qquad\qquad 4\big|+O_2$$

$$RCH_2CH=CH_2 \qquad RCH_2CH_2CH_2OOH +R\cdot \qquad RCHCH_2CH_2OOH$$

$$+\ HO_2^{\cdot} \qquad\qquad\qquad\qquad\qquad\qquad\qquad \big|$$

$$O\!\!-\!\!O\cdot$$

SCHEME 1

The step[1] in Scheme 1 is an intermolecular transfer which due to the fast reaction $R\cdot +$ $O_2 \rightarrow RO\cdot_2$ leads to the accumulation of alkylhydroperoxides and to degenerated chain branching. The intramolecular transfer reaction which, depending on conditions, may be preferred before its intermolecular equivalent is another possibility. Among several alternatives to intramolecular hydrogen abstraction, the most facile isomerizations are those which involve 1,5-H transfer where a 6-membered ring of the transition state is formed (step 2). In subsequent step 3 of the β-fragmentation reaction, γ-hydroperoxyalkyl radicals may decompose to carbonyl compounds, alkene, and hydroxy radicals, or undergo cyclization to oxetane (Equation 4).

$$\begin{array}{ccccc} R^1CHOOH & \cdot CHR^2 & & R^1CH\!-\!O\!-\!CHR^2 & \\ \diagdown & \diagup & \rightarrow & \diagdown \quad \diagup & +\ \cdot OH \\ & CH_2 & & CH_2 & \end{array} \qquad (4)$$

The role of the fast reaction of β-hydroperoxyalkyl radicals with oxygen (step 4, Scheme 1) is not frequently incorporated into the schemes of reaction mechanisms, even though it represents the efficient accumulation of chemical energy in one molecule. According to some authors[2] dihydroperoxides formed from peroxyhydroperoxide radicals, however, yield only two hydroxy radicals and carbonyl compounds in the decomposition reaction (Equation 5)

$$CH_2(CH(CH_3)OOH)_2 \rightarrow 2\ \cdot OH\ +\ CH_3COCH_3\ +\ CH_3CHO \qquad (5)$$

and cannot be considered as the source of further degenerated branching.

To complete the list of possible reaction pathways of peroxy radical transformations it is necessary to underline reaction 5 in Scheme 1, which may compete with the formation of peroxyalkyl radicals, namely, the disproportionation of alkyl radicals with oxygen which fits to the high yield of alkenes in the primary steps of hydrocarbon oxidation. It is difficult to distinguish the above mechanism and that based on 1,4-isomerization (Equation 6) of alkylperoxy radicals followed by splitting

$$RCH_2CH_2OO\cdot \rightarrow R\dot{C}HCH_2OOH \rightarrow HO_2^{\cdot}\ +\ RCH=CH_2 \qquad (6)$$

off $HO\cdot_2$ radicals. The marked decrease in the yield of alkenes with a pressure in the range of 20 to 50 kPa observed during the early and advanced stages of oxidation reaction supports the disproportionation mechanism. The above results were interpreted alternatively by the propagation reaction (Equation 7) which

$$2\ RO_2^{\cdot} \rightarrow 2\ RO\cdot\ +\ O_2 \qquad (7)$$

was shown that might be of importance at concentrations of about 10^{-8} mol dm^{-3} for such alkylperoxy radicals which are restricted in their ability to isomerize. Otherwise, the above reaction proceeds as an actual disproportionation step (Equation 8)

$$2 \, RO_2^{\cdot} \rightarrow \, >C=O \, + \, >CHOH \, + \, O_2 \tag{8}$$

giving carbonyl compounds, oxygen, and alcohol. All these apparently opposing hypotheses of elementary steps of further propagation of the oxidation reaction are more or less frequently used in the explanation of a negative temperature coefficient of reaction and cool flame periodicity. The passage from the stage of reaction where mainly hydroperoxides are formed to the regime of prevailing peroxy radical isomerization or alkene formation may cause the gradual decrease of the reaction rate regardless of the fact that the temperature increases. In this parametric region with negative temperature coefficient, which in hydrocarbon oxidation usually extends from about 300 to 450°C, and which varies with hydrocarbon structure and concentration, the stationary level of $HO_2^{\cdot}$ radicals becomes considerably higher than that of $RO_2^{\cdot}$ radicals. It was also found that in oxygen rich mixtures, the region of negative temperature coefficient was shifted to higher temperatures and the interpretation was put forward, namely, that at higher oxygen concentrations the acylperoxy $R(CO)OO^{\cdot}$ radical is the major peroxy radical in the system, whereas alkyl peroxy radicals are preponderant at high hydrocarbon concentrations. When we compare the yields of CO and CO_2 in the mixtures of 20% and 80% butane with oxygen determined at the maximum of light emission varying with the temperature, we see that at about 350°C, a sharp increase in the amount of CO and CO_2 occurs (CO concentration being by 1 order higher than that of CO_2). At the same time, the mixtures with 20% butane give 6 times more CO and 5 times more CO_2 than those with 80% butane.[3]

The formation of both CO and CO_2 may be understood on the basis of aldehyde oxidation (Equations 9 to 14) in the advanced

$$RCHO + ROO^{\cdot} \rightarrow ROOH + R\overset{\cdot}{C}O \tag{9}$$

$$R\overset{\cdot}{C}O \rightarrow R^{\cdot} + CO \tag{10}$$

$$R\overset{\cdot}{C}O + O_2^{\cdot} \rightarrow R(CO)OO^{\cdot} \tag{11}$$

$$R(CO)OO^{\cdot} + RCHO \rightarrow R(CO)OOH + R\overset{\cdot}{C}O \tag{12}$$

$$R(CO)OOH \rightarrow RCO_2^{\cdot} + {\cdot}OH \tag{13}$$

$$RCO_2^{\cdot} \rightarrow R^{\cdot} + CO_2 \tag{14}$$

stage of the reaction. This, however, does not explain the actual reasons for the appearance of cool flames in the oxidation of hydrocarbons, aldehydes, ketones, and other compounds. It should be recalled that in static systems the appearance of a cool flame is characterized by a pressure pulse which is accompanied by a moderate temperature rise (<200°C) and a pale blue weak light emission corresponding to the formaldehyde fluorescence spectrum. The propagation velocity of a cool flame which may be stabilized in the flow system is low (about 1 to 10 cm/sec).

The increase of temperature in a cool flame may lead to a "normal" ignition which in spark engines is premature and gives the unwanted knock behavior of the motor.

The question now is whether the cool flame may be interpreted as an accumulation of alkylhydroperoxides and peroxides, and as their subsequent quasi-explosive decomposition

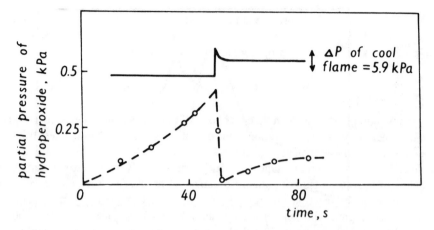

FIGURE 3.   The change in the yield of organic peroxides during the cool flame oxidation of heptane.[2] Initial temperature 242°C, initial pressure of heptane 6.5 kPa; initial pressure of O = 6.5 kPa. (From Burgess, A. R. and Loughlin, R. G., *Chem. Commun.*, 729, 1967. Copyright by Royal Society of Chemistry, London. With permission.)

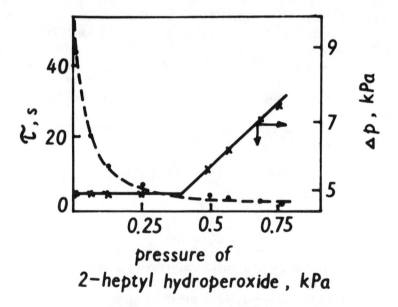

FIGURE 4.   The changes of induction period τ of cool flame appearance[2] characterized by the pressure increase ΔP on concentration of added 2-heptyl hydroperoxide at 242°C.

which is stopped by the negative temperature coefficient of reaction, into the zone of which the process is shifted by self-heating, or whether the explanation lies elsewhere? One thing is certain, namely, that the passage of the cool flame decreases the level of alkylhydroperoxides in the system. This may be demonstrated on cool flame oxidation of heptane where the partial pressure of organic peroxides increases up to a maximum immediately before a cool flame and drops practically to zero during its passage (Figure 3).[2] If a small amount of 2-heptyl hydroperoxide were added to the reaction mixture, a significant reduction of the induction period of the cool flame appearance would be observed while the pressure change in the reactor due to the cool flame would remain unchanged. A larger amount would have little further effect on the induction period, but the pressure change would increase (Figure 4).[2] These results suggest that some critical concentration of organic peroxides should be attained to make possible the propagation of a cool flame.

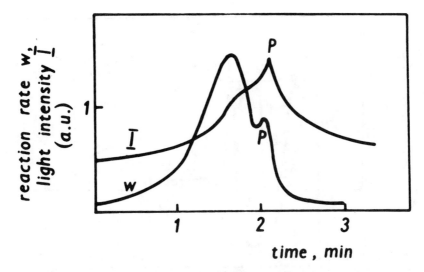

FIGURE 5.   Illustration[2] of the course of propane oxidation in closed reactor at 430°C.

The main light emitter of the cool flame, the excited formaldehyde, may be formed directly in decomposition of methyl peroxide or methyl hydroperoxide in highly exothermic termination reactions (Equations 15 to 17).

$$CH_3O\cdot + \cdot OCH_3 \rightarrow CH_2O + CH_3OH \tag{15}$$

$$CH_3O\cdot + \cdot OH \rightarrow |CH_2 + H_2O \tag{16}$$

$$CH_3O\cdot + \cdot CH_3 \rightarrow CH_2O + CH_4 \tag{17}$$

At a sufficiently high temperature, the concentration of peroxy radicals becomes low and the cross termination (Equation 18) can't be ignored any

$$RO_2^\cdot + R\cdot \rightarrow ROOR \tag{18}$$

more when compared with self-termination of peroxy radicals (Equation 8). Dialkyl peroxides are known to decompose with markedly higher rates than corresponding hydroperoxides and may be a source of additional branching. Such unexpected acceleration of the oxidation reaction represented by the peak ascribed to the effect of intermediately formed dialkyl peroxides was actually observed (Figure 5) during the final stages of oxidation of some hydrocarbons, when the oxygen concentration in the closed reactor falls to a very low value. It is characteristic with the sudden increase of the reaction rate just before the complete consumption of oxygen.

In the somewhat advanced stages of the oxidation reaction where aldehydes become one of the main reaction by-products, other reactions (Equation 19) may occur which also possibly bring about branching.

$$RCHO + O_2 \rightarrow HO_2^\cdot + R\dot{C}=O \tag{19}$$

After the reaction of carbonyl radicals with oxygen (Equation 11) and after the transfer reaction of corresponding acylperoxy radicals to the hydrocarbon (Equation 12), the formed peracids may also cause branching of (Equation 13) the oxidation reaction.

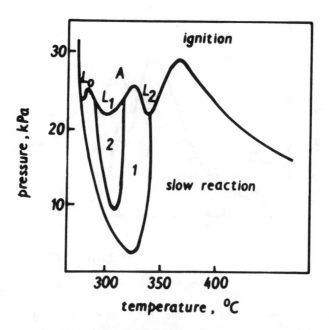

FIGURE 6.   The ignition[2] diagram for 3-ethyl pentane and oxygen (1:2), cylindrical pyrex vessel, volume 280 cm.[3] A — zone of multiple stage ignition composed of lobes L_0, L_1, and L_2; 1 — zone of the appearance of one cool flame; 2 — zone of the appearance of two cool flames. (From Barat, P., Cullis, C. F., and Pollard, R. T., *13th Symp. Int. on Combustion, The Combustion Institute*, Academic Press, London, 1971, 171. With permission.)

Intermolecular abstraction of hydrogen by acylperoxy radicals is known to be markedly faster than by corresponding alkylperoxy radicals. The decrease in the rate and appearance of a negative temperature coefficient with increasing temperature may thus also be due to fragmentation of carbonyl radicals to carbon monoxide, which obviously takes place at higher temperatures than the intermolecular hydrogen transfer by acylperoxy radicals. Furthermore, some alkenes are known to inhibit aldehyde oxidation. Their effect consists in efficient scavenging of acylperoxy radicals (Equation 20)

$$R^1C(O)OO\cdot + R^2CH{=}CHR^3 \rightarrow R^{1\cdot} + CO_2 + R_2CH\overset{\displaystyle O}{\overset{\displaystyle /\,\diagdown}{-}}CHR^3 \qquad (20)$$

and in the formation of oxiranes. How the inhibition of aldehydes, or if we want hydrocarbons where alkenes are formed preponderantly in the early stages of the reaction, contributes to the observation of a negative temperature coefficient is questionable.

One can expect that such a spectrum of possibilities which were proved to be real in independent experiments should lead to the more complicated shape of ignition diagrams, where the onset and decay of various branching reactions should produce some "fine structure". Indeed, as an example we may give the ignition diagram of 3-ethylpentane (Figure 6).[2] The individual lobes on the diagram which are characteristic for multiple-stage ignition were supposed to be brought about by the accumulation of various hydroperoxides and peracids. In the region of temperature above which alkene and $HO\cdot_2$ formation begins to prevail, direct peroxidation is usually called the high temperature oxidation region. Hydrogen peroxide produced by subsequent reactions of $HO\cdot_2$ radicals is decomposed here similarly as in the case of the third limit of hydrogen combustion. This may cause the further acceleration of the reaction, the temperature coefficient of which becomes positive and the normal explosion (or ignition) results.

The high temperature burning of all hydrocarbons has one important thing in common, namely, that most of the molecules are consumed by reaction with H atoms and that the reactions of the H_2-O_2 system play an important role in the flame from the very beginning of the process. The radicals H·, ·OH, ·O·, and HO·$_2$ are always present in the cool part of the flame where they can attack the fuel molecules. They are supplied there from the hotter zone of the flame, where they are usually present in overconcentration when compared to equilibrium.

Unsaturated hydrocarbons have two important pathways of oxidation reaction (Equations 21,22). Of interest is that the otherwise

$$RO_2^{\cdot} + RCH{=}CH_2 \rightarrow RO\cdot + RCH_2\overset{O}{\overset{/\ \ \backslash}{CH - CH_2}} \tag{21}$$

$$RO_2^{\cdot} + RCH_2CH{=}CH_2 \rightarrow ROOH + R\dot{C}HCH{=}CH_2 \tag{22}$$

rather inactive terminating allyl radical may in oxidation systems undergo rapid reaction with peroxy radicals (or HO·$_2$ radicals — Equation 23) to form much more reactive hydroxy

$$R\dot{C}HCH{=}CH_2 + HO_2^{\cdot} \rightarrow RCH{=}CHCH_2O\cdot + \cdot OH \tag{23}$$

and alkoxy radicals.[4] The sufficiently fast transfer which can convert nonreactive peroxy radicals into ·OH radicals would bring about a rapid oxidation chain.

The transfer reaction in alcohols obviously begins on the α-carbon (Equations 24 and 25) and chain branching occurs probably

$$RCH_2OH + HO_2^{\cdot} \rightarrow HOOH + R\dot{C}HOH \tag{24}$$

$$R\dot{C}HOH + O_2 \rightarrow RCHO + HOO\cdot \tag{25}$$

in the oxidation of the corresponding aldehyde. In spite of the fact that no pronounced negative temperature coefficient was observed, ethanol easily gives rise to cool flames.

Various compounds arise in the advanced stages of oxidation of hydrocarbons, and some particular features of the oxidation of each group may throw some light on the oxidation of saturated hydrocarbons. One may feel that while there is some agreement on many details of the mechanism of oxidation, no general consensus exists for the entire scheme.

III. OXIDATION OF HYDROGEN

A. General

Among different fuels, hydrogen has the highest heat of combustion per mass unit of fuel; its value is almost three times higher than that of hydrocarbons. Since it is potentially available from water and its combustion gives no toxic products, it belongs among perspective fuel sources in the future. Apart from its own intrinsic interest as the simplest combustion reaction used in rocket engines, it has long been assumed that the knowledge of hydrogen oxidation may contribute appreciably to understanding of the later stages of hydrocarbon oxidation. It is not, therefore, surprising that hydrogen oxidation has been studied very extensively and the general pattern of the process is fairly known.

The gas phase reaction of hydrogen and oxygen takes place at an elevated temperature ($>500°C$) if it is induced thermally and at ambient temperatures when it is induced catalytically or photochemically. Among the elementary steps of importance here are especially branching reactions (Equations 26 and 27) which multiply the amount of free radicals in the system.[5]

$$H· + O_2 \rightarrow ·O· + ·OH \tag{26}$$

$$·O· + H_2 \rightarrow ·OH + H· \tag{27}$$

The additional branching may be represented by radical decomposition of hydrogen peroxide which is the reaction intermediate. Primary radicals arise from the slow bimolecular reaction of oxygen and hydrogen molecules (Equation 28).

$$H_2 + O_2 \rightarrow 2 ·OH \tag{28}$$

The main reaction product is water formed in the propagation step of hydroxy radicals and hydrogen molecules (Equation 29).

$$HO· + H_2 \rightarrow H_2O + H· \tag{29}$$

The concentration of radicals in the hydrogen flame depends on the instantaneous occurrence of respective competition reactions. In the stoichiometric mixture of burning hydrogen ($2 H_2O + O_2$) with oxygen, the most numerous are hydrogen atoms (about 10^{16} cm^{-3}), by one order less are oxygen atoms, while the concentration of hydroxyl radicals is about three orders lower. At the same time, the overall concentration of radicals is by six orders higher than should correspond to the equilibrium value of the thermal dissociation of hydrogen molecules at the temperature of the flame.

Three explosion limits (Figure 2) may be interpreted as follows. The first limit at low pressures is connected with the compensation of branching processes by termination reactions of active particles on the walls of the reactor. Provided that the surface decay of reactive particles manages to eliminate the growth of their amount, the explosion does not occur. The latter reaction is of a heterogeneous character and the critical pressure depends significantly on the coating of the walls of the reaction vessel and on vessel sizes. As may be deduced from the ignition diagram, this heterogeneous reaction depends only weakly on the temperature.

The second explosion limit corresponds to somewhat higher pressures where the third body M, dissipating the reaction heat participating in reaction 30, competes successfully with the branching reaction (Equation 26). The value of the rate constant at 700 K of reaction 26 is about $1·10^6$ dm^3 mol^{-1} sec^{-1}, whereas that for reaction 30 reaches up to $4.4·10^9$

$$H· + O_2 + M \rightarrow HO_2^· + M \tag{30}$$

dm^3 mol^{-1} sec^{-1}. The branching reaction (Equation 26) may thus be realized only at concentrations of the third body (e.g., oxygen, nitrogen, hydrogen) lower than 0.0003 mol dm^{-3}.[5] Above these values, no explosion can occur, since practically all hydrogen atoms, as the most important chain carriers, are scavenged and transformed to hydrogen peroxy radicals which are less reactive under given conditions. Between the first and second explosion limits the explosion of the system occurs. At considerably higher pressures, there may finally be observed the third limit which corresponds either to the subsequent branching reactions of hydrogen peroxide decomposition or hydrogen peroxy radicals (Equation 31) as a consequence

$$HO_2\cdot \rightarrow HO\cdot + \cdot O\cdot \qquad (31)$$

of the energetic branching due to the reaction heat.

In the textbooks of combustion chemistry the analysis of presumed kinetic schemes is put forward under some simplifying assumptions of steady-state conditions. Taking into account the elementary steps (Equation 26 to 30) and monomolecular termination of hydrogen atoms, we receive the set of differential equations (32 to 34) for concentration of hydroxy

$$d[\cdot OH]/dt = 2\,k_{28}[H_2][O_2] - k_{29}[\cdot OH][H_2] + k_{26}[H\cdot][O_2] + k_{27}[\cdot O\cdot][H_2] \qquad (32)$$

$$d[\cdot O\cdot]/dt = k_{26}[H\cdot][O_2] - k_{27}[\cdot O\cdot][H_2] \qquad (33)$$

$$d[H\cdot]/dt = k_{29}[\cdot OH][H_2] - k_{26}[H\cdot][O_2] + k_{27}[\cdot O\cdot][H_2] - k_{30}[H][O_2][M] - k_t[H\cdot] \qquad (34)$$

radicals, oxygen and hydrogen atoms. Making the left sides of the first two equations (32 and 33) equal to zero and substituting the concentrations of hydroxy radicals and oxygen atoms, we have Equation 35

$$d[H\cdot]dt = \varphi[H\cdot] + 2\,k_{28}[H_2][O_2]$$

where

$$\varphi = 2\,k_{26}[O_2] - k_{30}[O_2][M] - k_t \qquad (35)$$

where $\varphi = 2\,k_{26}[O_2] - k_{30}[O_2][M] - k_t$ is called the branching factor, k_t is termination rate constant. If it is positive, the concentration of hydrogen atoms increases exponentially in time (Equation 36)

$$[H\cdot] = v/\varphi\,(e^{\varphi t} - 1) \qquad (36)$$

where v is the rate of initiation (v = $2\,k_{28}[H_2][O_2]$). For $\varphi<0$, the concentration of hydrogen atoms reaches steady value after some time $[H\cdot_{st}] = v/[\varphi]$, whereas at $\varphi = 0$ the concentration of hydrogen atoms increases linearly ([H] = vt). Regardless of the fact that we have omitted some elementary steps from the complex scheme of hydrogen oxidation (which are predominantly of termination and initiation character) and have used the steady-state assumption for concentrations of oxygen atoms and hydroxy radicals, the critical condition of hydrogen ignition, $\varphi = 0$ is quite obvious. Since in the equation for branching factor φ (Equation 35) the concentrations of O_2 and M are both the functions of overall pressure, we may show that, indeed, two limiting pressures correspond to the zero value of φ thus approximately determining the borderline in the ignition diagram.

The interpretation of the first and second pressure limit of hydrogen ignition in terms of the above analysis, which belongs to the classical knowledge of combustion chemistry, is very illustrative and comprehensible. Its shortcoming consists in the fact that no explosion in the mathematical sense of infinite value of the variable at finite time follows from it.

It is of interest that the solution of the complete set of differential equations, corresponding to 13 elementary steps from Reference 5 for temperatures and pressures above the critical conditions of the ignition, does not overcome it and only exponential growth of hydrogen atoms can be obtained with increasing time.

Table 1
VALUES OF ARRHENIUS PARAMETERS FOR REACTION OF H, O, AND OH RADICALS WITH PRIMARY, SECONDARY, AND TERTIARY C-H BOND

Reaction	C–H	A, dm^3 mol^{-1} sec^{-1}	E, kJ/mol
H·+RH → H$_2$+R·	Primary	2.2 · 10^{10}	41
	Secondary	5 · 10^{10}	35
	Tertiary	8.7 · 10^{10}	28
O·+RH → ·OH+R·	Primary	5 · 10^9	24
	Secondary	1.3 · 10^{10}	19
	Tertiary	1.6 · 10^{10}	14
·OH+RH → H$_2$O+R·	Primary	7 · 10^8	6.9
	Secondary	1.5 · 10^9	3.6
	Tertiary	1.3 · 10^9	0.6

B. The Influence of Additives of Hydrogen-Oxygen Reaction

The traces of NO$_2$ added to some hydrogen-oxygen mixtures lower the ignition temperature of the system by more than 200 K. A similar effect is also demonstrated by nitrosyl chloride, nitrous oxide, ammonia, cyanogen, chlorpicrin, and nitric oxide. The reaction has an induction period of several seconds between the dosage of reactants into the reactor and its onset.

The key to the sensitization effect may be revealed in the reaction of hydrogen peroxy radicals with NO (Equation 37)

$$HOO· + NO → ·OH + NO_2 \tag{37}$$

which has a rate constant on the order of 10^8 dm^3 mol^{-1} sec^{-1} at room temperature. This reaction converts less reactive HO$_2$· radicals to very reactive hydroxy radicals even at higher pressures of the system. In the case of NO$_2$ addition there exist its pressure limits, lower and upper, which determine the occurrence of ignition. When we follow the changes of NO$_2$ pressure from the beginning of the reaction, we may notice that during the induction period, NO$_2$ pressure gradually falls down from the initial value. Provided that ignition occurs, this decrease is sharply accelerated at the end of the induction period, whereas in the case of the slow reaction, the acceleration declines and NO$_2$ pressure reaches some stationary value. Nitrogen dioxide is depleted by the reaction 38.

$$·O· + NO_2 → NO + O_2 \tag{38}$$

When small amounts of hydrocarbons are added to the hydrogen-oxygen mixture, three effects may be observed, namely:

1. The first pressure limit is raised, second pressure limit is reduced (inhibition of explosion)
2. Increase of the maximum rate of reaction in the parametrical region of slow oxidation
3. Induced explosion occurring outside the normal H$_2$ + O$_2$ explosion peninsula

In principle, all hydrocarbons can suppress the low pressure ignition of the H$_2$ + O$_2$ system. There is, however, a large difference in the effect of individual hydrocarbons. This may be understood from Table 1, where Arrhenius parameters[2] are given for reactions potentially competing with branching steps. These reactions include the interaction of hydrogen and oxygen atoms and hydroxy radicals with hydrogen on primary, secondary, and tertiary carbon atoms of hydrocarbon.[6]

Since the rate constants of the reaction (Equation 39) with

$$H\cdot + RH \rightarrow H_2 + R\cdot \qquad (39)$$

primary hydrogens at 700 K is $2\cdot10^7$ dm^3 mol^{-1} sec^{-1}, with secondary hydrogens $7.3\cdot10^8$ dm^3 mol^{-1} sec^{-1}, and the rate constant of the branching reaction (Equation 26) is only $1\cdot10^6$ dm^3 mol^{-1} sec^{-1}, we may deduce which will be the inhibition effect of individual hydrocarbons on the oxidation course. The hydrocarbons with primary hydrogens, such as methane or neopentane, lower, e.g., the second limit only slightly with the increasing concentration until its critical value is reached at which the explosion is completely eliminated. This sudden inhibition is, however, due to subsequently formed formaldehyde which is a powerful inhibitor of the $H_2 + O_2$ low pressure explosion.

Hydrocarbons with secondary and tertiary hydrogens suppress the explosion region in a more pronounced way even at fairly low concentrations. Hydroperoxides from hydrocarbons are formed which enrich the system with other free radicals, and, consequently, in the slow oxidation the rate of the process may be higher. The induced explosion of nonisothermal character which occurs outside of the $H_2 + O_2$ explosion peninsula may be ascribed to reactions of hydrocarbon oxidation as was shown earlier.

It should be pointed out that small additions of hydrogen peroxide itself may also bring about the retardation of the reaction because of the deactivation reactions (Equations 40 and 41).

$$H\cdot + H_2O_2 \rightarrow H_2 + HO_2\cdot \qquad (40)$$

$$HO\cdot + H_2O_2 \rightarrow H_2O + HO_2\cdot \qquad (41)$$

IV. THE OXIDATION OF CARBON MONOXIDE

The behavior of carbon monoxide-oxygen mixtures may be described in terms of similar explosion diagrams as the hydrogen-oxygen system. There is, however, an additional region in coordinates of pressure and temperature (Figure 7), where a slower reaction occurs producing a weak blue chemiluminescence which reminds one of a feeble flame. This glow reaction which can put considerable uncertainty into the determination of explosion limits is interesting from the viewpoint of its possible oscillatory course.

The dry and impurity free system of $CO + O_2$ is difficult to ignite, and when ignited, burns with a lower velocity than that in the presence of moisture. Also, the chemistry of the dry $CO + O_2$ system and interpretation of explosion limits brings some problems which may be overcome by assumption of energetical branching. For pure $CO + O_2$ systems, it is assumed that the process proceeds as a sequence of elementary reactions (Equation 42 to 45).

$$CO + \cdot O\cdot \rightarrow CO_2^* \qquad (42)$$

$$CO_2^* + O_2 \rightarrow CO_2 + 2\cdot O\cdot \qquad (43)$$

$$\cdot O\cdot + O_2 + M \rightarrow O_3 + M \qquad (44)$$

$$CO + O_3 \rightarrow CO_2 + 2\cdot O\cdot \qquad (45)$$

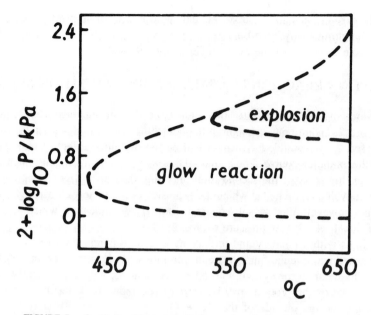

FIGURE 7.   Qualitative illustration of glow and explosion limit curves for $CO + O_2$ (2:1) mixture.[5] (From Hoare, D. E. and Walsh, A. D., *Trans. Faraday Soc.*, 50, 37, 1954. Copyright by Royal Society of Chemistry, London. With permission.)

There are, however, doubts whether CO reacts with ozone directly (Equation 45) or whether its reaction is preceded by decomposition of ozone into oxygen atoms.

Reactions 44 and 45 are, therefore, sometimes replaced by reactions 46 and 47. The termination reactions in the

$$CO_2^* + CO \rightarrow CO^* + CO_2 \tag{46}$$

$$CO^* + O_2 \rightarrow CO_2 + \cdot O \cdot \tag{47}$$

region of the lower explosion limit seem to be light emitting wall reactions (Equations 48 and 49).

$$walls + \cdot O \cdot \rightarrow products + h\nu \tag{48}$$

$$walls + CO_2^* \rightarrow CO_2 + h\nu \tag{49}$$

In the case of the second explosion limit, it is probably the reaction (Equation 50). Practically, it is difficult to exclude

$$CO + \cdot O \cdot + M \rightarrow CO_2 + M \tag{50}$$

from the $CO + O_2$ system the presence of hydrogeneous compounds and also reactions from the hydrogen-oxygen system (Equations 26 and 27) may be of great importance as the reactions of hydroxy radicals (Equation 51) expressed as summary reaction 52.

$$\cdot OH + CO \rightarrow CO_2 + H \cdot \tag{51}$$

$$CO + H_2O \rightarrow CO_2 + H_2 \tag{52}$$

The explosion limits depend strongly on the presence of traces of hydrogen-containing compounds which may turn the description of the system to that of hydrogen and oxygen mixtures.

V. OXIDATION REACTIONS CATALYZED BY TRANSITION METAL IONS

The effective way of acceleration or inhibition of a free radical reaction is the addition of some transition metal ions, which play either a direct or an indirect role at all oxidation reactions. Their trace quantities may significantly influence the overall course of the radical reaction. Even though some reactions concerning the generation of free radicals were dealt with, it should be recalled that transition metal ions may also enter the propagation and termination step of a chain reaction, and their reactivity may vary to a large extent from an accelerating to an inhibiting effect, bringing about quite different reaction products. This may well be documented in the case of Fenton's reagent,[7] which in an inert atmosphere and at only a mild excess of hdyrogen peroxide, reacts with tert-butyl alcohol, yielding 2,5-dimethyl-2,5-hexandiol as the main reaction product (Equations 53 and 54). The small

$$(CH_3)_3COH + HO\cdot \rightarrow (CH_3)_2C(OH)CH_2\cdot + H_2O \qquad (53)$$

$$2(CH_3)_2C(OH)CH_2\cdot \rightarrow (CH_3)_2C(OH)CH_2CH_2C(OH)(CH_3)_2 \qquad (54)$$

amount of Cu(II) ions, however, prevents dimerization and gives 2-methyl-2,3-propandiol (Equation 55).

$$(CH_3)_3CH(OH)\dot{C}H_2 + Cu(II) + H_2O \rightarrow (CH_3)_2C(OH)CH_2OH + Cu(I) + H^+ \qquad (55)$$

The character and number of surrounding ligands do not only influence the electron density on a respective ion, but may facilitate or retard the electron transfer from/to the substrate. During their catalytic action, the transition metal ions change the oxidation state as well as the character of the coordination field, which leads either to deactivation or activation of the catalytic effect.

As was pointed out, the generation of free radicals by transition metal ions is not conditioned by peroxides only, but may involve other substrates, too, such as hydrocarbons, halocarbons, halogens, etc., capable of accepting or donating electrons from/to the transition metal compound (ML_n).

As an example of oxidation by the electron transfer mechanism, we present the reaction of alkylbenzenes taking place in the presence of Co(III) acetate and oxygen[8] (Equations 56 to 64).

$$C_6H_5CH_3 + Co(III) \Leftrightarrow [C_6H_5CH_3^{\oplus}]\cdot + CO(II) \qquad (56)$$

$$[C_6H_5CH_3^{\oplus}]\cdot \rightarrow C_6H_5CH_2\cdot + H^+ \qquad (57)$$

$$C_6H_5CH_2\cdot + Co(III) \rightarrow C_6H_5CH_2^+ + Co(II) \qquad (58)$$

$$C_6H_5CH_2^+ + HOAc \rightarrow C_6H_5CH_2OAc + H^+ \qquad (59)$$

$$C_6H_5CH_2\cdot + O_2 \rightarrow C_6H_5CH_2OO\cdot \qquad (60)$$

$$C_6H_5CH_2OO\cdot + Co(II) \rightarrow C_6H_5CH_2OOCo(III) \qquad (61)$$

$$C_6H_5CH_2OO\cdot + C_6H_5CH_3 \rightarrow C_6H_5CH_2OOH + C_6H_5CH_2\cdot \tag{62}$$

$$C_6H_5CH_2OOCo(III) \rightarrow C_6H_5CHO + Co(III)OH \tag{63}$$

$$Co(III)(OH) + HOAc \rightarrow AcOCo(III) + H_2O \tag{64}$$

The above sequence of reactions, which via benzaldehyde leads to benzoic acid, may proceed even without the apparent formation of free cation radicals. The fast transfer of protons from it may occur in the cage of the solvent so that only benzyl radicals arise.

Aldehydes are much more susceptible to interaction with the transition metal ions than are hydrocarbons (Equation 65).

$$RCHO + Co(III) \rightarrow R\overset{\cdot}{C}O + H^+ + Co(II) \tag{65}$$

In the presence of oxygen the formed acyl radical is transformed to peracid with back regeneration of Co(III) (Equations 66 to 69).[9-11] Oxidation of alkylaromatic hydrocarbons taking

$$R\overset{\cdot}{C}O + O_2 \rightarrow RCO_3\cdot \tag{66}$$

$$RCO_3\cdot + RCHO \rightarrow RCO_3H + R\overset{\cdot}{C}O \tag{67}$$

$$RCO_3H + Co(III) \rightarrow Co(III)OH + RCO_2\cdot \tag{68}$$

$$RCO_2\cdot + RCHO \rightarrow RCO_2H + R\overset{\cdot}{C}O \tag{69}$$

place in the presence of Co or Mn salts is strongly accelerated by bromide anions. The reason for this is likely to consist in the formation of Co(II)(AcO)Br compound oxidized further by hydroperoxide to Co(III) (AcO)Br which eliminates bromide atom carrying the oxidation chain (Equations 70 and 71).

$$Co(III)(AcO)Br \rightarrow Co(III)(OH)(AcO) + Br\cdot \tag{70}$$

$$Br\cdot + C_6H_5CH_3 \rightarrow HBr + C_6H_5CH_2\cdot \tag{71}$$

One thing is not sure here, namely, whether the acceleration effect is due to the direct reaction of peroxy radicals with free HBr, the reservoir of which is Co(III)(AcO)Br, or if it is brought about by the instability of the latter only.

Another effect of transition metal ions may consist in the activation of molecular oxygen. The direct reaction of hydrocarbons with the triplet ground state of molecular oxygen is spin forbidden. If molecular oxygen is complexed with some transition metal ion, its spin state is changed and the reaction with hydrocarbon becomes allowed. Interaction of molecular oxygen with complexes of transition metal ions may occur either as one or two electron transfer.[12] The former case leads to superoxo complexes, the latter to peroxo complexes. Co(II), e.g., give with molecular oxygen superoxo complexes primarily (Equation 72) which in the subsequent reaction with

$$Co(II) + O_2 \rightarrow Co(III)OO\cdot \tag{72}$$

Co(II) yield a bicentric bridged peroxo complex (Equation 73).

$$Co(III)OO\cdot + Co(II) \Leftrightarrow Co(III)OOCo(III) \tag{73}$$

In dipolar proton donating solvents the superoxo complexes are of limited stability and undergo fast exchange reactions (Equation 74) leading to hydrogen peroxy radicals, which

$$M^{n+}OO\cdot + ROH \rightarrow M^{n+}OR + HOO\cdot \tag{74}$$

may initiate further radical reactions.[13] This example is one of the prerequisites of the start of thermo and photooxidation degradation of hydrocarbon polymers which may change their use properties considerably even during storage. Trace quantities of metal ions in cooperation with air humidity may bring about the gradual development of a chain oxidation reaction in the sense of the already quoted schemes. Moreover, there always exists some probability of back restoration of original $M^{(n-1)+}$ ions from $M^{n+}OR$ species consisting in the reaction (Equation 75) where HOOH is formed in the oxidation process.

$$HOOH + M^{n+} OR \rightarrow ROH + M^{(n-1)+} + HOO \tag{75}$$

The above reaction, which is one of the several possible alternatives of catalytic reaction cycles of transition metal ions, represents a real danger of stepwise development of an unwanted oxidation reaction. Because of the usually very low concentration of metal ions, this reaction is very slow in its initial stages; however, as the concentration of peroxides gradually increases, it may become important.

On the other hand, these reactions may play some positive role in peroxide formation in the living tissues. The question of the role of peroxides in normally functioning organisms was reexamined several times in recent years, and the first approach — that the peroxides are always harmful to living tissues — was rejected. It was ascertained that a certain level is even necessary for normal metabolism. It was also shown that lipid peroxides are intermediates for synthesis of biologically active compounds, such as prostaglandins, thromboxanes, and steroid hormones, and that their presence up to a certain level is a necessary condition of life.[14]

It is believed that peroxidation of lipids is, on the one hand, necessary for the regeneration of phospholipid membranes, while above some critical rate it leads to aging. Lipid peroxides can, e.g., activate or inhibit the effect of some enzymes.

Which mechanism controls the stationary course of lipid peroxidation? The ideal scheme for such control is forwarded by nonenzymatic oxidation of lipids with reactive free radical intermediates and with the participation of transition metal ions-oxygen complexes quoted in the preceding text.

Provided that $HO_2\cdot$ radicals or HO_2H are efficiently scavenged, the level of peroxides is low, and peroxidation is, in fact, a series of nonchain radical reactions and as such, may serve as an example of "desired" peroxidation. At the same time, the mosaic structure of the membrane bilayer with lipid islands can ensure that this process takes place under quasi-homogeneous conditions as in solution. The limit of turnover from nonchain to chain per-oxidation will be very subtle and will depend on peroxide concentration; the higher level of peroxides may start the sequence of hydroxy radical generation (Equation 76) which would be the harmful process in such an interpretation due to high nonselectivity of hydroxy radicals.

$$M^{n+}OO\cdot + H_2O_2 \rightarrow M^{n+}(OH) + O_2 + \cdot OH \tag{76}$$

An interesting chain reaction with respect to transition metal ions may be the insertion of molecular oxygen to the M-R bond where M is metal (Equations 77 to 79). The initiation step involves

$$[Cr(III)CH(CH_3)_2]^{2+} \rightarrow Cr(II) + \dot{C}H(CH_3)_2 \qquad (77)$$

$$(CH_3)_2\dot{C}H + O_2 \rightarrow (CH_3)_2CHOO\cdot \qquad (78)$$

$$(CH_3)_2CHOO\cdot + [Cr(III)CH(CH_3)_2]^{2+}$$

$$\rightarrow [Cr(III)OOCH(CH_3)_2]^{2+} + (CH_3)_2\dot{C}H \qquad (79)$$

here the homolysis of the metal-carbon bond, and, in propagation, alkyl radicals attached to metal are displaced by peroxy radicals.

Ion metal complexes, as e.g., those of Ir(I) and Rh(I) may catalyze the oxidation of dienes to polyperoxides (Equation 80) proceeding by a similar substitution mechanism.

$$(80)$$

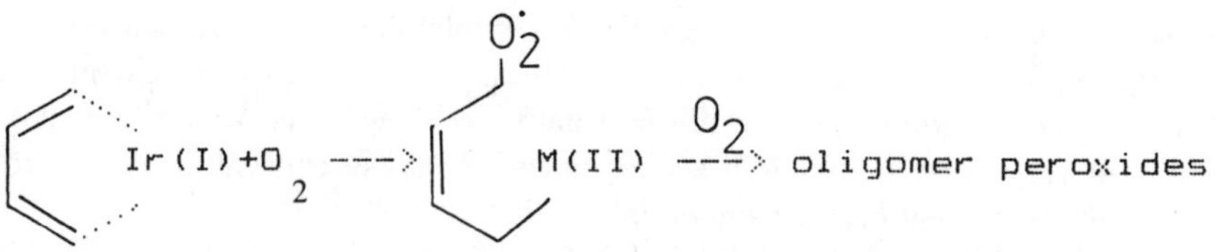

Until now, the role of free radicals at the heterogeneous oxidation of organic compounds taking place on oxometallic compounds such as OsO_4, RuO_4, MoO_3, and MnO_2 (which found wide application in organic synthesis due to their capability to transfer oxygen to many substrates under relatively mild conditions) has not been quite clear. At the catalytic oxidation of alkenes[15,16] allyl intermediates on the surface of oxo compounds are formed (Equation 81). Hydrolysis of the last surface product

$$CH_2{=}CH{-}\dot{C}H_2\cdots O{=}Mo(V){=}O \rightleftharpoons CH_2{=}CH{-}CH_2{-}\overset{\displaystyle OH}{\underset{\displaystyle |}{Mo}}(VI){=}O \qquad (81)$$

$$CH_2{=}CH{-}CH_2{-}\overset{\displaystyle OH}{\underset{\displaystyle |}{Mo}}(VI){=}O \rightarrow CH_2{=}CH{-}CH_2{-}O{-}\underset{\displaystyle \overset{\displaystyle |}{OH}}{Mo}(IV)$$

(Mo=O represents a part of the surface skelet.)

by water vapors gives allyl alcohol as the primary product of oxidation, which is further oxidized to acrolein.

π-Allyl complexes of heterogeneous oxometallic compounds may be understood to be precursors of allyl free radicals which may be released from the complex to the system quite easily.

$$Mo(VI)CH_2CH{=}CH_2 \;\rightleftharpoons\; Mo(V) + \cdot CH_2CH{=}CH_2$$

SCHEME 2

Biochemical oxidation may well be demonstrated on metalloporphyrins which represent a large group of biological substrates responsible for activation, transport, and storage of molecular oxygen in living tissue.[17]

The term porphyrin involves porphin and phtalocyanine and their derivatives which have the cyclic structure composed of pyrol and benzpyrol units mutually linked by methine or azine groups (Scheme 3).

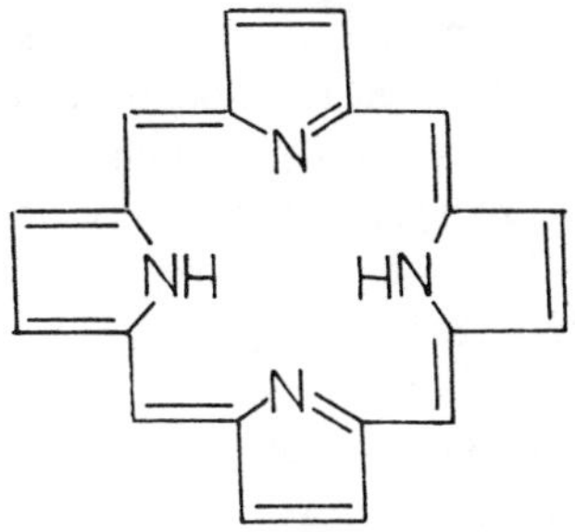

porphine

SCHEME 3

The metal ion is complexed here among four nitrogen atoms. High conjugation within the porphyrin cycle makes the transfer of electron density possible from the periphery of the cycle to the metal center by changing the peripheral substituents. The complexed metal ions, such as Mg in the case of chlorophyll or Fe in heme, have two free coordination sites in a *trans* position which facilitates the transfer of electrons between the axial ligands through the metal center. Thus, the formation of the metal-oxygen bond and its strength may be controlled by both the substituents and axially coordinated ligands.

Depending on the number of π-electrons at the metal center, metalloporhyrins, due to their conjugated electron system, may undergo reduction or oxidation either of the central atom or ligand.[18]

VI. OXIDATION OF POLYMERS

Organic polymers, like other organic compounds, if subjected to the action of oxidizing gas and/or temperature and light, undergo more or less extensive chemical reaction which ultimately leads to the degradation, cross-linking, and the change (deterioration) of use properties. The mechanism of this type of process is of a free radical nature predominantly including degenerated branching via hydroperoxides. It may have features in common with other low molecular organic substrates, but it may also be substantially different. The difference may be due to the macromolecular character of polymer molecules which usually do not have the ideal structure but involve some concentration of defects and anomalous centers on which the initiation of the free radical reaction starts. These anomalous structures, which are incorporated into macromolecules during its synthesis, storage, or processing may include both the structural defects and low molecular additives, impurities, residues of polymerization catalysts, transition metal ions, etc.; their concentration in the polymer is sometimes so low that it cannot be detected by routine analytical techniques. Structural defects cannot also be removed by any arbitrary method of polymer purification. The other peculiarity of the oxidation reaction of polymers consists in the more difficult diffusion conditions for the oxidant in the condensed medium of the polymer material. The chain-centered free radicals, once formed in the solid polymer, propagate only to the close surroundings of their formation in the presence of oxygen, and microheterogeneous domains of different composition from the parent polymer usually arise. The radical reaction in the condensed phase of the polymer to higher distances are more likely to be mediated by low-molecular free radicals or compounds which arise from the macromolecular precursor by fragmentation reactions. The concentration heterogeneity phenomenon is particularly important in the case of semicrystalline polymers, where all low molecular compounds (oxygen, oxidation products, inhibitors, plasticizers, etc.) accumulate in the amorphous regions of a polymer.

In the presence of oxygen, the primary reaction centers on the macromolecule formed there by the action of light, heat, different initiators of radical reactions (peroxides, ozone, halogens, oxygen, azocompounds, carbonyl compounds), electric field, etc., are transformed to peroxy radicals POO· which propagate the oxidation reaction. In accordance with the above-mentioned schemes of hydrocarbon oxidation, the overall mechanism may be illustrated very simply (Equations 82 to 88).

$$P\cdot + O_2 \rightarrow POO\cdot \tag{82}$$

$$POO\cdot + PH \rightarrow POOH + P\cdot \tag{83}$$

$$POOH \rightarrow PO\cdot + \cdot OH \tag{84}$$

$$PO\cdot(P\cdot) \rightarrow P\cdot' + RH \tag{85}$$

$$POO\cdot + POO\cdot \rightarrow \tag{86}$$

$$POO\cdot + P\cdot \rightarrow \quad\quad \text{termination} \tag{87}$$

$$P\cdot + P\cdot \rightarrow \tag{88}$$

(P denotes the macromolecular C-fragment)

The system becomes more complicated by fragmentation reactions of PO· or P· radicals (Equation 85) from which lower molecular fragments are formed, and the molecular weight of the polymer decreases. Provided that on the macromolecule backbone there are either inherently present or subsequently formed some reactive groups such as double bonds, some reactive side groups etc., the decrease of the molecular weight may be slowed down or even changed due to the advanced cross-linking reaction.

We have already pointed out that reactions 86 and 87 are not, in fact, termination reactions as such since they may generate alkoxy radicals either directly or via dialkylperoxides. The nondissipated heat of reaction 87 is sufficient to induce the immediate decomposition of POOP molecule on its O-O bond. Since alkoxy radicals are more reactive in transfer reactions than alkyl peroxy radicals, the course of the oxidation reaction may thus be considerably faster when compared to that with pure alkoxy peroxy radical propagation. This may also be the reason that the rate constants referred to in the reaction (Equation 83) of qualitatively identical peroxy radicals differ to the extent of several orders.[19] (For peroxy radicals of polymethyl methacrylate at 273 K $k_{max}[PH] = 1\cdot10^{-4}$ sec^{-1}, whereas k_{min} [PH] is only $3.9\cdot10^{-6}$ sec^{-1}. Also, the activation energy and preexponential factor show relatively large scatter.

It should be recalled here, that even though much effort has been devoted to the explanation of the fate of polymer alkyl and peroxy radicals in the macromolecular medium, only partial success has been achieved. The commonly accepted approach to the interpretation of some nivelization in reactivity on one side and the difference when compared to low molecular medium on the second side is based upon the idea of kinetic nonequivalency of respective reaction sites in different polymer domains. So-called polychromatic kinetics, the conception of the migration of the radical center along and across macromolecular chains and nivelization of the reactivity of low molecular additives (antioxidants and stabilizers) in polymers are only a few examples of this relatively large research field.

VII. COMBUSTION OF POLYMERS

A burning polymer is a highly complex system involving the radical reactions preponderantly occurring in three interdependent regions, namely, in the condensed phase, at the interphase between the condensed phase and the gas, and in the gas phase. The polymer first decomposes in the condensed phase to give gaseous products which are usually combustible and enter the flame zone above its surface. Here they burn as gas which leads to the final products (CO_2 and water) and to the heat release. Under steady-state conditions of burning, part of the liberated heat is conducted back to the polymer surface where it brings about further reaction and yields more combustible gaseous products. For the subsequent course of the process, it is important whether the character of the polymer surface changes significantly when compared with the original sample or not. In the former case, the formed carbonaceous residue usually makes the propagation of the flame more difficult and the smoldering or glowing of the surface layer controls the development of the fire. The latter case represents the burning of the polymer combined with evaporation of the polymer sample preceded by its degradation with more or less significant participation of oxygen.

If we compare the qualitative pattern of ignitability or tendency to burning for low molecular hydrocarbons and their polymer analogs, our attention will be attracted by two important interrelated processes.

1. Degradation and volatilization of polymer preceding the gas phase reactions
2. Possible ejection of reactive free radicals from the condensed phase and its influence on the initiation of free radical reactions in the gaseouse phase[20,21]

If the second item is of importance, the separate stages of the polymer combustion depicted in the following Scheme 4, where P is the organic polymer, ΔH is the heat, C_1 and C_2 are combustible gaseous and final combustion products, could involve the participation of radicals R· released from the condensed phase in the ignition of the combustible gas and both processes, namely, the ignition of gaseous and polymer hydrocarbon will be considerably different.

$$P + \Delta H + (O_2) \rightarrow C_1(+R\cdot)$$
$$C_1 + O_2 + R\cdot \rightarrow C_2 + \Delta H$$

SCHEME 4

Even though it was unequivocally proved that there exists the release of radical particles from the condensed phase of thermally exposed polymers to the gaseous phase dependent, of course, on the nature of the respective polymer, we feel that they play a minor role in the polymer ignition. The yield of free radicals escaping to the gaseous phase will be considerably higher for polymers undergoing statistical degradation, such as polyolefines, when compared to those decomposing by an unzipping mechanism (polyoxymethylene, polymethyl methacrylate). The footprint of the former case on the surface of the molten polymer can be demonstrated on the simultaneous thermogravimetry and DTA records of

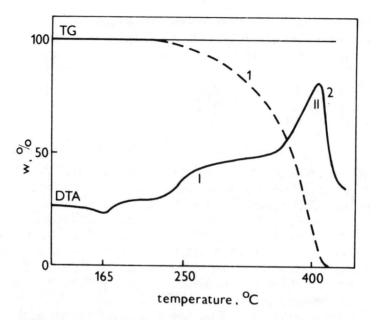

FIGURE 8. Simultaneous record of the loss of volatile products (1) and differential thermal analysis (2) at the decomposition of polypropylene (50 mg) in the air (the rate of heating 10°/min W is the residual weight of sample.)

polypropylene in air. The DTA curve demonstrates two exotherm maxima (Figure 8). The first maximum can be easily cancelled by current phenolic antioxidants, while the second cannot. The first maximum represents the thermooxidation of a polymer, while the second is closely connected with a cool flame phenomenon appearing in the gas phase. We recall that under some conditions, the cool flame may ignite the hot flame, too, but in a variety of different fuels this is not a necessary condition. Polymethyl methacrylate, e.g., supplies monomeric methylmethacrylate to the gas phase without noticeable cool flame combustion, the polymer being as easily ignited as, e.g., polypropylene.

The simplest approach to polymer combustion, therefore, considers the polymer only as a reservoir of the fuel, the rate of its supply being checked by polymer decomposition and its peculiarities. The flame itself is then the matter of the nature of the released fuel. The predominant decomposition of the polymer to the original monomer involves only a few cases (polymethyl methacrylate, polymethacrylonitrile, poly-α-methylstyrene, polytetrafluoroethylene, and polyoxymethylene) and generally, a mixture of different volatile products is formed and the description of the burning process is rather complex.

As was pointed out, the polymers yielding the char residue in their degradation tend to glow along with flaming combustion. Glowing can be understood as a heterogeneous exothermic process of a carbonaceous residue oxidation. Let us image that the carbonaceous residue can be represented by a polyaromatic conjugated system. Its reaction is likely to start on the most reactive superficial polyconjugated double bonds so that the addition of $HO_2\cdot$ radicals takes place here and oxirane rings are primarily formed (Equation 89). At

$$+ HO_2\cdot \quad \text{---} \rightarrow \quad HO\cdot \ +$$

(89)

elevated temperatures the decomposition of these oxiranes yields biradicals (Equation 90), which in the presence of oxygen, react

$$(90)$$

further to cyclic peroxides which decompose and undergo β-scission with the elimination of CO_2, CO, and the skeleton of a carbon residue thus loses one aromatic ring. Because of the strong exothermicity of reaction, enhanced moreover by the subsequent oxidation of CO, the process may go on in the presence of oxygenated radicals and the carbon residue is "evaporated" in a stepwise manner. Provided that the concentration of oxygen is not sufficient, only smoldering occurs, which is, in fact, incomplete glowing combustion. Smoldering combustion has been studied very little from a fundamental point of view. It takes place when cellulosic materials or certain thermosetting polymers with a high surface/volume ratio, such as polyurethane foams, are thermally isolated because of, e.g., the large thickness of the material, which impedes the efficient heat decrease and thus sustains the exothermic process.

The cross-linking reaction and char formation has, thus, an apparently controversial effect in polymer combustion; on one hand, it decreases the tendency of the polymer to combust, since the supply of the fuel to the gaseous phase becomes restricted; on the other hand, it may support the occurrence of heterogeneous processes of nonflaming combustion.[22]

VIII. INHIBITION OF SLOW CHAIN REACTIONS

A large variety of pathways exist which slow down free radical reaction intervening in the initiation, propagation, and termination stages of the process. The additives which interrupt the chain radical reaction, decreasing the rate of initiation and propagation and increasing the rate of termination, are generally called stabilizers. According to the mode of their effect we recognize preventive stabilizers and chain-breaking stabilizers. Preventive stabilizers take part in the initiation step of the free radical reaction by slowing down free radical generation. As an example, we may show the decomposition of hydroperoxides by dilaurylthiodipropionate ($R_2'S$) (Equation 91) which occurs in a nonradical way.

$$ROOH + R_2'S \rightarrow ROH + R_2'SO \qquad (91)$$

Metal deactivators which bind the transition metal ions into the complex and impede, thus, the free radical decomposition of hydroperoxides or the direct formation of free radicals (Equations 2 and 3) belong to another group of stabilizers. The deactivation of metal ions consists in the formation of nonsoluble complexes, in the change of the redox potential by chelating hydroxy acids or ethylenediaminotetraacetic acid derivatives.

The initiation reaction may be efficiently suppressed also by competitive scavenging of other initiating molecules such as ozone, singlet oxygen, nitrogen oxides, etc. This is the case of antiozonants which protect rubbers and other unsaturated polymers and compounds against the harmful effect of ozone. Ozone reacts with C=C double bonds in rubbers very fast (for polybutadiene, at 20°C in CCl_4 the rate constant is $6 \cdot 10^4$ dm^3 mol^{-1} sec^{-1}; for polyisoprene $4.4 \cdot 10^5$ dm^3 mol^{-1} sec^{-1}), giving ozonide, cyclic peroxides, and hydroperoxy groups.[23] The efficient antiozone protection requires the presence of compound which may successfully compete with it. Such a demand is met by some diamines, as, e.g., *N,N*-dialkyl-*p*-phenylene-diamine and other compounds (the rate constant of aromatic diamines with ozone is of the order of $10^6 \cdot dm^3 \, mol^{-1} \, sec^{-1}$) which are assumed to enter the reaction by the secondary amino group (Equation 92).

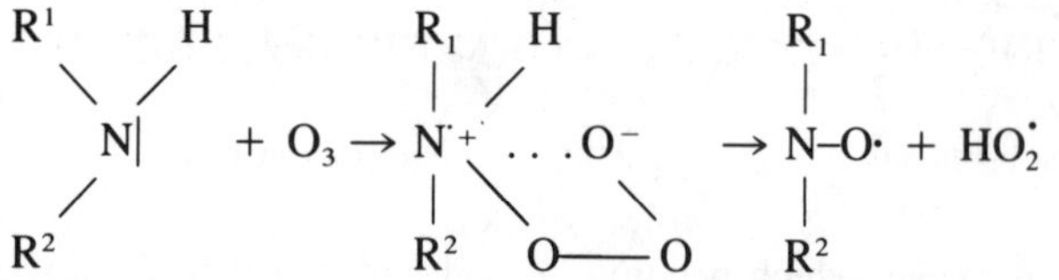

$$
\begin{array}{c}
R^1 \quad\ H \\
\diagdown \diagup \\
N| \\
\diagup \\
R^2
\end{array}
+ O_3 \rightarrow
\begin{array}{c}
R_1 \quad H \\
| \diagup \\
N^+ \cdots O^- \\
| \diagdown \quad \diagdown \\
R^2 \quad O\!-\!\!-\!O
\end{array}
\rightarrow
\begin{array}{c}
R_1 \\
| \\
N\!-\!O\cdot \\
| \\
R^2
\end{array}
+ HO_2^{\cdot}
\tag{92}
$$

In the case of polymers, the effect of ozone is focused particularly on the surface layers of the sample, the periodical mechanical stress having a pronounced effect on the sample fatigue. Many questions on the antiozone protection concerning the mechanism remain unanswered. The crucial point lies directly in the mechanism of the ozone effect itself which, despite many uncertainties, seems to demonstrate the mutual interplay of radical and ionic schemes.

Another way of suppressing the initiation phase of a free radical reaction is advanced by ultraviolet absorbers and quenchers of excited states. The filtration effect of these compounds (derivatives of *o*-hydroxybenzophenone, benzotriazol, carbon black, etc.) does not allow the excitation of chromophores by ultraviolet light (carbonyl groups) in some materials and transforms the excess of absorbed energy to heat. Similarly acting are the quenchers of exicted states (as, e.g., Ni salts of some organic acids).

Chain-breaking stabilizers interfere with the propagation phase of chain reactions. Hindered phenols and aromatic amines are used as antioxidants. Their effect is based upon easy reaction with peroxy radicals where the formed phenoxy or amino radical is considerably less reactive in propagation.

As was already pointed out, the reactivity of these intermediate phenoxy or amine radicals is determined to a large extent by sterical shielding of a formed radical site, by the degree of its delocalization, and not least, by the effect of substituents having opposite electron activities. This may be, e.g., seen on derivatives of 3-aniline-1,5-diphenyl pyrazoles (Scheme 5) where X, Y, Z are substituents.

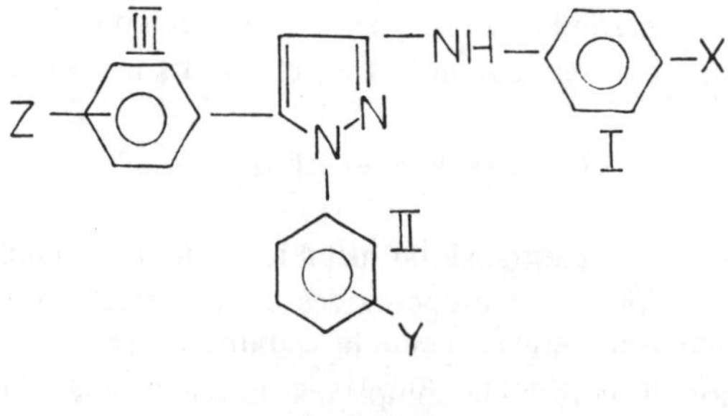

SCHEME 5

Estimating the antioxidant efficiency of such compounds in polypropylene by the chemiluminiscence method and relating it to the nonsubstituted compound (X = Y = Z = H), it may be seen that there exists the correlation between the antioxidant efficiency and the quality of the respective substituent on the I, II, and III phenyl ring, the most efficient being those which have X and Y substituents influencing the reactivity of the radical site by a push-pull mechanism. The correlation may be expressed in terms of dependence of antiox-

Table 2

ANTIOXIDANT EFFICIENCY (I) OF X,Y,Z-SUBSTITUTED *N*-PHENYL *N*-PYRAZYL AMINES IN POLYPROPYLENE ESTIMATED FROM THE INHIBITION PERIODS OF CHEMILUMINESCENCE CURVES, HAMMET'S CONSTANTS OF THE SUBSTITUENTS X,Y,Z AND THEIR DIFFERENCE M

Substituents			I	Hammet's constants			Diff.
X	Y	Z	Rel.u.	X	Y	Z	m
Cl	OCH$_3$	H	0.47	0.23	−0.27	0	0.50
Cl	H	CH$_3$	0.79	0.23	0	−0.17	0.40
Cl	H	H	0.95	0.23	0	0	0.23
H	H	H	1.0	0	0	0	0
H	CH$_3$	H	1.1	0	−0.17	0	0.17
H	OCH$_3$	H	1.23	0	−0.27	0	0.27
H	H	Cl	1.55	0	0	0.23	−0.23
CH$_3$	OCH$_3$	H	1.68	−0.17	−0.27	0	0.10
CH$_3$	H	H	1.87	0	0.23	0	−0.23
H	Br	H	2.13	−0.17	0.23	0	−0.40
H	NO$_2$	H	2.53	0	1.27	0	−1.27
CH$_3$	NO$_2$	H	3.75	−0.17	1.27	0	−1.44

idant efficiency and of the difference m between Hammet's constants of the respective substitutent on the phenyl ring, the tendency of which is clear from Table 2.

The more negative the value of the difference m between Hammet's constant of substituents X on the phenyl ring I and that on the rings II or III of substituents Y and Z, the more antioxidative is the effect of the respective compound.[24] The most pronounced is the case of X = CH$_3$ and Y = NO$_2$ when the substituents have the most distinct opposite (push-pull) effect on electron density of the potentially formed radical site on nitrogen.

In some cases, the observed tendency may be somewhat diminished because of unknown antioxidant efficiencies of the transformation product of a parent antioxidant.

Bioantioxidants such as tocopherols (Scheme 6)

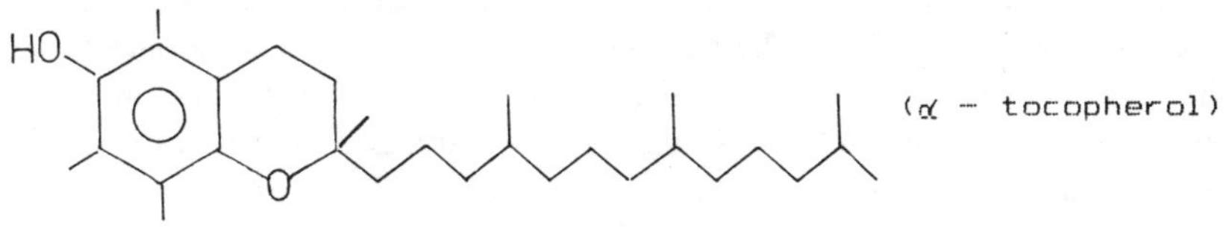

(α - tocopherol)

SCHEME 6

ubiquinones, steroid hormones, etc. are products of living tissues inhibiting the reactions of oxidation. These compounds are polyfunctional in living oragnisms, i.e., despite antioxidant activity they play some other specific roles. It is assumed that tocopherols can stabilize membranes owing to specific interactions of a side fytyl group of tocopherol and a fatty acid chain of phospholipids. This interaction is performed by trapping the CH$_3$ group of a side chain of tocopherol in holes formed by double bonds of fatty acid moieties, which is of van der Waal's character and increases the density of packing in the phospholipid membrane and impedes the diffusion of oxygen to double bonds. In such a way the effect of tocopherol on the decrease of the oxidation rate of lipids can be only structural. On the other hand, the scavenging effect of tocopherols and other bioantioxidants cannot be ex-

Table 3
RATE CONSTANTS OF INTERACTION OF PEROXY RADICALS WITH DIFFERENT FORMS OF BIOANTIOXIDANTS[14]

		k, dm^3 mol^{-1} sec^{-1}		
Antioxidant		Quinone form	Hydroquinone form	Cyclic form oxychromane
α-Tocopherol	(60°C)	2.5 10^3	4.7 10^6	3.1 10^6
Ubiquinone Q-O	(50°C)		3.3 10^5	
Ubiquinone Q-2			3.1 10^5	
Ubiquinone Q-6		3.0 10^2	3.1 10^5	1.4 10^5
Ubiquinone Q-9		2.8 10^2	3.6 10^5	
Vitamine K$_1$[a]	(37°C)	2.1 10^2	5.8 10^6	

[a]   Side hydrocarbon group of this vitamine has 20 carbon atoms.

cluded, either, since the rate constant of the tranfser reaction with peroxy radicals is relatively high (Table 3).[14] From this viewpoint, it may be expected that a large part of antiradical activity of bioantioxidants belongs to their reduced hydroquinone form which has a rate constant of transfer reaction with peroxy radicals on the order of 10^6 dm^3 mol^{-1} sec^{-1}. This value is at least by one order higher than that for synthetic phenolic antioxidants such as 2,6-di-*tert*-butyl-4-methyl phenol or butoxyanizol. One should, therefore, expect that such compounds in a model system of ethyl benzene will be more effective. The opposite is, however, true.

Some years ago it was shown that the mixture of fat soluble vitamin E (tocopherol) and water soluble vitamin C had a very powerful antioxidant activity in the oxidation of edible oils.[25] Each of them alone, however, does not behave as an antioxidant. The mixture of both vitamins is not efficient in the stabilization of dried foods such as precooked cereals, potato flakes, etc. The mutual interaction of vitamins in vivo autooxidation of polyunsaturated lipids of cellular membranes is the demonstration of effective synergism of both components. It was ascertained that radicals generated during autooxidation of cell membrane lipids[26] are very easily intercepted by the tocopherol present there. Radicals of tocopherol regenerate the tocopherol in reaction with ascorbic acid which approaches the cell walls as the hydrosoluble component of cell fluids. The transfer reaction takes place at the interface between hydrophilic and hydrophobic sites. With this regeneration scheme we can comprehend why no marked characteristic picture of a deficiency of vitamin E is known in living organisms. Vitamin E, even in small concentrations undergoes a one electron oxidation and can be regenerated by vitamin C or other reducing agent (glutathione, etc.).

IX. FLAME RETARDANCY

Flame retardants are chemical compounds which in a direct or indirect way decrease the flammability of some material. Their effect may be either physical (heat loss by different cooling systems or additives or the mechanical prevention of oxygen or fuel from approaching the flame zone) or chemical. In this latter case, the flame retardant reacts with reactive atoms and radicals in the flame and suppresses the branching reactions referred to earlier. Hydrogen bromide released from many brominated aliphatic hydrocarbons reacts with hydrogen atoms and hydroxy radicals (Equations 93 and 94) so that

$$H \cdot + HBr \rightarrow H_2 + Br \cdot \tag{93}$$

$$HO \cdot + HBr \rightarrow H_2O + Br \cdot \tag{94}$$

less reactive bromine atoms are formed. They abstract hydrogen atoms from the fuel and the hydrogen bromide is regenerated (Equation 95).

$$Br \cdot + RH \rightarrow HBr + R \cdot \tag{95}$$

The main elements used in flame retardant compounds are those of odd groups of the Periodical Table, particularly from Group VII (Cl, Br), V (P, Sb), and III (B, Al). Current phenolic or amine antioxidants which break the oxidation chains efficiently at temperatures up to 250°C are inefficient and are usually very flammable. On the other hand, in the case of perhalogenated aromatic compounds, the release of bromine atoms may occur only after preceding addition of reactive radicals to the benzene ring (Equation 96).

The ratio of the effect of individual halogens corresponds approximately to their atomic weights, namely, F:Cl:Br:I = 1:1.9:4.2:6.7. Particularly efficient is antimony trihalogenide which in the gaseous phase reacts with hydrogen atoms and hydroxy radicals in a sequence of reactions (Equations 97 to 101).

$$SbX_3 + H \cdot \rightarrow HX + SbX_2 \tag{97}$$

$$SbX_2 + H \cdot \rightarrow HX + SbX \tag{98}$$

$$SbX + H \cdot \rightarrow HX + Sb \tag{99}$$

$$Sb + \cdot OH + SbOH \tag{100}$$

$$SbOH + H \cdot \rightarrow SbO + H_2 \tag{101}$$

The efficiency of SbX and Sb is based upon their higher residence time in the flame zone when compared to simple hydrogen halides. Phosphorus-containing flame retardants may function in a similar way, namely, via PO· and HPO radicals which scavenge ·OH and H· radicals. The introduction of finely divided metal or metal oxide particles to the flame, usually formed *in situ,* is one of the most effective ways of inhibiting flame reactions.[27] It was shown that flammable iron pentacarbonyl and tetraethyl lead and nonflammable chromyl chloride are better by at least one order of magnitude than the best inhibitors in decreasing the flame velocity of a stoichiometric hexane-air flame by 30%.

Even though the effect declines with decreasing pressure, iron pentacarbonyl is still about 25 times more effective than CCl_4 in extinguishing hydrocarbon diffusion flame at 10 kPa.

Metal oxides formed from organometallic compounds in the flame are assumed to catalyze the recombination of radicals (Equation 102).

$$H\cdot + \cdot OH + \text{oxide surface} \rightarrow H_2O + \text{oxide surface} \tag{102}$$

The specificity of polymer flame retardancy consists in the possibility of preventing the formation of flammable fuel from the condensed phase by additives which form on the polymer surface glassy or carbonaceous protecting layer. Additives which are required to influence gaseous phase reactions should have appropriate characteristics of their decomposition and should be supplied to the flame zone at a proper time. The weak point in the strategy of the flame retardancy of polymers is the necessity of using a relatively large amount of the additive flame retardant which may lead to deterioration of some useful properties of the polymer product. Attention is, therefore, gradually directed to the effect of some inorganic additives, some of them being classical ones such as trihydrate of aluminum trioxide, which bring about the decline of reaction heat by endothermic processes of water release. Inorganic phosphates and polyphosphates which support carbonizations found relatively good application in flame retardancy of cellulosic materials.[28] The protective effect of carbonized residue is particularly enhanced in the case of intumescent additives, which are composed of a carbon forming compound (polyol, starch, etc.), the gas releasing component (melamine, guanidine, etc.), and acid forming component (P_2O_5, inorganic phosphates, borates, etc.). The synchronization of the carbonization procedure with the release of the gas initiated by the present acid leads to the carbonaceous foam which more or less isolates the flammable material from the heat source. The additional source of free radicals, such as, e.g., peroxides, halogenated flame retardants, etc., markedly reduce the amount of carbonaceous foam intervening obviously with built-up reaction of carbonaceous residue as was indicated at the scheme of glowing combustion of carbon residue.[29]

In the past few years, empirical research has gone into designing better fire retardants for specific polymers, but fundamentally no new systems have emerged from these studies.

X. OTHER EXAMPLES OF BRANCHED CHAIN REACTIONS

Besides oxygen, the substrates may burn in N_2O, fluorine, chlorine, sulfur vapors, etc. At the concentration of hydrogen in fluorine considerably higher reaction heat (535 kJ/mol) is released than for oxygen (242 kJ/mol), and consequently the flame has the higher temperature. Oxygen is, however, preferred since it is much cheaper and nontoxic and noncorrosive products are formed. The burning of hydrogen in fluorine is therefore used only in special cases of high temperature flame at the cutting of some materials.

Another branched chain reaction may be the nitration of alkanes which is initiated by the thermal decomposition of nitric acid (Equation 103).[30]

$$HNO_3 \rightarrow HO\cdot + \cdot NO_2 \quad (k = 1.6 \ 10^{15}\text{sec}^{-1}) \tag{103}$$

Alkyl radicals which are formed by the transfer of peroxy radicals to hydrocarbon RH form with NO_2 alkyl nitrites which play the role of a degenerated branching agent like hydroperoxides at the oxidation of hydrocarbons by oxygen. Nitration of alkanes which attracted considerable attention from research workers is used now mainly for synthetic purposes, such as the production of tetranitromethane, nitroethane, and nitroparaffins. The mechanism of this process is of importance from the viewpoint of smog formation which involves the oxidation of alkanes in the presence of NO and NO_2. The most irritating components of the smog, peroxyacetyl nitrate and peroxypropionyl nitrate, which are somewhat more stable than ozone, are present in the polluted atmosphere at concentrations corresponding to 1.0

to 3% of w. They decompose to reactive free radicals and may initiate the further generation of harmful compounds. The amount of formed smog was decreased by free radical scavengers, the most active being diethylhydroxyl amine which rapidly reacts with all reactive free radicals.

REFERENCES

1. **Johnson, P. R.,** A general correlation of the flammability of natural and synthetic polymers, *J. Appl. Polym. Sci.,* 18, 491, 1974.
2. **Pollard, R. T.,** Hydrocarbons, in *Comprehensive Chemical Kinetics, Gas Phase Combustion,* Bamford, C. H. and Tipper, C. F. H., Eds., Elsevier, Amsterdam, 1977, 249.
3. **Dechaux, J. C., Flament, J. L., and Lucquin, M.,** Negative temperature coefficient in the oxidation of butane and other hydrocarbons, *Combust. Flame,* 17, 205, 1971.
4. **Benson, S. W.,** Cool flames and oxidation: mechanism, thermochemistry and kinetics, *Oxid. Commun.,* 2, 169-188, 1982.
5. **Dixon-Lewis, G. and Williams, D. J.,** The oxidation of hydrogen and carbon monoxide, in *Gas-Phase Combustion, Comprehensive Chemical Kinetics,* Bamford, C. H. and Tipper, C. F. H., Eds., Elsevier, Amsterdam, 1977, 1.
6. **Hoyermann, K. and Wagner, H. G.,** Elementary reactions in the high temperature oxidation of hydrocarbons, *Oxid. Commun.,* 2, 259, 1982.
7. **Walling, C.,** Fenton's reagent revisited, *Acc. Chem. Res.,* 8, 125, 1975.
8. **Hendriks, C. F., vanBeck, H. C. A., and Heertjes, P. M.,** Oxidation of substituted toluenes by Co(III) acetate in acetic acid solutions, *Ind. Eng. Chem. Prod. Res. Dev.,* 17, 256, 1978.
9. **Schwab, A. W.,** Catalytic autooxidation of 9,10-formyl-stearic acid, *J. Am. Oil Chem. Soc.,* 50, 74, 1973.
10. **Allen, G. C. and Aquilo, A.,** Metal-ion catalyzed oxidation of acetaldehyde, *Adv. Chem. Series,* 76, 363, 1986.
11. **Marta, F., Boga, E., and Matok, M.,** Interpretation of limiting rate and of induction period in oxidation of benzaldehyde catalyzed by cobaltous acetate, *Discuss. Faraday Soc.,* 46, 173, 1968.
12. **Wong, C. L., Switzer, J. A., Balakrishnan, K. P., and Endincott, J. F.,** Oxidation reduction reactions of complexes with macrocyclic Co(II) (NH) system, *J. Am. Chem. Soc.,* 102, 5511, 1980.
13. **Sheldon, R. A. and Kochi, J. K.,** *Metal Catalyzed Oxidation of Organic Compounds,* Academic Press, New York, 1981.
14. **Burlakova, E. B. and Khrapova, N. G.,** Peroxidation of lipid membranes and natural antioxidants (Russian), *Usp. Khim.,* 54, 1540, 1985.
15. **Trifiro, F. and Pasquon, I.,** Classification of oxidation catalysts according to type of metal-oxygen bond, *J. Catal.,* 12, 412, 1968.
16. **Dai, P. S. E. and Lunsford, J. H.,** Catalytic properties of molybdenum containing zeolites in epoxidation reactions, *J. Catal.,* 64, 173-184, 1980.
17. **Mlodnicka, T.,** Metalloporphyrins as catalysts in autooxidation process, *J. Mol. Catal.,* 36, 205, 1986.
18. **Golubchikov, O. A. and Berezin, B. D.,** Some aspects of porphyrine chemistry (Russian), *Usp. Khim.,* 55, 1361, 1986.
19. **Emanuel, N. M. and Buchachenko, A. L.,** *Chemical Physics of Aging and Stabilization of Polymers* (Russian), Publishing House Nauka, Moscow, 1982.
20. **Tkac, A.,** Radical processes in polymer burning and its retardation. I. ESR methods for studying the thermal decomposition of polymers in the preflame and flame zones, *J. Polym. Sci., Polym. Chem. Ed.,* 19, 1475, 1981.
21. **Tkac, A. and Spilda, I.,** Radical processes in polymer burning and its retardation. II. An ESR study of flame retardation of polypropylene, *J. Polym. Sci., Polym. Chem. Ed.,* 19, 1496, 1981.
22. **Cullis, C. F. and Hirschler, M. M.,** *The Combustion of Organic Polymers,* Clarendon Press, Oxford, 1981.
23. **Razumovskii, S. D., Rakovski, S. K., Shopov, D. M., and Zaikov, G. E.,** *Ozone and Its Reactions with Organic Compounds,* Publishing House Bulgarian Academy of Science, Sofia, 1983.
24. **Rychla, L., Rychly, J., Ambrovic, P., Mogel, L., and Schulz, M.,** Thermooxidative stability of polypropylene in the presence of substituted 3-anilino-1,5-diphenyl pyrazoles, *Polym. Degrad. Stab.,* 14, 147, 1986.
25. **Lambelet, P., Saucy, F., and Loliger, J.,** Chemical evidence for interaction between Vitamin E and C, *Experientia,* 41, 1384, 1985.

26. **Lambelet, P., Ducret, F., Saucy, F., Savoy, M. C., and Loliger, J.,** Generation of radicals from antioxidant type molecules by polyunsaturated lipids, *J. Chem. Soc. Trans. Faraday I,* 83, 141, 1987.
27. **Warren, P. C.,** Stabilization against burning, in *Polymer Stabilization,* Hawkins, W. C., Ed., Wiley Interscience, New York, 1972, 314.
28. **Aseeva, R. M. and Zaikov, G. E.,** *Combustion of Polymers,* Hanser Publishers, New York, 1986.
29. **Rychly, J., Matisova-Rychla, L., and Vavrekova, M.,** Carbonization of Intumescence Fire Retardant in Polypropylene, *J. Fire Ret. Chem.,* 8, 82-92, 1981.
30. **Ballod, A. P. and Shtern, V. Ya.,** Reactions of gas phase nitration of alkanes (Russian), *Usp. Khim.,* 45, 1428, 1976.

Chapter 8

OSCILLATING REACTIONS

I. INTRODUCTION

From the complex processes of both ionic and radical mechanisms not fully understood yet but highly fascinating in their performance, oscillatory reactions have challenged the effort of many research workers in recent years. Even though no practical application is available from the study of such systems to date, the analogous features in the dynamics of many biological and chemical oscillators seem to be very promising. All living organisms exhibit processes such as heart beats or nerve impulses that have the character of oscillatory changes, the origin of which is ultimately chemical in nature.[1] The study of oscillatory systems may thus be the way to understand some vital functions of biological objects.

It is obvious that many oscillatory processes have a trivial origin and can be explained on the basis of a heterogeneous character of the system or, consequently, by the transport delay to the reaction site induced by some regulatory resistance of the surrounding medium. The observation of this type of phenomenon in chemistry or biochemistry is, however, still rather exciting and at first sight is contrary to the common sense of classical chemical kinetic laws. Now, it is known that the necessary condition of the appearance of oscillatory behavior is the mutual coupling of at least two system variables such as:

a. Concentration — concentration
b. Concentration — temperature
c. Concentration — pressure
d. Concentration — coordinates
e. Temperature — coordinates
f. Pressure — coordinates
g. Temperature — pressure

(Temperature may be sometimes replaced by light intensity.)[2] This coupling is performed via the corresponding rate equations by the existence of the feed-back which positively or negatively influences the rate of some preceding or subsequent reaction. The dependence of the reaction rate on the concentration, temperature, or pressure should be nonlinear. Simultaneously, with a permanent supply of reactant or energy the first derivation according to time in some concentration, temperature, or pressure interval, should be positive. This is a characteristic feature of chain processes which may change the reaction rate suddenly by relatively little intervention. Besides the typical chain reactions this sudden autoacceleration or deceleration is substantial for some enzymatic or nonisothermal processes. The points d, e, and f involve coupling of the chemical reaction with transport processes which are of highest importance, especially in biological systems, at reactions of enzymes attached to a particular site of the membrane.

What is the role of free radical processes in the oscillatory mechanism? The chain character of some free radical reactions sometimes brings the system to the limit of its stability. Since one radical site may propagate over thousands of molecules of substrate, the radical chain reactions as such may be considered as some kind of chemical "amplifier". Eliminating the radical center, we substantially decrease the overall rate of reaction; the generation of new radical centers leads to a subsequent considerable increase of reaction rate. The latter case may, thus, bring about the explosion of the system. In the formal description it is expressed as a singular point in the corresponding system of differential equations. The process accelerated towards explosion may, however, be turned back by the appearance of the inhibitor and kept at a small rate until its consumption and repetition of the quasiexplosive course. Because of the high reactivity of reactive free radicals, their stationary concentrations in the system are by several orders lower than those of other reacting intermediates, and the Bodenstein rule of steady state may sometimes be used for formulation of the rates of their appearance and decay. This makes possible the easy appearance of nonlinear members in corresponding rate equations.

It seems thus that in experiments, oscillations may be obtained very easily. The contrary is, however, true. Only little is known about the mechanisms on a molecular level in many cases of complex reactions, and the mutual coupling of intermediates may bring about cancellation of the phenomenon.

Chain free radical reactions, especially in the gaseous phase, are known to be accompanied by a considerable release of heat; the effect of the temperature change cannot be ignored which complicates the matter considerably. Generally we may only say that chain radical reactions satisfy all requirements necessary for observation of oscillatory behavior in chemistry. Since it is difficult to estimate now, to what extent free radical mechanisms are significant in individual oscillating systems, in the following text, we present a description of the most important oscillatory reactions and allow the reader to draw his own conclusions. Our attention will be mainly focused on points a and b, which are the most interesting from the viewpoint of chemistry. It should be borne in mind, however, that in some examples of oscillating systems presented under these points the transport phenomena may also play an important role.

II. SURVEY OF THE MOST IMPORTANT OSCILLATING SYSTEMS

Even though the number of newly discovered oscillatory systems steadily increases, no systematic classification is available. According to the field of chemistry in which the phenomenon was observed, we speak about chemical oscillatory systems as such: thermo-kinetic, biochemical, electrochemical oscillations, etc.

A. Chemical Oscillatory Systems

Considerable attention has been paid especially to the Belousov-Zhabotinskii (B-Z) reaction, which represents a system with certain concentrations of malonic acid and potassium bromate with cerium or manganese ions in diluted sulfuric acid. As we ourselves have done (Figure 1), anyone of average skill as a chemist can reproduce the experiment very easily and trace the oscillations in the concentration of cerium (or bromide) ions; the reaction is also accompanied by small periodical changes of temperature or the rate of heat release.

The oscillatory course is characterized by some preoscillatory stage (induction period A); then the concentration of Ce^{4+} suddenly decreases and the periodical process begins. The sharp increase of Ce^{4+} concentration in the oscillatory course is followed by the steady decrease (B) and sharp drop (C) of Ce^{4+} which turns out to continuously decrease until a new sharp increase of Ce^{4+} will appear again. According to the reaction conditions, the oscillations may have a "flip-flop" or "saw-tooth" like shape.

The following overall reactions are involved (Equations 1 to 3).

$$BrO_3^- + 4Ce^{3+} + CH_2(COOH)_2 + 5H^+ \rightarrow BrCH(COOH)_2 + 4Ce^{4+} + 3H_2O \tag{1}$$

$$4Ce^{4+} + BrCH(COOH)_2 + 2H_2O \rightarrow Br^- + 4Ce^{3+} + HCOOH + 2CO_2 + 5H^+ \tag{2}$$

$$BrO_3^- + 2Br^- + 3CH_2(COOH)_2 + 3H^+ \rightarrow 3BrCH(COOH)_2 + 3H_2O \tag{3}$$

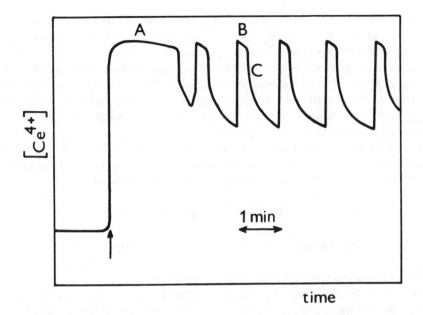

FIGURE 1. Potentiometric trace of cerium ion concentrations (without absolute calibration) for the system of 41 mℓ of Ce(SO₄)₂·4H₂O· (0.003 M), (M = mol·dm⁻³, KBrO₃ (0.02 M), malonic acid (0.02 M) in 0.6 M H₂SO₄ at 28.5°C (air, electromagnetic mixing). Arrow indicates the addition of Ce^{4+} to solution of the bromate and malonic acid.

These stoichiometric equations cannot, of course, give any information about the possibility of oscillations. A detailed study has revealed that reaction 3 consists of four elementary reactions (Equations 4 to 7). The rate determining step turns

$$BrO_3^- + Br^- + 2\,H^+ \rightarrow HBrO_2 + HOBr \tag{4}$$

$$HBrO_2 + Br^- + H^+ \rightarrow 2\,HOBr \tag{5}$$

$$HOBr + Br^- + H^+ \rightarrow Br_2 + H_2O \tag{6}$$

$$Br_2 + CH_2(COOH)_2 \rightarrow BrCH(COOH)_2 + Br^- + H^+ \tag{7}$$

out to be reaction 4. Provided that the concentration of bromide ions is low, reaction 3 proceeds slowly, and bromate ions can react with Ce^{3+} and malonic acid according to the overall reaction 1, whose elementary reactions have been shown to be Equations 8 to 12. Here, the rate determining step is reaction 8.

$$BrO_3^- + HBrO_2 + H^+ \rightarrow 2\,BrO_2{}^\cdot + H_2O \tag{8}$$

$$BrO_2{}^\cdot + Ce^{3+} + H^+ \rightarrow HBrO_2 + Ce^{4+} \tag{9}$$

$$BrO_2{}^\cdot + Ce^{4+} + H_2O \rightarrow BrO_3^- + Ce^{3+} + 2\,H^+ \tag{10}$$

$$2HBrO_2 \rightarrow BrO_3^- + HOBr + H^+ \tag{11}$$

$$HOBr + CH_2(COOH)_2 \rightarrow BrCH(COOH)_2 + H_2O \tag{12}$$

The bromination of malonic acid proceeds via step 7 and 12 and its rate is apparently determined by enolization of malonic acid catalyzed by sulfuric acid. The presence or absence of bromomalonic acid[3] is essential for the appearance of an induction period of oscillations. Initially present bromomalonic acid shortens the period significantly since it reacts according to reaction 2 with Ce^{4+} ions and initiates the considerable increase of Br^- concentration. The increase in the concentration of bromide ions inhibits the oxidation of Ce^{3+} ions to Ce^{4+} ions since bromination of malonic acid prevails. The consumption of Br^- in reaction 3 is necessary for reaction 1 to start again.[4]

Less attention has been devoted to radical bromination, which obviously should also be involved because of the reaction of malonic acid with Ce^{4+} and the formation of malonyl radicals (Equations 13 and 14). This reaction proceeds as a chain radical

$$Ce^{4+} + CH_2(COOH)_2 \rightarrow Ce^{3+} + \cdot CH(COOH)_2 + H^+ \tag{13}$$

$$\cdot CH(COOH)_2 + Br_2 \rightarrow Br\cdot + CHBr(COOH)_2 \tag{14}$$

process with bromine atoms as the main chain carriers.

In the case of the B-Z reaction, malonyl radicals are assumed to be depleted by the subsequent process with cerium (IV) ions. The idea was expressed[5] that the details of those subsequent fast processes are not important in the kinetics of an oscillating reaction. There are, however, indications that the role of malonyl radicals in the raction mechanisms was underestimated; Ce^{4+} ions and malonic acid in diluted sulfuric acid medium are, e.g., efficient initiators of free radical polymerization of acrylamide as well as of the oxidation of ethyl alcohol. At the same time, both acrylamide and ethyl alcohol[6] strongly damp the oscillatory behavior of the B-Z reaction and at certain concentrations completely inhibit it. The oscillatory course is, moreover, inhibited also by hydrogen peroxide and oxygen[7] which obviously interact with malonyl radicals.[8] It was found[9] that mesoxalic, glyoxalic, and oxalic acid, which are by-products of malonic acid oxidation react with Ce^{4+} ions so rapidly that the rate can not be followed. These reactions which also proceed by a radical mechanism were not considered as important, too.

It is, however, of interest that oxalic acid, cerium ions, and potassium bromate cause the oscillations in the rates of CO_2 and Br_2 evolution, provided that the flow of carrier gas (H_2) passes through the solution.[10,11]

It was shown that the continuous removal of bromine is necessary to receive "well behaved" oscillatory systems. Instead of bromide ions, elementary bromine was proposed to play the role of the key intermediate here. This new approach was applied also to the original B-Z reaction with malonic acid and the following free radical sequence was introduced into the mechanism (Equations 15 to 19). Bromine inhibits the autocatalytic regeneration

$$BrO_2\cdot + Br_2 \rightarrow Br_2O_2 + Br\cdot \tag{15}$$

$$Br_2O_2 \rightarrow 2BrO\cdot \tag{16}$$

$$Ce^{3+} + Br\cdot \rightarrow Ce^{4+} + Br^- \tag{17}$$

$$H^+ + Ce^{3+} + BrO\cdot \rightarrow Ce^{4+} + HOBr \tag{18}$$

$$3H^+ + BrO_2\cdot + 3Ce^{3+} \rightarrow 3Ce^{4+} + HOBr + H_2O \tag{19}$$

of $HBrO_2$ because step 15 competes with steps 9 and 10 of the originally proposed mechanism. Due to the physical removal of Br_2 by a carrier gas, the autocatalytic production of $HBrO_2$ gradually accelerates again. The formation of Br_2 has, however, a certain induction period because some amount of HOBr is necessary to appear from $HBrO_2$ and the oscillatory cycle can repeat again.

According to References 5 to 12, the most important radical species involved in the mechanism of B-Z reaction are $BrO_2 \cdot$ radicals. Their main role consists in both the oxidation and reduction of cerium ions according to steps 9 and 10.

The additional mechanistic observation is the suppression of oscillatory behavior in perchloric acid. No plausible explanation was also developed on the inhibition effect of chloride ions.

After a detailed study of the B-Z reaction with cerium or manganese ions, it was found that oscillations will occur also in modified systems. Cerium, manganese, or ferrous ions were, e.g., replaced by complex ions of ruthenium,[13] attached to the polymer carrier of poly-4-vinyl-4-methyl bipyridine.[14] Instead of malonic acid, other organic compounds, such as succinic acid, acetone, pentane-2,4-dione, acetyl acetone,[15] cyclopentanone, cyclohexanone[16], etc., were used. The only compound retained from the original B-Z reaction was potassium bromate. Surprisingly, no experiments were referred to until now[14] on the B-Z system, where the mutual coexistence of a separate liquid and gaseous phase, making possible free escape of gaseous components from the liquid, was excluded.

It was ascertained that in some systems oscillations proceed even without the presence of ions of the catalyst, and the systems of only potassium bromate and oxidized compounds, such as phenols and polyphenols (gallic acid, pyrogallol, resorcinol, etc.), aromatic amines, and aminophenols in sulfuric acid may oscillate, too.[17]

The frequency of oscillations increases and the amplitude and induction period decreases with the increasing concentration of sulfuric acid in such systems. The other oxidizing agents used instead of bromate (such as MnO_4^-, $Cr_2O_7^{2-}$, ClO_3^- and IO_3^-) do not bring about chemical oscillations. Also aromatic amines or phenols which produce bromine immediately by reaction with bromate, do not give the oscillatory course. No effect of inert carrier gas and of the continuous removal of bromine was examined.

The oscillation process which attracted fairly large attention is the Bray-Liebhafsky reaction between hydrogen peroxide and potassium iodate in diluted sulfuric acid (Equations 20 and 21).

$$2\ HIO_3 + 5\ H_2O_2 \rightarrow I_2 + 5\ O_2 + 6\ H_2O \tag{20}$$

$$I_2 + 5\ H_2O_2 \rightarrow 2\ HIO_3 + 4\ H_2O \tag{21}$$

The reaction was initially thought to be driven by the heterogeneous processes such as oxygen nucleation and supersaturation effects, but after the detailed elaboration of the Belousov-Zhabotinskii system it was accepted that the reaction is, in fact, homogeneous.[18] It was, however, found that the rate of oxygen evolution has its maximum in the phase with decreasing concentration of iodine. According to the above stoichiometric equations, oxygen appears only in reaction 20; so it was assumed that process 21 is accompanied by dismutation of hydrogen peroxide[19] of uncertain stoichiometry (Equation 22). This was also the reason why the mechanism of

$$2n\ H_2O_2 \rightarrow 2n\ H_2O + n\ O_2 \tag{22}$$

this reaction was considered to be a branched free radical chain reaction. Unambiguous identification of reaction intermediates is, however, lacking and the detailed mechanism is still an open question as is the cause of the oscillations themselves.

Addition of manganese ions and malonic acid[12] to a Bray-Liebhafsky system of acidic iodate and hydrogen peroxide leads to the enhanced oscillations with a period of only a few seconds. Based on the results of Reference 12, the systematic search for other oscillators has started. The attack for designing chemical oscillators included the following approaches:

1. Find an autocatalytic system.
2. Run the reaction in a continuous flow reactor which makes it possible to keep the reaction far from its equilibrium.
3. Vary the conditions until a region of bistability of the autocatalytic system is found.
4. Introduce another substance capable of affecting both branches of bistability differently and inducing the oscillations.

Systems such as iodate and arsenite ions, iodide, and chlorite ions were thus found which seem to be very promising from the view point of further study of oscillating chemical reactions.

The chlorite systems especially give oscillators of a much broader range than bromates or iodates. Chlorite oscillators, however, like the majority of systems found by such an approach show periodical behavior only under continuous flow conditions.

Another interesting oscillatory system is the decomposition of formic acid with concentrated sulfuric acid.[20] The mechanism of the decomposition reaction (Equation 23) is more

$$HCOOH \rightarrow CO + H_2O \tag{23}$$

complicated as an overall reaction and involves the direct participation of CO and radical processes in the feed-back step.[21-23] The oscillations are obviously connected with supersaturation effects and with the release of CO molecules out of the liquid, which is indicated by the strong influence of the mixing speed on their appearance. The bursts of bubbles of CO are so spectacular that the tentative model called the "bubblator" was developed for this system. Initiation of the reaction apparently consists in the effect of iron salts and perhaps traces of nitrates always present in sulfuric acid.

Worth mentioning is also the air oxidation of benzaldehyde catalyzed by cobalt and bromide ions,[24] which performs the oscillatory behavior in the temperature range of 55 to 100°C; the reaction is likely to proceed at least partially by a radical mechanism.

Although much has been done in the interpretation of so-called purely chemical homogeneous oscillation systems, there still exist doubts whether the driving force of the process is of an only chemical nature. One of the reasons of such uncertainty is the fact that in the vast majority of chemical oscillation systems known until now, gas (CO_2, O_2, CO, NO_2, etc.) is released or absorbed. Even in the B-Z reaction case it has been said that supersaturation of bromine[25] allows bromate to shift from the reaction step with bromide to the reaction of Ce^{3+} ions. These opinions are supported mainly by the dependence of the period of oscillation and the duration time of the oscillation process on the rate of mixing of the system.

B. Thermokinetic (Thermochemical) Oscillations

The oscillatory reaction course was also observed in a number of gas reactions. A typical example is the oxidation of propane[26] which, under certain conditions, is accompanied by the appearance of several cool flames. As was already mentioned earlier, cool flames have been observed in the oxidation of methane, ethane, acetaldehyde, formaldehyde, butane, 2-methyl-pentane, acetone, pentane, heptane, propionaldehyde, butyraldehyde, ethers, side chain aromatics and amines, etc.[27] They are strongly nonisothermal and considerable movement of the gas in the nonstirred reaction vessel (due to convection currents) occurs.

By the continuously fed and stirred tank reactor, the thermochemical oscillations are easily reproducible and regular. It is believed that the main reason for the oscillatory character of cool flames consists in the self-heating of the reaction system due to the accelerated formation of hydroperoxides which brings it to the interval of the negative temperature coefficient; the reaction rate decreases as well as the temperature. The decrease of the temperature out of the interval negative temperature coefficient leads to the repetition of the process.

Oscillations were also observed in the oxidation of CO by oxygen. The reaction proceeds in the gas phase at a pressure of 1 to 4 kPa and in the temperature range of 550 to 580°C. This system indicates some degree of product inhibition which probably brings the reaction to the oscillatory state with the glow course from the excited CO_2. Carbon suboxide radicals were proposed as an inhibitor.

Typical thermochemical oscillation may be demonstrated on the oxidation of ethyl alcohol by H_2O_2 in the presence of Fe^{3+} catalyst. Here, the variations in temperature are of the extent of 30°C and exceed the boiling point of ethyl alcohol.[28] The loss of heat in the preparatory stage of the phase transition seems to be the decisive factor in observing oscillatory behavior. This was demonstrated on the effect of an additional source of heat (heat spiral) immersed in the reaction mixture. Provided that its temperature was 20°C higher than the initial temperature of the reactor, the oscillations disappeared.

Interesting oscillation phenomena may be observed during the oxidation of heavy hydrocarbon fuels, including hexadecane and higher hydrocarbon paraffin waxes, stearic acid, polypropylene,[29] etc. At temperatures above 280°C and up to 330°C and certain concentrations of oxygen supplied to the surface of the molten fuel placed, e.g., in the tube, periodic bursts of fuel located at the sample surface appear. They propagate to the gas phase through the cool flame stabilized above the sample surface and give periodical changes of temperature (Figure 2) and intensity of luminiscence emission. The temperature changes in the gas are to the extent of several tens of degrees, whereas those at the surface are much lower ($\pm$ 0.2°C). The process has been called secondary multiple cool flame.[30]

This periodic reaction, with a period of several seconds which may be treated as an isothermal reaction from the side of the surface and as thermokinetic oscillations from the side of the gas phase, is a nice demonstration of the mutual coupling of both chemical (accumulation of hydroperoxides and peroxyradicals on the surface) and physical processes (flow of fuel out of the surface and oxidizing gas at surface, exchange of heat between gaseous and liquid phase, etc.).

The period and amplitude of oscillations depend strongly on the conditions and the geometry of the reactor, and the oscillatory behavior may be completely suppressed by the presence of an antioxidant.[31] The impression of fuel breathing may well be obtained observing this phenomenon.

C. Biochemical Oscillations

The most easily observable biochemical oscillations at the metabolic level may be encountered in different glycolytic reactions. The first reported oscillations of this kind were observed in suspensions of intact yeast cells in cell-free extracts. In these systems, all metabolites with a concentration of 10^{-3} to 10^{-5} mol dm^{-3} oscillate as does the pH and the rate of CO_2 release.[32] It was assumed that the important role in glycolytic oscillations is played by an allosteric enzyme, phosphofructokinase, which is inhibited by one of its substrates, adenosine triphosphate, whereas its products, adenosine diphosphate and fructose 1,6-diphosphate have an activating effect. The chemical feedback necessary for oscillations is thus ensured.

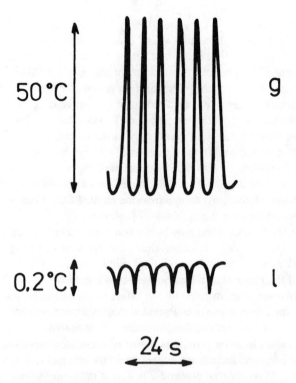

FIGURE 2.   Periodical changes of the temperature in the gaseous (g) phase and on the surface (1) for a parafin wax obtained
at the temperature of tubular reactor T = 306°C, flow rate of
the mixture of N_2 and O_2 = 6.6 dm³/hr and oxygen concentration 9.1%.

The allosteric nature of phosphofructokinase and corresponding cooperative allosteric
transitions seem, thus, to be an essential property of periodic phenomena in biochemical
nonequilibrium systems. It has been suggested that this property observed at the genetic,
enzymatic, and membrane level might well represent the common molecular basis for many
periodic phenomena in biology.

Quantitative agreement found for the oscillations analyzed in biochemical experiments
and those in simple phosphofructokinase, pyruvate kinase, or adenylate kinase models leads
to the conclusion that the dynamic behavior of the complex systems can be reduced to the
molecular property of a simple protein species operating as a master enzyme.

Models of glycolytic oscillations for the anaerobic degradation of glucose (G) in living
cells to pyruvate involve, e.g., the next essential quasi-elementary steps (Equation 24 to
29).

$$G \rightarrow FP \tag{24}$$

$$FP + {}^*E_1 \rightarrow {}^*E_1FP \tag{25}$$

$${}^*E_1FP \rightarrow {}^*E_1 + FDP \tag{26}$$

$$FDP + E_1 \rightarrow {}^*E_1 \tag{27}$$

$$FDP + E_2 \rightarrow E_2FDP \tag{28}$$

$$E_2FDP \rightarrow E_2 + GAP \tag{29}$$

Here, $*E_1$ denotes activated phosphofructokinase, E_2 is a combination of aldolase and triose phosphate isomerase, GAP is glyceraldehyde phosphate, FP is fructose 6-phosphate, FDP is fructose diphosphate, and E_1 is an inactive form of phosphofructokinase.

Another alternative scheme (Equations 30 and 31) is:

$$ATP + *E_1 \leftrightarrow ATP^*E_1 \tag{30}$$

$$ATP^*E_1 \rightarrow *E_1 + ADP \rightarrow \alpha ADP + E_1 \leftrightarrow *E_1 \tag{31}$$

ATP is adenosine triphosphate, ADP is adenosine diphosphate, and α is the number of product molecules necessary for the activation of enzymes E_1.

Both alternatives involving autocatalytic steps (Equations 25, 26, and 31) were designed to fulfill the Lotka-Volterra scheme.

D. Oscillations in Heterogeneous Electrochemical Reactions

Although none of the above mentioned oscillation systems can be called purely homogeneous, we should differentiate them from the oscillatory reactions which are strictly linked with the structure of interface and are heterogeneous as such. The longest known chemical oscillations are periodic electrode processes. Most of them were observed at electrodes on which an anodic or cathodic surface layer can form. Its removal and new formation, i.e., the passivation and activation of an electrode, is one of the prerequisites for the occurrence of oscillations. This above mechanism is of much general validity since oscillating catalytic or enzymatic systems may also work on a similar principle. Provided that the product formed by some coupled reaction deactivates the catalytic centers and subsequently this deactivator is removed by another process, we get the type of kinetic oscillatory scheme which is based on the product end inhibition.

Electrochemical decomposition of H_2O_2 on Cu_5FeS_4 (bornite) is an interesting example of a relatively simple case of a heterogeneous oscillatory process.[33] Three reactants were considered to participate there, namely, metal atoms (Fe, Cu) in the sulfide compound, the hydroxide groups bound to the metal atom on the surface, and H_2O_2.

The stationary electrochemical reduction of hydrogen peroxide involves one electron and results in one hydroxyl anion and hydroxyl radical (Equation 32).

$$H_2O_2 + e^- \rightarrow {}^-OH + \cdot OH \tag{32}$$

Provided that the electron is taken from the metal (M), the hydroxyl anion remains bound to the surface (Equation 33).

$$H_2O_2 + M \rightarrow \cdot OH + MOH \tag{33}$$

Hydroxy radicals may react with hydrogen peroxide with the formation of $HO_2\cdot$ radicals (Equation 34). It was suggested that hydrogen peroxy radicals are important particles in the oscillatory process because of

$$\cdot OH + H_2O_2 \rightarrow H_2O + HO_2\cdot \tag{34}$$

their participation in surface reactions (Equations 35 to 37).

$$M + HO_2 \cdot \rightarrow MO + HO \cdot \tag{35}$$

$$M + MOH + HO_2 \cdot \rightarrow 2\,M + O_2 + H_2O \tag{36}$$

$$MO + MOH + HO_2 \cdot \rightarrow 2\,MOH + O_2 \tag{37}$$

Reaction 36 is an autocatalytic radical reaction which leads to the cleaning of the electrode surface from OH groups. On the other hand, reaction 37 is the second autocatalytic reaction tending to increase the surface concentration of OH groups. A similar mechanism is obviously valid for oscillating reactions of hydrogen with oxygen and oxidation of CO or hydrocarbons on the surface of platinum.[34]

"Beating mercury heart" belongs to the class of electrochemical and mechanical oscillators. It may well be demonstrated by a mercury drop placed in a watch glass and covered with a dilute aqueous solution of strong acid such as H_2SO_4 or HNO_3 containing a few crystals of $K_2Cr_2O_7$, $Na_2S_2O_8$, or $KMnO_4$.

These solutions produce insoluble films on the mercury surface. An iron nail or sewing needle, when brought up to touch the mercury drop from the side, will start the rhythmical change of its shape with a frequency of 2 or 3 per second. The oscillations may easily sustain more than 1 hr.[35]

An interesting cross-catalytic effect involving the electrocapillarity of mercury was proposed to be the origin of the oscillations in such systems. Similarly, as at electron-conducting electrodes, electrochemical oscillations can occur at ion-conducting membranes. The electrophysical oscillations of nerves, muscles, and sensory organ cells of living organism are known to take place at such membranes.

III. CLASSIFICATION OF THE MODELS OF OSCILLATORY REACTIONS

As far as we know, most of the oscillatory systems may be constructed from the so-called element of instability defined by differential Equation 38

$$dX/dt = a \cdot f(X) + b \cdot X + c \tag{38}$$

where $f(X)$ is a nonlinear growing function of variable X, and a, b, c are coefficients. Provided that $a, c > 0$, the variable X may at certain times increase to infinity. The driving force of oscillatory behavior here is the tendency of the system to undergo a reaction ending in explosion, phase transition, or other singular state of the system.

This final stage, which functions as an attractor of the reaction system by which the rate of the process changes suddenly, may, however, be avoided by the existence of the negative feedback which adapts Equation 38 to the form of Equation 39, where $\beta \ll 1$, and $g(X,Y..)$ is another growing function of variables X,Y,Z.

$$dX/dt = af(X) + bX + c - \beta\, g(X,Y..)(a_1 f(X) + b_1 X + c_1) \tag{39}$$

$$I. \qquad\qquad\qquad\qquad II.$$

With increasing X, term II of Equation 39 becomes more significant and it slows down and stops the further increase. The variable X in Equation 39 thus performs two steady states; the first for $I = 0$, the second for $I - II = 0$.

The system may oscillate if coupling with other variables exists, say Y, Z, etc., which modulate the values of coefficients a, b, c, β, a_1, a_2, a_3 in Equation 39, and provide the switching on between these two steady states. The models of oscillation reactions may well be classified assuming the system of only two variables (say X and Y); the other variables when coupled with X or Y will oscillate, too. If we take into consideration only absolute members of Equation 38 (in Equation 39), i.e., $a = b = 0$, $a_1 = b_1 = 0$, and g (X,Y..) $\sim$Y and combine it with the corresponding differential equation for Y, we shall obtain the systems of equations (40 and 41) describing the harmonic oscillator.

$$dX/dt = c - \beta c_1 Y \tag{40}$$

$$dY/dt = \beta c_2 X + c_1 \tag{41}$$

Linear and absolute members of Equation 39, $a = 0$, $b = 0$, $c = 0$, $b > 0$, and $g(X,Y..)$ $\sim$Y, in combination with the corresponding equation for Y, lead to the Lotka-Volterra model; nonzero values of coefficients at nonlinear members give models such as the brusselator, oregonator for steady value of Y, the model by Iwamoto and Seno, thermokinetic oscillations, etc.

Let us, e.g., consider the model[36] represented by Equations 42 to 45 of two intermediates, X and Y, at constant

$$P \rightarrow Y \qquad\qquad k_1 \tag{42}$$

$$X + Y \rightarrow X + R \qquad k_2 \tag{43}$$

$$A + 2X + Y \leftrightarrow 3X + Y \qquad k_3 \rightarrow, k_4 \leftarrow \tag{44}$$

$$B + X \leftrightarrow C \qquad k_5 \rightarrow, k_6 \leftarrow \tag{45}$$

concentrations of compounds A, B, C, and P in an open system, and by two differential Equations 46 and 47 in which the

$$dY/dt = k_1 P - k_2 XY \tag{46}$$

$$dX/dt = k_3 AYX^2 - k_4 YX^3 - k_5 BX + k_6 C \tag{47}$$

concentrations are denoted in the same manner as compounds. The reader may realize that step 44 is questionable because of higher molecularity than 2, but the model illustrates well all the essential features of oscillation reactions. The numerical solution of the system of Equations 46 and 47 for arbitrarily chosen parameters is in Figure 3. It may be shown that the period of oscillations is sensitive mainly to the rates of supply and delivery of compounds P, B, and C into or out of the system. It is generally valid that the period of oscillations decreases with increasing concentrations or the rate of supply of the initial reacting compounds and vice versa.

In the terms of the above mentioned classification of models the polynom of the third degree in differential Equation 47 may well be replaced by function 48.

$$dX/dt = aX^2 + bX + c - \beta g(X,Y) (a_1 X^2 + b_1 X + c_1) \tag{48}$$

Provided that $g(X,Y) = X$, the equations 48 and 47 are identical for $\beta = 1$, $a_1 = k_4 Y$, $b_1 = 0$, $c_1 = k_5 B$, $a = k_3 AY$, $b = 0$, and $c = k_6 C$. If the process starts from initial conditions $t = 0$, $X = 0$, and $Y = 0$ then at the beginning, X is relatively small and because $\beta g(X,Y) \ll 1$, the second term II of Equation 48 may be neglected when compared to term I. The latter

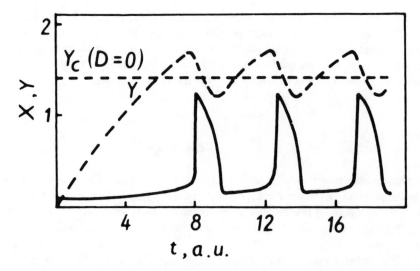

FIGURE 3.   The concentration of intermediate products X and Y as a function of time for the system of Equations 42 to 45. The values of parameters: $k_1P = 0.291$, $k_2 = 0.476$, $k_4 = 48.64$, $k_5B = 50.20$ and $k_6C = 5.35$; Y_c denotes the critical values of Y, if $D = 0$. (From Rychlý, J. and Rychlá, L., *Oxid. Commun.*, 7, 123, 1984. With permission.)

thus determines the course of the solution. Provided that the discriminant of term I becomes negative, X tends to increase to infinity. Under such conditions, term II cannot be ignored, and the increase of X is thus stopped and turned out.

With increasing X, the values of dY/dt may change the sign and become negative. The values of Y then decrease, and simultaneously modulate coefficients a and a_1. The decrease of Y after the value of X has attained its maximum leads to the change of the sign of dX/dt and X will start to decrease, too. The value of X in the vicinity of its maximum may be so large that the values of c in the quadratic polynom I may now be ignored and the decrease of X after maximum may be described by Equation 49.

$$dX/dt = -X[\beta a_1 X^2 + (b_1\beta - a)X + \beta c_1] \qquad (49)$$

In the expression a new quadratic polynom appears which at a parallel change of $a_1 \sim Y$ may have its discriminant negative. The system may then exhibit a sudden decrease of X until the value of X is again considerably less than 1 and the process may repeat again.

In the formalism of isothermal elementary reactions of reactant A, product P, and intermediate X in a homogeneous medium, the element of instability may be expressed by Equations 50 to 52 or 53 to 55 and described by differential Equations 56 or 57.

$$A \rightarrow X \qquad\qquad k_1 \qquad\qquad (50)$$

$$X \rightarrow P \qquad\qquad k_2 \qquad\qquad (51)$$

$$B + 2X \rightarrow 3X \qquad\qquad k_3 \qquad\qquad (52)$$

$$A \rightarrow X \qquad\qquad k_1 \qquad\qquad (53)$$

$$X \rightarrow P \qquad\qquad k_2 \qquad\qquad (54)$$

$$C + X \rightarrow 2X \qquad\qquad k_3 \qquad\qquad (55)$$

$$dX/dt = k_1A - k_2X + k_3BX^2 \tag{56}$$

$$dX/dt = k_1A + (k_3C - k_2)X \tag{57}$$

For $k^2 - 4 k_1k_3B < O$ or $k_3C - k_2 > O$, X in Equations 56 and 57 will always go to infinity with increasing time.

Autoinhibition (in terms of free radical chemistry we may speak also about termination) coupled with an explosion-like increase of variable X, may be formulated in different oscillations models thus arisen. Taking into consideration the kinetic element of instability (Equations 53 to 55), the easiest turnover of reaction is autocatalysis coupled with another variable Y (Equations 58 to 60) by which X is depleted to some very low

$$X \rightarrow Y \qquad (\text{or } B \rightarrow Y) \tag{58}$$

$$X + Y \rightarrow 2Y \tag{59}$$

$$Y \rightarrow P \tag{60}$$

value, and after the consumption of Y in reaction 60 the process may start again. The scheme (Equations 53 to 55, 58 to 60) is called the Lotka-Volterra model and is very popular in the description of predator-prey relations in nature and of all consequences which follow from competing populations, and intervene in the biosphere, human societies, or even economics.[37] It takes place as soon as the resources necessary for survival are limited or exhausted and may give rise to an evolution of the competing entities.

Provided that the inhibition manifests itself in a more complicated way, some other models may be derived (Equations 61 to 66).

$$A + Y \rightarrow X \qquad k_1 \tag{61}$$

$$X + Y \rightarrow P \qquad k_2 \tag{62}$$

$$B + X \rightarrow 2X \qquad k_3 \tag{63}$$

$$2X \qquad \rightarrow Q \qquad k_4 \tag{64}$$

$$Z \qquad \rightarrow fY \qquad k_5 \tag{65}$$

$$C + X \rightarrow Z \qquad k_6 \tag{66}$$

If X, Y, Z are variable (intermediates), A, B, C are initial reactants, P, Q are products, and f is the stoichiometric factor, we obtain the model called the oregonator. In the case of homogeneous systems other models exist such as the brusselator, explodator,[38,39] berlinator, etc.[40] by which their authors try to interpret the behavior of particular oscillatory systems. It seems that the Lotka-Volterra scheme and oregonator unify the most important properties of all homogeneous chemical oscillators. The modeling of chemical oscillations as such is an interesting part of the theory, but its importance cannot be overestimated; otherwise applied mathematics will dominate chemistry.

At this place we should remind the reader that the limit cycle approach discovered by Poincare or the stability theory by Lyapounoff is suitable for the description and analysis of all known types of oscillatory systems including 2 variables. The limit cycle is a closed

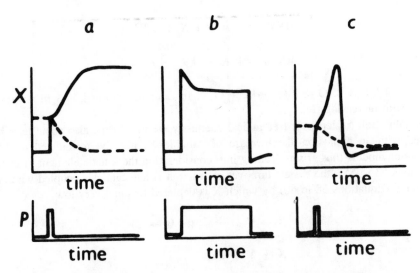

FIGURE 4.   Nonperiodic patterns of systems which potentially may perform oscillatory behavior: a — excitable bistability, b — overshoot behavior, c — excitable pulse formation; P is external fluctuation acting on the system. Broken lines perform subthreshold behavior. (From Franck, U. F., *Angew. Chem.*, 17, 1, 1978. With permission.)

curve in the plane of variables X and Y derived from the original differential equations for the rate changes of X and Y. It is stable if all other trajectories in its neighborhood are spirals winding themselves onto it. If the trajectories wind themselves away from it, the limit cycle is unstable. From simple topological considerations, the limit cycle for realistic types of oscillations, which have the period and amplitude independent of the initial condition variables, must contain at least one singular point inside it; if the limit cycle is stable this must be an unstable one. Only this unstable node or unstable focus can emanate trajectories which could wind onto a limit cycle. From this viewpoint, the Lotka-Volterra model which has a stable center, is considered as an unrealistic one.

Many oscillatory models theoretically derived from Equations 38 and 39 may be analyzed in such a way. Replacing X, e.g., by coordinate, temperature, or by other variable, we may obtain mechanochemical, thermokinetic, and other oscillatory models giving either the sustained oscillations if the system is open or damped oscillations if the system is closed.

IV. NONOSCILLATORY PHENOMENA CHARACTERISTIC FOR SYSTEMS CAPABLE OF OSCILLATIONS

The special features of oscillatory systems, namely, nonlinear relationships, kinetic coupling, and instability lead to typical temporal behavior when conditions do not allow the oscillations to occur. A system no longer in oscillation usually still possesses bistability (Figure 4); it may perform the overshoot phenomenon or excitable pulse formation.[40,41] Interconversion of the two steady states may be initiated by external perturbation P. Individual pulse appearance is a characteristic phenomenon accompanying the end of each oscillatory process.

Excitability of the oregonator model (Equations 61 to 66) is illustrated in Figure 5 for a particular combination of stoichiometric factor f and constant k_5. If Y is decreased by 6% from the steady value, the perturbation grows to 10% and then decays to the stable steady state. Decrease of Y by 6.5%, however, initiates the abrupt change of Y until a value 1000 times higher than it corresponds to the steady state to which it returns with damped oscillations (at the time 450 rel. u. under given conditions).

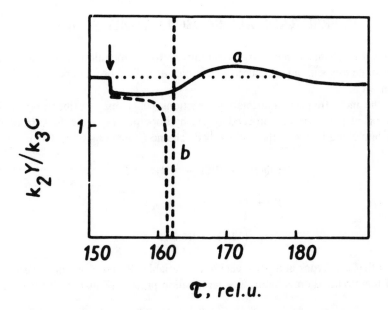

FIGURE 5. Effect of perturbation of Y of an oregonator in a stable steady state. (Parameters corresponding to Schemes 61 to 66 are: k_5/k_1 A = 24.88, k_3C/k_1A = 5970, f = 1.5, B = C, k_6 = k_3, $2k_1k_4/k_2k_3C$ = 8.375·10^{-6}). In curve a, Y was reduced by 6% at a time 153 of rel. u., in curve b, Y was reduced by 6.5%. Reprinted with permission from Field, R. J. and Noyes, R. M., *Acc. Chem. Res.*, 10, 214, 1977. Copyright 1977 by the American Chemical Society.)

Of special interest are the systems which form optically active states as dissipative structures. A dissipative structure may be understood as self organization of the system which exists only during the exchange of energy and matter between the system and its surrounding medium; i.e., only under nonequilibrium conditions.[42]

In unstirred reagents, the spatial patterns which depend on surface/volume ratio of the reacting system can develop which may serve as heuristic models for processes such as morphogenesis and nerve impulse transmission. Such traveling waves initiated, e.g., in thin layer of B-Z reagent by a pacemaker around which concentric rings appear are very spectacular. The experiments cited in Reference 43, e.g., describe the so-called spirals of chemical activity which may be generated in a shallow dish with ferroin as a catalyst. A blue ring was induced by touching the surface of the solution with a hot filament and the dish rocked gently to break the ring. The free ends of the fragmented circular wave curl around a pivot near each point winding up into spirals that have uniform spacing between waves. Wherever two waves collide head on, both vanish.

The passage of traveling waves is obviously connected with the excitability of the system. In an oscillatory B-Z reagent it was experimentally demonstrated that phase and trigger waves simultaneously exist. A trigger wave is a disturbance which moves undamped at constant velocity through a uniform medium. Unlike a phase wave, propagation of a trigger wave can be blocked by the physical barrier. Its passage establishes a continuous phase gradient and initiates a set of phase waves traveling behind, at the velocity of the initial trigger wave and appearing at each point with the period of temporal oscillation.

The phenomenon of excitability may be more important than oscillatory behavior per se. A steady state may be stable to infinitesimal fluctuations due to random molecular motion but very minor external perturbations may still generate pulses of large chemical change. These pulses may then migrate to transmit signals. This may be the probable mechanism of the functioning of a biological clock and seems to be of considerable importance in interpreting some basic processes of life itself.

V. FREE RADICALS IN OSCILLATORY SCHEMES

Free radical mechanisms have large flexibility in the formulation of oscillatory schemes. We demonstrate this with a few tentative examples of radical oscillatory models based upon the oregonator.

One of the most frequent autocatalytic reactions which may be incorporated into it is autooxidation of hydrocarbons proceeding via hydroperoxides (variable X in oregonator), where the branching factor of the undisturbed scheme (Equations 67 to 69) is, equal to 2.

$$ROOH \rightarrow (RO\cdot + \cdot OH) \rightarrow 2R\cdot \qquad (67)$$

$$R\cdot + O_2 \rightarrow RO_2\cdot \qquad (68)$$

$$RO_2\cdot + RH \rightarrow ROOH \qquad (69)$$

Provided that aldehydes and peroxyacids are variables Z and Y we may easily construct the rest of the modified oregonator. The termination process (Equation 70) is equivalent to

$$2 RO_2\cdot \rightarrow products \qquad (70)$$

reaction step 64, and monomolecular fragmentation of peroxy radicals to aldehydes and hydroxy radicals (Equation 71) is

$$RO_2\cdot \rightarrow RCHO + \cdot OH \qquad (71)$$

equivalent to the formation of compound Z (step 66). Aldehydes in a subsequent step are oxidized to peroxy acids (Equation 72)

$$RCHO + O_2 \rightarrow RCOOOH \qquad (72)$$

(equivalent of Y). Peroxy acids may react with hydroperoxides to alcohol, acid, and oxygen (Equation 73) and through steps 74, 68,

$$RCOOOH + ROOH \rightarrow ROH + O_2 + RCOOH \qquad (73)$$

and 69 may simultaneously initiate the further formation of

$$RCOOOH \rightarrow 2 R\cdot \qquad (74)$$

hydroperoxides. The scheme (Equations 67 to 74) is an isothermal alternative of cool flame periodicity; the negative temperature coefficient being only the consequence and not the cause of the phenomenon.

Autooxidation and bromination proceeding in parallel is another example of mutually coupled reactions with potential oscillatory behavior. Since it is known that alkyl radicals react with molecular bromine in preference to oxygen, the formation of hydroperoxides may be temporarily stopped when bromine appears in the reaction mixture. Its formation from some reservoir such as $HBrO_3$ by coupled reaction with hydroperoxides ensures the conditions of the potential oscillatory course.

Let us consider the set of reactions 75 to 82.

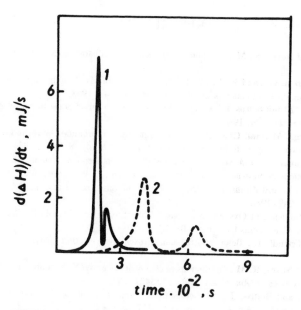

FIGURE 6. The time course of the rate of heat release for the mixture of polypropylene with 20% of w. of N-bromosuccinimide. Temperature 390 K (1), 380 K (2).

$$ROOH + HBrO_3 \rightarrow ROH + HBrO_2 + O_2 \tag{75}$$

$$ROOH + HBrO_2 \rightarrow HBrO + O_2 \tag{76}$$

$$2\ HBrO \rightarrow HBr + HBrO_2 \tag{77}$$

$$HBr + HBrO \rightarrow Br_2 + H_2O \tag{78}$$

$$HBr + ROOH \rightarrow HBrO + ROH \tag{79}$$

$$R\cdot + Br_2 \rightarrow RBr + Br\cdot \tag{80}$$

$$Br\cdot + Br\cdot \rightarrow Br_2 \tag{81}$$

$$Br\cdot + R\cdot \rightarrow BrR \tag{82}$$

Provided that a sufficient concentration of hydroperoxide is formed, it brings about the delayed formation of HBr. In the presence of hydroperoxides, reaction step 78, which competes with step 79 and produces elementary bromine, is suppressed. Below some level of hydroperoxides, however, molecular bromine will appear since the rate of step 78 will dominate that of step 79 and by reactions 80 and 82 will inhibit the further formation of hydroperoxides. At a low level of hydroperoxides, bromine is consumed (Equations 80 to 82) and autooxidation may start again.

The indication of such behavior was observed in a reaction of polypropylene with N-bromosuccinimide in the presence of air, investigated by differential scanning calorimetry (Figure 6).[44] In closed aluminum pans the rate of the released reaction heat gives two damped waves whose distance increases with decreasing temperature. The organic substrate here is polypropylene, the reservoir of bromine, N-bromosuccinimide; bromine is formed by Equation 119 in Chapter 6.

REFERENCES

1. **Field, R. J. and Noyes, R. M.,** Mechanisms of chemical oscillators: conceptual basis, *Acc. Chem. Res.,* 10, 214, 1977.
2. **Gaspar, V., Bazsa, G., and Beck, M. T.,** The influence of visible light on the Belousov-Zhabotinskii reactions applying different catalysts, *Z. Phys. Chem.,* 264, 43, 1983.
3. **Ungvarai-Nagy, Z. and Koros, E.,** Study of the preoscillatory period of the Belousov-Zhabotinskii system, *Acta Chim. Hung.,* 117, 29, 1984.
4. **Koros, E., Varga, M., and Gyorgyi, L.,** Thorough study of bromide control in bromate oscillators. 1. The effect of bromo complex forming metal ions, *J. Phys. Chem.,* 88, 4116, 1984.
5. **Field, R. J., Koros, E., and Noyes, R. M.,** Oscillations in chemical systems. II. Thorough analysis of temporal oscillations in bromate-cerium-malonic acid system, *J. Am. Chem. Soc.,* 94, 8649, 1972.
6. **Ungvarai-Nagy, Z. and Zimanyi, I.,** On the alcohol inhibition of the reacting BZ system, *React. Kinet. Catal. Lett.,* 31, 249, 1986.
7. **Tkac, I.,** The Influence of Oxygen on Belousov-Zhabotinskii Reaction and Oscillation of Enol Form of Substrate, thesis, Commenius University, Bratislava, 1984.
8. **Tkac, I. and Triendl, L.,** Belousov-Zhabotinskii oscillating reaction (Slovak), *Chem. Listy,* 77, 698, 1983.
9. **Jer Jwo, J. and Noyes, R. M.,** Oscillations in chemical systems. IX. Reaction of Ce^{4+} ions with malonic acid and its derivatives, *J. Am. Chem. Soc.,* 97, 5422, 1975.
10. **Noszticzius, Z. and Bodiss, J.,** A heterogeneous chemical oscillator. The Belousov-Zhabotinskii type reaction of oxalic acid, *J. Am. Chem. Soc.,* 101, 3171, 1979.
11. **Sevcik, P. and Dubrovska, J.,** Oscillations of bromine in Belousov-Zhabotinskii type reaction with oxalic acid, *Collect. Czech. Chem. Commun.,* 50, 1450, 1985.
12. **Ariese, F. and Ungvarai-Nagy, Z.,** The disproportionation of $HBrO_2$, key species of the Belousov-Zhabotinskii oscillating reaction, *J. Phys. Chem.,* 90, 4, 1986.
13. **Koros, E., Ludanyi, L., Friedrich, V., Nagy, Z., and Kis, A.,** The $Ru(dipy)_3^{2+}$-bromate-malonic acid oscillation system, *React. Kinet. Catal. Lett.,* 1, 455, 1974.
14. **Ishiwatari, T., Kawagishi, M., and Mitzuishi, M.,** Oscillating reactions in polymer systems, *J. Polym. Chem., Polym. Chem. Ed.,* 22, 2699, 1984.
15. **Stroot, P. and Janjic, D.,** Oscillations chimiques lors de l'oxydation de dicetones par de bromate en presence d'une catalyseur, *Helv. Chim. Acta,* 58, 116, 1975.
16. **Farage, V. J. and Janjic, D.,** Reactions chimique oscillantes, *Helv. Chim. Acta,* 61, 1539, 1978.
17. **Orban, M. and Koros, E.,** Chemical oscillations during the uncatalyzed reaction of aromatic compounds with bromate. 1. Search for chemical oscillations, *J. Phys. Chem.,* 82, 1672, 1978.
18. **Degn, H.,** Oscillating chemical reactions in homogeneous phase, *J. Chem. Educ.,* 49, 302, 1972.
19. **Noyes, R. M. and Field, R. J.,** Mechanisms of chemical oscillations. Experimental examples, *Acc. Chem. Res.,* 10, 273, 1977.
20. **Showalter, K. and Noyes, R. M.,** Oscillations in chemical systems. 24. Oscillatory decomposition of formic acid in sulfuric acid, *J. Am. Chem. Soc.,* 100, 1042, 1978.
21. **Kaushik, S. M., Shi Yuan, and Noyes, R. M.,** A simple demonstration of a gas evolution oscillator, *J. Chem. Educ.,* 63, 76, 1986.
22. **Zhi Yuan, Ruoff, P., and Noyes, R. M.,** Gas Evolution Oscillators. 7. A quantitative modeling test for Morgan reaction, *J. Phys. Chem.,* 89, 5726, 1985.
23. **Kaushik, S. M., Rich, L. R., and Noyes, R. M.,** Gas evolution oscillators. 6. Quantitative characterization of an experimental system, 89, 5722, 1985.
24. **Jensen, J. H.,** A new type of oscillation reactions. Air oxidation of benzaldehyde, *J. Am. Chem. Soc.,* 105, 2639, 1983.
25. **Di Lorenzo, S. and d'Alba, F.,** Oscillatory phenomena in the Belousov-Zhabotinskii system, *J. Chem. Soc. Faraday Trans.,* 81, 421, 1985.
26. **Franck, U. F.,** Chemical Oscillations, *Angew. Chem. Int. Ed. Engl.,* 17, 1, 1978.
27. **Gray, B. F.,** *Kinetics of Oscillatory Reactions, Reaction Kinetics.* Vol. 1, *A Specialist Periodical Report, the Chemistry,* Burlington House, London, 1975.
28. **Vidal, C. and Noyau, A.,** Oscillations chimiques et oscillations thermochimique, *Nouv. J. Chim.,* 3, 83, 1979.
29. **Delfosse, L., Baillet, C., Lucquin, M., and Rychly, J.,** Pulsating combustion of liquid hydrocarbons, *Comb. Flame,* 54, 203, 1983.
30. **Baillet, C., Delfosse, L., and Lucquin, M.,** Nouveau type de reactivite avec l'oxygene de certain hydrocarbures lourds vers 300°C en presence de leur phase liquide, *Rev. Inst. Francais Petrole,* 36, 793, 1981.

31. **Rychly, J. and Rychla, L.,** The influence of additives on the oscillation course of surface oxidation of polypropylene at higher temperatures, *Oxid. Commun.,* 6, 125, 1984.
32. **Hess, B. and Chance, B.,** Oscillatory enzyme reactions, in *Theoretical Chemistry and Biology,* Eyring, H. and Henderson, D., Eds., Academic Press, New York, 1978, 159.
33. **Tributsch, H.,** Sustained oscillations during catalytic reduction of hydrogen peroxide copper-iron-sulfide electrodes, *Ber. Bunsenges. Phys. Chem.,* 79, 570, 1975.
34. **Slinko, M. G. and Slinko, M. M.,** Autooscillations in the rate of heterogeneous catalytic reactions (Russian), *Usp. Khim.,* 49, 561, 1980.
35. **Shu Wai Lin, Keizer, J., Rock, P. A., and Stenschke, H.,** On the mechanism of oscillations in the "beating mercury heart", *Proc. Natl. Acad. Sci. U.S.A.,* 71, 4477, 1974.
36. **Iwamoto, K. and Seno, M.,** Effects of fluctuations on dissipative structures. Part 1. Effect of fluctuations on chemical oscillatory systems, *J. Chem. Phys.,* 70, 5851, 1979.
37. **Nicolis, G. and Portnow, J.,** Chemical oscillations, *Chem. Rev.,* 73, 365, 1973.
38. **Noszticzius, Z., Farkas, H., and Schelly, Z. A.,** Explodator and oregonator. Parallel and serial oscillatory networks. A comparison, *React. Kinet. Catal. Lett.,* 25, 305, 1984.
39. **Noszticzius, Z., Farkas, H., and Schelly, Z. A.,** Explodator: a new skeleton mechanism for the halate driven chemical oscillators, *J. Chem. Phys.,* 80, 6062, 1984.
40. **Schmitz, G.,** Oscillations entretenue dans un systeme chimique homogene, *J. Chim. Phys.,* 71, 689, 1974.
41. **Ruoff, P. and Switters, B.,** Oscillations, excitability and stirring effects in closed methylmalonic acid Delousov-Zhabotinskii System, *Z. Physik. Chem. Neue Folge,* 135, 171, 1983.
42. **Glansdorff, P. and Prigogine, I.,** *Thermodynamical Theory of Structure Stability and Fluctuations,* Wiley Interscience, New York, 1971.
43. **Winfree, A. T.,** Rotating chemical reactions, *Nature (London),* 82, 1974.
44. **Janigova, I.,** The Decomposition of N-Halogen Compounds in Polymers (Slovak), thesis Polymer Institute, Slovak Academy of Science, Bratislava, 1986.

Chapter 9

CHEMICAL LASERS

I. INTRODUCTION

In chemical reactions, the products are frequently formed in different excited states. Optimization of reaction conditions with respect to the maximum yield of excited state generation led to the discovery of chemical lasers. The chemical laser is a tool for the transformation of chemical energy into the form of powerful laser radiation. Its principle includes the outcomes from many scientific regions (dynamics of molecular collisions, chemical kinetics, spectroscopy, gas dynamics, optics, etc.). To illustrate it, we describe in more detail the laser emission of infrared light from the vibrational levels of HF molecule.[1]

If the low pressure mixture of UF_6 and H_2 (1:1.5) is photolyzed with UV flashes of a xenon flash tube with a duration of $5 \cdot 10^{-5}$ sec, the pulse of infrared light starts about $4 \cdot 10^{-6}$ sec after each flash initiation and lasts for about $9 \cdot 10^{-6}$ sec. The processes taking place in such a short time interval are rather complicated.

The initiating reaction is photolysis of UF_6 (Equation 1)

$$UF_6 \rightarrow .UF_5 + F \cdot \qquad (1)$$

which generates atomic fluorine. $F \cdot$ atoms can be obtained also from other fluorine-containing molecules (e.g., from F_2). The electron beam, heat (temperatures 800 to 1500 K), electrical or microwave discharge are also used as alternative sources of initiation energy.[2-4] Especially in electrical discharge the rise time and total duration of the pulse can be easily varied, the repetition rate can be largely increased, and by the choice of various circuit elements, one can ensure a maximum energy deposit to the reactive medium. Very efficient generation of fluorine atoms is obtained when utilizing the chain reaction in $H_2 + F_2$ mixture (Equations 2 and 3).

$$F \cdot + H_2 \rightarrow HF (v) + H \cdot \qquad \Delta H_0 = -132 \text{ kJ/mol} \qquad (2)$$

$$H \cdot + F_2 \rightarrow HF (v') + F \cdot \qquad \Delta H_0 = -410 \text{ kJ/mol} \qquad (3)$$

As an important criterion of efficiency of chain reactions in chemical lasers, the term of effective or laser chain length is frequently used, which means only that number of molecules produced by one $F \cdot$ atom (obtained by action of outer initiation source) which gives the deposit into the energy of laser radiation. For example[3], the laser chain length is ~ 30 to 100 for the H_2-F_2 system, ~ 100-400 for the D_2-F_2-CO_2 system and ~ 10 for the H_2-ClF system.

At a sufficiently high concentration of fluorine atoms, the actual chemical reaction with lasing action can start. Reactions 2 and 3 are not only chain generators of $F \cdot$ atoms but produce also hydrogen fluoride molecules with high vibrational state quantum numbers v and v'. The state-selective rate constants of reaction 2 leading to the separate vibrational states with v = 0,1,2,3 are in the relations:[5] $k(0):k(1):k(2):k(3) = 0.06:0.29:1.00:0.63$. In reaction 3 the maximum population of vibrational levels is shifted higher:[6]

$$k(0):k(1):k(2):k(3):k(4):k(5):k(6):k(7):k(8):k(9) =$$

$$0.025:0.052:0.072:0.075:0.334:0.76:1.00:0.114:0.052:0.075$$

Nevertheless, the main lasing chemical reaction in H_2-F_2 laser is reaction 2 with maximum contribution ($\sim$0.6) to the total energy of laser radiation from the emission of transition v $= 2 \longrightarrow$ v $= 1$ (wavelength $\sim$2800 nm). This unusual order of corresponding rate constants leads to the inversion in the population of individual vibrational levels. This is the main condition for the existence of an effective chemical laser. This type of population inversion can be created not only on vibrational but also on electronic or rotational energy levels. The powerful iodine laser based on a photodissociation reaction (Equation 4) provides 100% population inversion because only

$$CF_3I \xrightarrow{\ h\nu\ } \cdot CF_3 + \cdot I(^2P_{1/2}) \tag{4}$$

the $^2P_{1/2}$ iodine atoms are formed. The emission $^2P_{1/2} \longrightarrow {}^2P_{3/2}$ (wavelength 1300 nm) gives us an example of a laser working on electronic energy levels. The photodissociation lasers are not pure chemical lasers. Although the lasing particles have an origin in chemical reaction, the energy of the laser radiation is not obtained from the chemical energy but mainly from the energy of the outer initiation source.

The wavelength of chemical lasers working on rotational energy levels goes up to the far IR region. As an example,[7] the chemical reaction of photoelimination of HF from the CH_2CF_2 provides the HF molecules with high rotational numbers J. For example, one of the major components of such laser radiation with wavelength 19,370 nm can be ascribed to the transition v $= 0$, J $= 13 \longrightarrow$ v $= 0$, J $= 12$. The lasers based on pure rotational transitions cannot be expected to provide laser radiation of very high power. Very fast nonradiative relaxation of rotationally excited states prevents the population inversion from occurring for a sufficiently long time necessary for this purpose.

Population inversion between two energetical levels makes an intensive emission of photons possible then. When this emission proceeds in a laser tube with mirrors, each spontaneously emitted photon goes many times through the gaseous content of the tube and stimulates an emission of further photons from other excited molecules. The cascade of photons produced by this process has a character of coherent radiation, since stimulated emission synchronizes the radiation of excited molecules. The laser gain, therefore, depends also on the Einstein coefficient for stimulated emission between two levels of a given lasing molecule. Besides this desirable process, the collisional quenching and parasitic chemical reactions take place in this stage of the overall process. If the rate of decrease of the concentration of excited HF molecules through these channels is large in comparison with the rate of stimulated emission, the intensity of the laser beam decreases.

II. CLASSIFICATION OF CHEMICAL LASERS

The transfer of vibrational energy from the molecule excited in a chemical reaction to different molecules is generally a nondesirable type of collisional quenching. However, in a particular class of chemical lasers this effect plays a positive role. The D_2-F_2-CO_2 laser serves as an example. Here the chemical chain reactions of 2 and 3 types produce vibrationally excited DF molecules. The vibrational energy is then transferred by collisions to the specific vibrational mode of CO_2 molecules, which are the actual source of laser radiation in such a system. What is the advantage of this transfer? On one hand, it is obvious that the DF molecule alone can be the source of laser radiation and a large loss of vibrational energy can be expected during this transfer. On the other hand, the relaxation processes in the case of vibrationally activated CO_2 molecules are not so fast as in the case of hydrogen halides. Also, the population inversion in DF molecules is not a necessary condition for obtaining the population inversion of CO_2 molecules. This fact significantly prolongs the laser chain length. As a result, the addition of CO_2 to the D_2-F_2 mixture significantly increases the energy of radiation and allows it to work at atmospheric pressure.

Besides these two groups of chemical lasers (with the direct or indirect generation of lasing molecules in chemical reaction), we have mentioned the possibility of classifying chemical lasers on the basis of their initiation sources or on the energetical levels serving for the emission of photons (vibrational, electronic, rotational).

One important characteristic of lasers is also the duration of the radiation. The continuous-working (CW) and pulsed chemical lasers utilize the same basic principles but they significantly differ in technical design. The simple pulsed laser has been described at the beginning of this chapter. The operation of CW chemical lasers is based on the mixing of reagents in streams with sub- or supersonic velocity. The fast flow of reactants is one of the necessary conditions which guarantees fast and continuous generation of population inversion in a reactor. The problems which have to be also solved are the necessity for homogeneous mixing of reagents, the extreme heating of the active medium, and the corrosive activity of gases. Take as an illustration a purely chemical CW laser based on a $NO-F_2-D_2-CO_2$ mixture which fulfills the main goal of construction of chemical lasers — the transformation of chemical energy to the laser radiation without the presence of other physical souces of energy.[8] In the preliminary stage, the $NO\cdot$, F_2 and CO_2 gases are mixed and the generation of atomic fluorine can start (Equation 5).

$$NO\cdot + F_2 \rightarrow NOF + F\cdot \tag{5}$$

In the next part of the reactor the D_2 is injected and previously described indirectly generated population inversion in CO_2 evolves. Of major importance here, and mainly in the case of supersonic CW chemical lasers, are the shape of the reactor and especially the shape of the nozzles which significantly influence the gas dynamics and consequently the total energy output of the laser.

The chemical laser can serve not only as generator but also as an amplifier of radiation from the previous laser unit. The high density of photons of the input laser beam remove the population inversion from lasing molecules of the amplifier more efficiently than its own photons and an optical resonator is not necessary in such an arrangement. Large power beams are generated in this way.

Chemical lasers can be classified also on the basis of other characteristic parameters. Spectra of laser radiation can have a single or multi-line character. A broad range of wavelengths is important in laser chemistry. However, the most followed parameters of chemical lasers is their power and its relation to the input initiation energy. The range of chemical lasers varies from miniaturized, low-power devices to lasers projected for nuclear fusion, with excellent beam quality which can provide extreme energies in very short pulses. However, greater increases in power are still desirable and large groups of scientists and technicians are working in this direction.

III. IMPORTANT FACTORS

As follows from the previous discussion, the main goal of chemical lasers is an effective transformation of chemical energy to the radiation energy concentrated in a narrow space interval and sometimes also in a narrow time interval (pulsed lasers). Many effects, which must be suppressed to a minimum, act against this effort. From many factors which can be utilized for the efficient control of processes taking place in the overall mechanism, the chemical reaction producing a population inversion seems to play a decisive role.

The demands put on the suitable type of chemical reaction are of different character. First of all, the reaction must be fast in order to accumulate a sufficient number of lasing molecules and thus prevail over the rate of different unavoidable deactivation processes and losses of energy. This means that the activation energy of such a reaction should be small (e.g., for reaction 2 the value of $E_A \sim 4$ kJ/mol has been measured).[9,10] In the case of reaction sequence 6 and 7, not

$$\cdot CH_3 + \cdot NF_2 \rightarrow CH_3NF_2 \text{ (v)} \tag{6}$$

$$CH_3NF_2 \text{ (v)} \rightarrow HF \text{ (v')} + CH_2=NF \tag{7}$$

only the recombination reaction 6 but also the unimolecular elimination reaction 7 should be sufficiently fast. On the other hand, the rate of reaction 7 should not exceed the rate of redistribution of vibrational energy in the chemically activated CH_3NF_2 molecule. These conditions are necessary for effective channeling of kinetic and potential energy of $\cdot CH_3$ and $\cdot NF_2$ species into the vibrational mode of HF molecules. Because the reactions of free radicals mostly fulfill the conditions of small activation energy, they are therefore main candidates for application in chemical lasers.

Besides the necessity of large total rate constants, the state-selective rate constants leading from the ground state of reactants to the different vibrational states of products and their mutual relations are also important. For obtaining population inversion it is necessary that some rate constants for higher vibrational states are larger than the constant for the vibrational ground state of products. The main presumption for this effect seems to be the exothermicity of the reaction; for three-atomic exchange reactions it is, e.g., sufficient to have a sum of reaction exothermicity and activation energy larger than 60 kJ/mol.[2] In practice, the highest populated vibrational level always has an energy lower or equal to the sum of reaction exothermicity and activation energy. This also means that the potential electronic energy-chemical energy (and not the kinetic energy of reactants) is a reason for population inversion. For a three-atomic exchange reaction (Equation 8) it is then desirable that BC molecules would

$$A + BC \rightarrow AB + C \tag{8}$$

have lower dissociation energy than AB molecules. This is valid if AB dissociation energy is very large (HF, LiF, and CO molecules have the values of 565, 574, and 1072 kJ/mol, respectively). Because the vibrational energy of products has its upper limit as the sum of exothermicity and activation energy, the increased reaction exothermicity enlarges the multiline character of laser radiation as a consequence of the increased number of transitions between different vibrational levels and hence extends the region of its wavelengths.

Theoretical analysis of reaction dynamics has shown that reactions with an activated complex more similar to the reactants than to the products are a proper source of vibrational population inversion.[11,12] Quantum chemical calculations[13] give for $F\cdot + H_2$ reaction (Equation 2) the following structure of the activated complex: the collinear geometrical arrangement with internuclear distance $R_{HH} = 0.76$ and $R_{HF} = 1.62$ (in 10^{-10} m). These values together with the $R_{HH} = 0.74$ in H_2 molecule and $R_{HF} = 0.92$ in the HF molecule illustrate the previous statement. The existence of a long-lived reaction intermediate during a given reaction has a negative influence on population inversion. Here the excess of vibrational energy has enough time for redistribution between different internal degrees of freedom, and channeling into the one vibrational mode is not so effective.

The next important factor, mainly in the case of chain-reaction lasers, is the overall character of complicated reaction mechanisms comprising a large number of initiation, propagation, branching, and termination reactions as well as different secondary processes. In large pulsed lasers it is necessary to have a stable reaction mixture at approximately atmospheric and higher pressures. But exothermic chain reactions in H_2-F_2 mixture at such pressure can also lead to spontaneous detonation. The branched reactions of vibrationally activated H and HF molecules (Equations 9 and 10) seem to be responsible for such an undesirable effect.[2,3]

$$H_2 (v) + F_2 \rightarrow HF + H\cdot + F\cdot \tag{9}$$

$$HF (v) + F_2 \rightarrow HF + 2 F\cdot \tag{10}$$

Molecular oxygen can be used as a stabilizer at such conditions according to reactions 11 and 12, where the radicals with significantly

$$H\cdot + O_2 + M \rightarrow HO_2 + M \tag{11}$$

$$F\cdot + O_2 + M \rightarrow FO_2\cdot + M \tag{12}$$

lower reactivity than hydrogen or fluorine atoms are formed. The controlled branched mechanism on the other hand can significantly increase a rate of initiation. An interesting possibility of increased number of active centers provides the photon branching of chain reactions.[3] In the system CH_3F- D_2 -F_2 -CO_2 -He, radiation of CO_2 with the wavelength of 9600 nm can be used in the initiation stage of the multiphoton vibrational excitation of CH_3F molecules (Equation 13). A secondary reaction (Equation 14)

$$CH_3F + m\ h\nu \rightarrow CH_3F (v) \tag{13}$$

$$CH_3F (v) + F_2 \rightarrow \cdot CH_2F + HF + F\cdot \tag{14}$$

provides the source of F· atoms without the necessity for external initiation radiation. This process can be called photon-branched only if one F· center is able to generate more photons than is necessary for its own generation. An optimal solution in the future will probably combine the photon and chemical or energetical branching into the common chain reaction mechanism.

As soon as the population inversion exists, different deactivation processes start to compete with the spontaneous or stimulated emission of photons. The collisions of lasing molecules with all types of molecules present in the reaction medium must be then taken into account. These collisions can lead to the chemical reactions (e.g., opposite reactions (2) and (3)) or to the nonreactive transfer of excess of vibrational energy. The second mentioned channel leads to the increased relative translational energy of colliding partners (V → T process) and hence to the increased temperature, to the increased rotational energy of one or both partners, (V → R process) and/or to the increased vibrational energy of the second nonlasing partner (V → V process). Because lasing molecules are usually diatomic, the fast process of intramolecular redistribution of vibrational energy between different vibrational modes is not present. The actual mechanism of these complicated relaxation processes is not fully understood to date, and only some qualitative conclusions can be postulated. Two of them are illustrated by the data in Table 1. Here the Z stands for the mean number of collisions necessary for one deactivation transfer of energy from vibrationally activated HF or HCl molecule. The ΔE_v value indicates the energy difference between the vibrational transition of a HF or HCl molecule of given frequency and the vibrational transition of the colliding partner nearest to this frequency ($\Delta E_v \rightarrow 0$ indicates the possibility of resonance between vibrational modes of both partners).

The first effect documented by these results is the influence of attractive intermolecular forces on the mean collision time and hence also on the efficiency of the vibrational energy

Table 1
DEACTIVATION OF VIBRATIONALLY ACTIVATED HF AND HCl MOLECULES IN COLLISIONS WITH DIFFERENT PARTNERS[2]

	HF		HCl	
	Z Ref. 14	$\Delta E_v(cm^{-1})$	Z Ref. 15,16	$\Delta E_v(cm^{-1})$
Ar	$>10^5$	—	$>3 \cdot 10^6$	—
N_2	$5.7 \cdot 10^4$	1630	$9.1 \cdot 10^3$	552
D_2	$2.7 \cdot 10^3$	971	143	-108
H_2	570	-198	$1.1 \cdot 10^5$	-1274
CH_4	160	941	106	-30
H_2O	1.3	205	10	-871

Note: Z — mean number of collisions; ΔE_v — difference in frequencies of colliding partners (see text).

transfer. The possibility of hydrogen bonding in the case of H_2O molecules significantly increases the mean lifetime of the collision complex in comparison with, e.g., inert Ar atoms, and the deactivation ability of such molecules is generally large. The second specific effect can be followed by comparison of rows for H_2 and D_2 molecules. The harmonic vibrational frequency in HF is closer to the value for the H_2 than for the D_2 molecule and the quasiresonance character of V → V transfer to the H_2 molecules significantly increases the total deactivation rate. On the other hand, the opposite situation in harmonic frequencies for D_2 and H_2 in collision with HCl stimulate a fast deactivation by the D_2 molecule.

There are also other important factors which influence the properties of chemical lasers. By the type and energy of initiation the initial concentration of free radicals may be controlled. The rate of reagent delivery significantly influences the rate of generation of lasing molecules at CW lasers. The decrease in initial temperature of the gas mixture can increase the energy output of the DF-CO_2 laser.[3] Increasing the pressure generally has a positive effect on the concentration of lasing molecules, but simultaneously increases the deactivation rate and some optimum value must be searched for. Further components can be added to the reaction mixture for its stabilizing (O_2), for better absorption of initiation radiation, and as an additional source of F atoms (SF_6) as well as for the increase of heat capacity and the decrease of deactivation rate (He). All these factors together with many other technical parameters allow efficient adaptation of chemical lasers for actual purposes.

IV. TYPES OF CHEMICAL REACTIONS

There are a number of types of chemical reactions on the basis of which the emission of laser radiation has been observed.[2] Attention in this section is concentrated on reactions with free radicals as reactants and/or as products of these reactions.

The simple three-atomic exchange reaction (8) has been illustrated and discussed on the example of the F· + H_2 reaction (2). As a further example the following reactions can be presented (Equations 15 to 21). This type of reaction obviously generates

$$F· + HX \rightarrow HF (v) + X· \qquad X = I, Br, Cl \qquad (15)$$

$$H· + Cl_2 \rightarrow HCl (v) + Cl· \qquad (16)$$

$$Br· + HI \rightarrow HBr (v) + I· \qquad (17)$$

$$\cdot O \cdot + \cdot CH \rightarrow CO \ (v) + H\cdot \tag{18}$$

$$\cdot O \cdot + \cdot CF \rightarrow CO \ (v) + F\cdot \tag{19}$$

$$\cdot O \cdot + \cdot CN \rightarrow CO \ (v) + \cdot \dot{N}\cdot \tag{20}$$

$$\cdot O \cdot + CS \rightarrow CO \ (v) + \cdot S\cdot \tag{21}$$

a significant population inversion and a large part of the electronic potential energy transforms effectively into the enhanced vibrational motion of created chemical bonds. Besides major frequencies of radiation from low lying vibrational states, also the lines having their origin in emission from high vibrational states can be observed.

Another important class of chemical lasers is based on abstraction chemical reactions. Due to the large energy of the HF bond, the $F\cdot$ atoms are able to abstract the $H\cdot$ atoms exothermically from different species according to the general scheme (Equation 22).

$$F\cdot + RH \rightarrow HF \ (v) + R\cdot \tag{22}$$

Not only hydrocarbons (CH_4, C_2H_6, C_3H_8, C_4H_{10}, $(CH_3)_3 \ CH$) but also inorganic hydrides as SiH_4, GeH_4, AsH_3, SbH_3, B_2H_6 can serve as donors of $H\cdot$ atoms. In this second group the weak chemical bond M-H allows one to obtain emissions also from high vibrational states of HF molecules. Also, the $H\cdot$ and $\cdot O\cdot$ atoms can take part in abstractions with a laser effect, e.g., Equation 23.

$$H\cdot + XeF_4 \rightarrow HF \ (v) + \cdot XeF_3 \tag{23}$$

However, the real reaction scheme of these reactions is much more complex, mainly in the case of initiation by electrical discharge.

As a product of many biomolecular exothermic reactions a single energized molecule can be obtained which, at convenient conditions (low pressure), is not stabilized by collision relaxation. The subsequent unimolecular elimination of a small molecule (usually HF) can proceed by a mechanism which directs the excess of internal energy into the vibrational mode of the eliminated molecule and a lasing effect is observed. The recombination radical reactions can be first mentioned in this context. The recombination of $\cdot CH_3$ and $\cdot CF_3$ radicals (Equation 24) leads to the vibrationally excited 1,1,1-trifluoroethane.

$$\cdot CH_3 + \cdot CF_3 \rightarrow CH_3CF_3 \ (v) \qquad \Delta H_0 = -420 \ \text{kJ/mol}; \ E_A \sim 0 \tag{24}$$

The energy released from the interaction of both radicals is probably statistically distributed among different vibrational modes of this molecule and, following elimination (Equation 25),

$$CH_3CF_3(v) \rightarrow HF \ (v') + CH_2{=}CF_2 \qquad \Delta H = 130 \ \text{kJ/mol}; \ E_A \sim 297 \ \text{kJ/mol} \tag{25}$$

produces the energized HF molecules, although the reaction (25) for $v = 0$, $v' = 0$ is endothermic. Statistical redistribution of vibrational energy significantly decreases the efficiency of energy transformation and population inversion is usually not reached in this type of reaction (the relation of occupancies of levels with $v' = 1$ and $v' = 0$ lies between the limits 0.58 to 0.9).[2] Nevertheless, the Boltzmann distribution is strongly perturbed and laser radiation is observed.

Some insertion reactions are also of similar character. The oxygen in excited electronic state[1] D is able to insert into the chemical bond (Equation 26).

$$\cdot O \cdot (^1D) + CHF_3 \rightarrow HOCF_3 \text{ (v)} \tag{26}$$

The energized $HOCF_3$ molecules split off (Equation 27), again with lasing HF molecules as a product.

$$HOCF_3 \text{ (v)} \rightarrow HF \text{ (v}') + F_2C=O \tag{27}$$

Such insertion properties also have $\cdot CH_2 \cdot$ and $\cdot NH \cdot$ species in excited states of similar singlet biradical character as $\cdot O \cdot$ (^1D) atoms.

The third type of this reaction starts as an addition reaction. Free radicals are able to add to the double bond according to the reaction which can, for simplicity, be written as summary Equation 28.

$$CH_2=CH_2 + 2 F \cdot \rightarrow CH_2FCH_2F \text{ (v)} \tag{28}$$

This reaction can also be followed by elimination of HF (Equation 29)

$$CH_2FCH_2F \text{ (v)} \rightarrow HF \text{ (v}') + CH_2=CHF \tag{29}$$

In comparison with the three-atomic reactions, the whole group of elimination reactions has a remarkably lower efficiency of transformation of chemical energy into the form of enhanced energy of a single vibrational mode. So these reactions are not suitable for projects of chemical lasers with very high power.

A large class of lasers uses UV radiation for excitation of reactants to the higher electronic states. This type of laser lies on the border between chemical lasers and other laser types. Although the chemical reaction is used and its products are generators of laser radiation, the energy for this radiation is consumed mainly from initiating UV radiation. In photoelimination reactions, the excess of energy can be channeled from the excited electronic state of a given molecule to the vibrational mode of the eliminated molecule at the proper arrangement of electronic potential energy surfaces (Equations 30 and 31).

$$CH_2=CHF + h\nu \rightarrow CH_2=CHF^* \tag{30}$$

$$CH_2=CHF^* \rightarrow HF \text{ (v)} + CH\equiv CH \tag{31}$$

As in the case of previously mentioned dehydrohalogenation reactions, the population inversion is usually not observed because of the large number of possibilities for energy redistribution to the different vibrational degrees of freedom in the polyatomic reaction intermediate. As a result, the emission from low-lying vibrational levels is only observed.

The products of photodissociation reactions in excited electronic or vibrational states are very frequent sources of laser radiation. The best known and most powerful iodine laser represented by reaction 4 has been discussed previously. A similar bromine laser based on reaction 32 provides emission

$$IBr + h\nu \rightarrow Br \cdot (^2P_{1/2}) + I \cdot \tag{32}$$

from transition between electronic states $Br\cdot(^2P_{1/2} \rightarrow {}^2P_{3/2})$ at a larger wavelength (2700 nm) than the iodine laser. The products in higher vibrational states can be obtained by photolysis of ClNO or C_2N_2 molecules. As a lasing molecule $NO\cdot$ and $\cdot CN$ radicals may be functioning in this case. The interesting possibility of simple regeneration of starting compounds by radical recombination increases the practical value of similar chemical lasers.

A very important group of lasers is based on exciplex association reactions.[17] Radiation from these lasers covers the visible and UV regions of light spectrum. Their principle can be explained on the reaction (Equation 33) of inert gas atoms in

$$Xe^* + Xe \rightarrow Xe_2^*$$ (33)

ground and excited electronic states. Minima on potential energy curves of excited $Xe_2{}^*$ dimers and the practically repulsive curve for ground electronic state of Xe_2 provide ideal conditions for the generation of strong population inversion, because the lower energy level is rapidly depleted after emission of photons by dissociation to inert Xe atoms. As exciplex lasing molecules the KrF, XeCl, PbXe, CdAr, XeO, Kr_2F, and many other systems can operate.

V. APPLICATIONS

The chemical laser is an efficient tool not only in science but also in industry.[4] In comparison with other types of lasers, it has the disadvantage of limited tunability. On the other hand, it is very efficient in the activation of molecules identical with lasing molecules or molecules with coinciding spectral lines. In chemical research, chemical lasers are able to give information about the state-selective behavior of many chemical reactions.[18] The specific excitation of reactants to the different vibrational-rotational levels can significantly change the rate of unimolecular or bimolecular chemical reactions. The mechanism of laser multiphoton dissociation has been discussed in the second chapter. Also, isomerization reactions can be significantly influenced by laser IR radiation. The enhanced rate of bimolecular reactions can be demonstrated by Equation 34 where the rate of reaction is found

$$Sr + HF \rightarrow \cdot SrF + H\cdot$$ (34)

to be by several orders of magnitude higher when the HF gas is irradiated by pulsed HF laser and HF at $v = 1$ state takes part in the reaction. New interesting possibilities of the detailed study of steric factors that control reactivity should be opened up now.[19] By the choice of angle between the laser beam and the beam of molecules, we may preferably excite molecules of specific space orientation. Such polarized molecular beams can be crossed and the orientation dependence on reactivity can be studied at single-collision conditions.

Promising advances are being made at isotope separation by laser multiphoton dissociation. The reason lies in the fact that from two isotopical analogues of a given molecule, only one can efficiently absorb a specific IR wavelength and dissociate to products significantly enriched by one of both isotopes. However, with increasing pressure, which is desirable for practical use, the process becomes less selective. As examples of parent molecules, H_2CO can be used for separation of H and D isotopes, CF_3I for ^{12}C and ^{13}C separation, SF_6 for ^{32}S and ^{34}S separation, and UF_6 for ^{235}U and ^{238}U separation.

The use of HX lasers is of importance also for the effective control of pollutants in the atmosphere, since they combine high selectivity with the possibility of long-distance monitoring. The beam of a HF laser can be oriented in a direction perpendicular to the exhaust stream of, e.g., an aluminum factory. The emitted and the retroreflected beams are then compared, and the concentration of HF can be monitored in the exhaust gases. The DF laser lines coincide with absorption lines of SO_2, $\cdot NO_2$, N_2O, and some other gases; therefore, the technique may be generally applied to pollution monitoring.

Chemical lasers are perspective tools also for the initiation of nuclear fusion of hydrogen isotopes D and T (Equation 35).

$$D + T \rightarrow {}^4He + n + 1.7 \; 10^9 \; kJ/mol \tag{35}$$

For successful thermonuclear fusion it is necessary to overcome the large repulsive electrostatic force between nuclei. The kinetic energy of nuclei sufficient for this purpose is reached only at temperatures higher than $5 \cdot 10^7$ K. This fact places large demands on possible laser sources of heat. It is expected that the power of some 100 TW must be concentrated in a laser pulse with a duration of less than 30 nsec. The iodine laser and the HF laser operating on the H_2 - F_2 reaction are also among candidates which are expected to fulfill these strong requirements. However, large technical problems must be solved, such as, e.g., the recycling of HF produced in the course of a laser reaction. Approximately $5 \cdot 10^5$ kg of F_2 should be regenerated every day by electrolysis of HF which is 100 times larger than the average F_2 production of an industrial plant. In spite of these problems, the development and wide use of new powerful chemical lasers is of major importance for the future of our energetics.

REFERENCES

1. **Kompa, K. L. and Pimentel, G. C.,** Hydrofluoric acid chemical laser, *J. Chem. Phys.*, 47, 857, 1967.
2. **Gross, R. W. F. and Bott, J. F.,** Eds., *Handbook of Chemical Lasers*, John Wiley & Sons, New York, 1976.
3. **Basov, N. G.,** Ed., *Chemical Lasers*, Nauka, Moscow, 1982.
4. **Telle, H. H.,** Chemical lasers — a tool or a toy?, *Acta Phys. Polonica*, A66, 337, 1984.
5. **Berry, M. J.,** $F + H_2$, D_2, HD reactions: chemical laser determination of the product vibrational state populations and the $F + HD$ intramolecular kinetic isotope effect, *J. Chem. Phys.*, 59, 6229, 1973.
6. **Herbelin, J. H. and Emanuel, G.,** Einstein coefficients for diatomic molecules, *J. Chem. Phys.*, 60, 689, 1974.
7. **Cuellar, E., Parker, J. H., and Pimentel, G. C.,** Rotational chemical lasers from hydrogen fluoride elimination reactions, *J. Chem. Phys.*, 61, 422, 1974.
8. **Cool, T. A. and Stephens, R. R.,** Chemical lasers by fluid mixing, *J. Chem. Phys.*, 51, 5175, 1969.
9. **Wurtzberg, E. and Houston, P. L.,** The temperature dependence of absolute rate constants for the $F + H_2$ and $F + D_2$ reactions, *J. Chem. Phys.*, 72, 4811, 1980.
10. **Heidner, R. F., Bott, J. F., Gardner, C. E., and Melzer, J. E.,** Absolute rate coefficients for $F + H_2$ and $F + D_2$ at T = 295-765 K, *J. Chem. Phys.*, 72, 4815, 1980.
11. **Polanyi, J. C. and Wong, W. H.,** Location of energy barriers. I. Effect on the dynamics of reactions $A + BC$, *J. Chem. Phys.*, 51, 1439, 1969.
12. **Mok, M. H. and Polanyi, J. C.,** Location of energy barriers. II. Correlation with barrier height, *J. Chem. Phys.*, 51, 1451, 1969.
13. **Steckler, R., Schwenke, D. W., Brown, F. B., and Truhlar, D. G.,** An improved calculation of the transition state for the $F + H_2$ reaction, *Chem. Phys. Lett.*, 121, 475, 1985.
14. **Hancock, J. K. and Green, W. H.,** Vibrational deactivation of HF (v = 1) in pure HF and in HF-additive mixtures, *J. Chem. Phys.*, 57, 4516, 1972.
15. **Chen, H. L. and Moore, C. B.,** Vibration $\rightarrow$ rotation energy transfer in hydrogen chloride, *J. Chem. Phys.*, 54, 4072, 1971.
16. **Chen, H. L. and Moore, C. B.,** Vibration $\rightarrow$ vibration energy transfer in hydrogen chloride mixtures, *J. Chem. Phys.*, 54, 4080, 1971.
17. **Rhodes, Ch. K.,** Ed., *Excimer Lasers*, Springer Verlag, Berlin, 1979.
18. **Letokhov, V. S.,** *Nonlinear Selective Photoprocesses in Atoms and Molecules*, Nauka, Moscow, 1983.
19. **Zare, R. N. and Bernstein, R. B.,** State-to-state reaction dynamics, *Phys. Today*, 33(11), 1, 1980.

Chapter 10

FREE RADICALS AND BIOCHEMISTRY

The idea of free radical participation in normally functioning biological tissue is more the subject of speculations than of experimentally founded reality. A much more complete view exists on the possible role of free radicals in pathological processes, aging, carcinogenesis, etc., but also much uncertainty exists here relating to the succession of individual radical steps and to their ratio in the complex of all possible reaction pathways. That is why in the excellent textbook on fundamental principles and mechanisms in biological chemistry published in 1978 no mention of free radicals was included in the subject index and all chemical mechanisms were presented there as ionic or molecular in nature.[1]

We are aware that a uniform theory of free radical mechanisms in biological organisms is still not possible considering the available experiments. Throughout the preceding chapters we have, however, attempted to indicate the free radical alternatives or examples of mechanisms in biochemistry, whenever it was rationalized by experimental facts. The rest of the facts related to this topic summarized here are aimed at emphasizing that some common language in this interdisciplinary approach is necessary.

Biochemists, e.g., frequently assume that at catabolic oxidation reactions, electrons and protons are released which are subsequently transferred to anabolic reduction reaction steps mediated by $NADP^+$ (nicotinamide adenine dinucleotide phosphate). The release of electrons suggests the existence of some carrier molecule and thus we get into the field of free radical chemistry. One thing is certain, namely, that a well-functioning living organism does not like reactive free radical intermediates moving freely in its tissue and eliminates them by all possible ways. This is natural because of the high reactivity of many free radicals and their harmful effect on all genuinely constructed and precisely equilibrated mechanism cycles.

This, however, does not exclude the existence of free radical intermediates in active stages of catalytic oxidative enzymatic reactions, where the radical center remains attached to the enzyme moiety and the mechanisms of the process are deduced from the free radical character of metabolites. Free radical centers are latently hidden in trace elements which are essential for living organisms, where they are present as central atoms of prosthetic groups of many enzymes.[2,3] The release of free radicals is then only a question of interaction with a proper copartner.

Copartners, which in interaction with trace elements may produce free radicals, usually arise from reactions of oxygen which is the strongest oxidation reactant in living matter. By its stepwise reduction superoxoanion radicals $\cdot O_2^-$, peroxyanions, and oxide ions may be formed. Superoxoanion radicals forming in living tissues by the action of some flavinoxidases (as e.g., xanthinoxidase) are especially reactive. Superoxide dismutase, which is structurally identical with blood protein erythrocuprein, is the enzyme which facilitates the disproportionation of $\cdot O_2^-$ to oxygen and more stable H_2O_2. Subsequent reaction of either $\cdot O_2^-$ anion radicals or H_2O_2 catalyzed with Fe ions in the nonenzymatic Haber-Weiss reaction (Scheme 1) gives hydroxyl radicals which

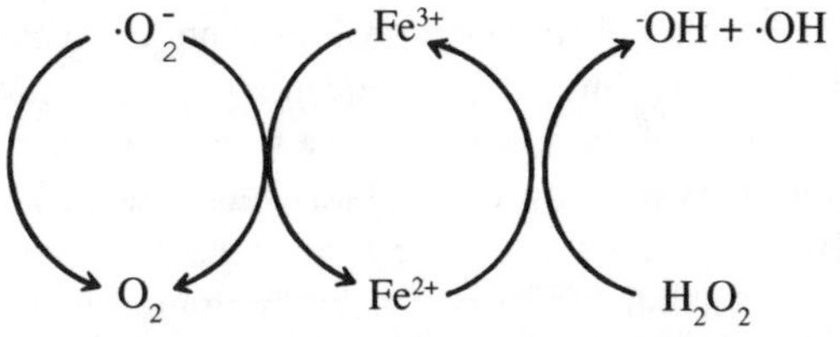

SCHEME 1

are very strong oxidizing particles and in some way are responsible for the toxicity of overconcentrations of oxygen.

Hydroxyl radicals have been implicated in a number of biological phenomena, generally from the viewpoint of their cytotoxic action. For example, cellular damage induced by ionizing radiation and the destruction of microorganisms by phagocytes have been attributed, in part, to the effect of hydroxyl radicals. Alloxan, which destroys the beta cells of the pancreas, and 6-hydroxy- and 6-aminodopamine, which destroy sympathetic nerves, are examples of cytotoxic agents whose mechanism of action are thought to involve hydroxyl radicals.[4] The scavengers for very reactive oxidant hydroxyl radicals are formic acid, thiourea, cystein, and ergotionein, the only mercaptoimidazole derivative in nature.[5] ·OH radicals may be scavenged there by the formation of disulfides (Equations 1 and 2).

$$RS^- + \cdot OH \rightarrow RS\cdot + {}^-OH \tag{1}$$

$$RS\cdot + RS\cdot \rightarrow RSSR \tag{2}$$

The harmful effect of oxygen on living tissue may be demonstrated on already quoted peroxidation of lipids (Scheme 2) which is the nonenzymatic radical chain reaction of polyconjugated fatty acids present as the integral part of cell membranes.[6]

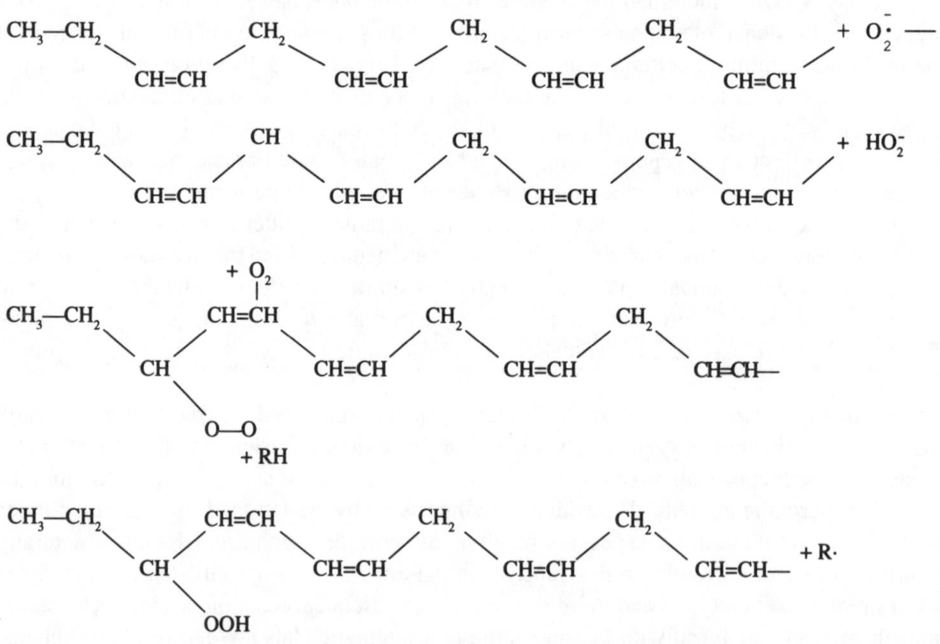

SCHEME 2

The mechanism of decomposition of formed peroxides is not quite clear, but as the stable product dialdehyde of malonic acid is formed. The sequence of peroxidation reactions disrupts the structure and the permeability of the membrane; the formed dialdehyde may, moreover, take part in cross-linking reactions with proteins and denaturate them. The lipid peroxidation is suspected to be responsible for aging and some serious diseases, such as arteriosclerosis and cancer.[7,8] In prooxidant states the intracellular concentration of the activated form of oxygen ($\cdot O_2^-$, HO_2, $ROO\cdot$, $\cdot OH$, $HOOH$) is increased, presumably because cells either overproduce these species, or are deficient in their ability to destroy them. In carcinogenesis, free radicals appear to play a role mostly in the promotion phase, during which gene expression of initiated cells is modulated by affecting genes (inducing alterations in DNA structure) that regulate cell differentiation and growth.[9] The detailed mechanism in carcinogenesis, however, remains unclear. Prooxidant states can be caused by different classes of agents, including hyperbaric oxygen, radiation, xenobiotic metabolites, modulators of the cytochrome P-450 electron-transport chain, and membrane-active agents. The support for the role of radicals in carcinogenesis derives from the observation that many radical precursors are tumor promotors; in contrast, many antioxidants are anticarcinogens.

Another aspect of lipid peroxidation in diverse biological systems consists in interesting and unique properties of intermediate peroxides. Hydroperoxides derived from arachidonic acid, e.g., serve as modulators of enzymes taking part in biosynthesis of prostaglandin. Since a relationship exists between the presence of prostaglandins and blood platelet functions, it was alternatively assumed that a heart attack is a lipid peroxidation disease. 5-Hydroperoxyeicosatetraenoic acid was again proposed as an intermediate in the biosynthesis of leukotrienes which were identified in leukocytes. The lipoxygenase enzyme pathway leading to it has been shown quite persuasively.[10]

The reaction of trace metal ions producing reactive free radicals with an excess of hydroperoxides is obviously harmful and the level of peroxide intermediates is therefore controlled by glutathione peroxidase (GSH),[11] which decomposes them in a nonradical way (Equation 3).

$$ROOH + 2\ GSH \rightarrow ROH + GSSG + H_2O \tag{3}$$

Glutathione peroxidase is regenerated from its disulfidic analog by reaction with reduced form of nicotinamide adenine dinucleotide phosphate NADPH (Equation 4).

$$GSSG + NADPH + H^+ \rightarrow NADP^+ + 2\ GSH \tag{4}$$

Less reactive free radicals may also be formed in the interaction of trace elements or subsequently formed reactive free radicals with different vitamins, especially with those that are fat soluble. The reaction behavior of vitamin A (retinol) (Scheme 3),

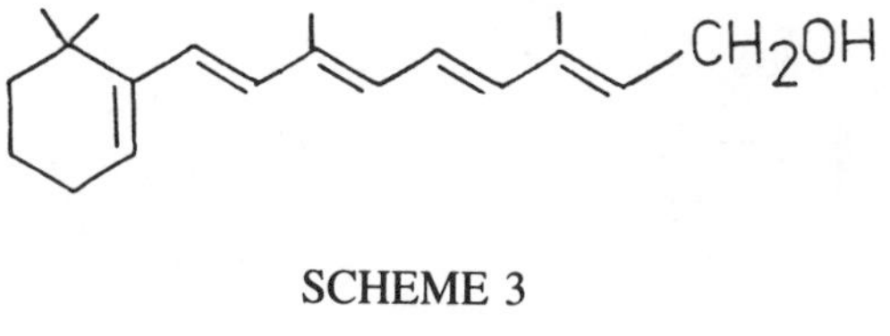

SCHEME 3

which is an unsaturated polyconjugated alcohol produced in the liver from its provitamin β-carotene and which is necessary for healthy mucous membranes and eyes, will be similar to that of unsaturated fatty acids. Vitamin D, occurring as cholecalciferol (Scheme 4)

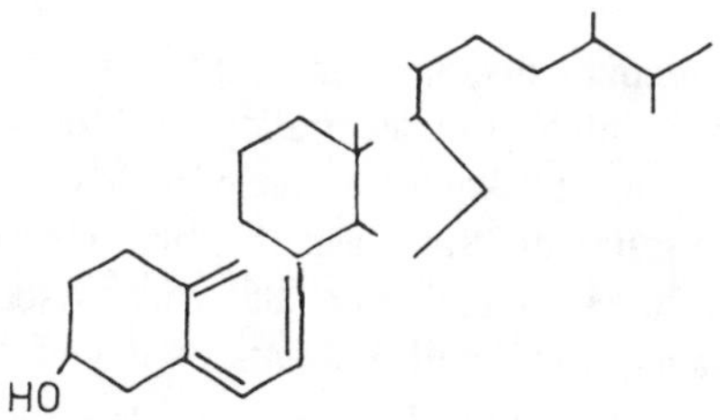

SCHEME 4

or ergocalciferol, is required for development of bones and teeth since it promotes the use of calcium and phosphates.

Much attention has been paid to vitamin E which is a mixture of various tocopherols and is known as a fat stabilizer. Its effect probably consists in antioxidant action, i.e., in scavenging reactive free radicals formed during oxidation. This fact has raised the question on the role of free radicals in pathological processes in living organisms; but what the tocopherols actually do is still the subject of controversy and details of the antioxidant mechanism are not fully understood.[12] Low levels of vitamin E in the blood, e.g., result in a higher susceptibility of red cells to hemolysis which might explain anemia. There are also studies on rabbits and rats demonstrating that a lack of vitamin E leads to muscular dystrophy.

The activity of vitamin K (Scheme 5 — vitamin K_2-phyloquinone), another substrate of a free radical trap which has an

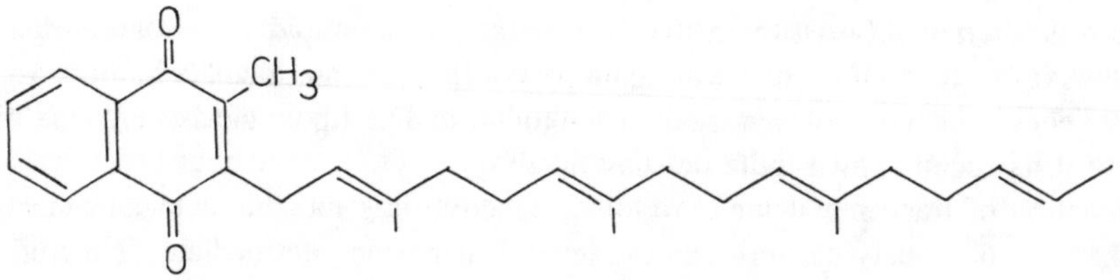

SCHEME 5

antihemorrhagic function, consists in quinoid groups which may easily be reduced to hydroquinone and reversibly oxidized and thus mediate the electron transfer. The quinone-hydroquinone electron shuttle provided by coenzyme Q (ubiquinone) is an important part of the respiratory chain from which oxidizing molecules obtain energy for living.

Vitamin C or L-ascorbic acid (Scheme 6) and its anions

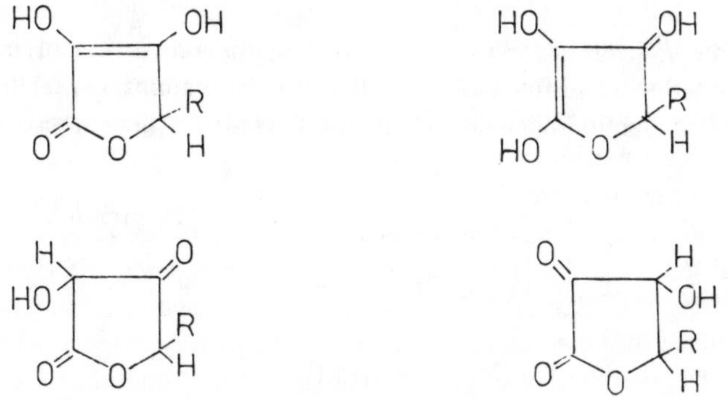

$$R = CH(OH)CH_2(OH)$$

SCHEME 6. Tautomers of L-ascorbic acid

and radicals play a vital role in plants, animals, and humans.[13] Ascorbic acid is omnipresent in blood and tissues, but the highest concentrations are found in the adrenal and pituitary glands, liver, spleen, and brain. It is known as one of the most important reducing substances occurring naturally in living tissues. Since ascorbic acid represents a powerful antioxidant, it is used as an additive to bread, meat, beer, and wine. Ascorbic acid acts synergistically with other food antioxidants, and small amounts of it yield a great gain in efficiency.[14] Ascorbic acid may provide some protection against free radicals and blocks the formation of N-nitrosocompounds which are carcinogenic.[15,16]

Cholesterol is an ingredient of cell membranes of vital importance for cell growth. However, its excessive presence leads to sedimentation on the walls of arteries and consequently to atherosclerosis. Ascorbic acid is necessary for the transformation of cholesterol into bile acids.[17]

It seems that ascorbic acid has an effect also on immunocompetence, since autooxidation of lipids facilitates penetration through the cell membranes. Then all direct and indirect stabilizers of membrane lipids influence the antigen-antibody reaction.[18]

Living organisms are frequently exposed to radiation and chemicals that can cause alterations in the structure of DNA. Evidence indicates that damage to DNA plays a role in mutagenesis, carcinogenesis, and aging.[19] The most widely studied radiation-induced modification of DNA is the formation of pyrimidine dimers by ultraviolet light. If these dimers are not repaired, they create blocks to the synthesis of normal DNA and a high rate of mutation. The helical axes of the models of DNA show substantial kinking and unwinding at the sites of damage, which may have long-range as well as local effects arising from the concomitant changes in the supercoiling and overall structure of the DNA.[20] Damage introduced to DNA by reactive radicals creates a number of structural defects including modified bases, base-free sites, and single and double strand breaks. Structural characterization of such defects is necessary for an understanding of their biological consequences and enzymatic repairability.[21]

Another DNA defect is caused by psoralen or its derivatives in conjunction with ultraviolet light. Psoralens are a class of heterocyclic compounds and anthralin derivatives (antipsoriatic drugs) which are used in the treatment of various skin diseases. For example, during the interaction of anthralin with psoriatic skin, free radicals from psoralen are formed, by oxidation with a polyunsaturated hydroperoxide.[22]

When compared to reactive and unstable free radicals which nonselectively enter a large variety of chemical reactions, the stable radicals in biological systems are usually the final stage of a complex transformation sequence and their correspondence to respective process may be more or less indirect.[23,24] ESR technique has led to a number of observations of free radicals in virtually all types of animal and plant preparations in both normal and pathological states. Because of the complexity of the cells and the relatively nonspecific information contained in ESR spectra of many bioradicals, however, estimation of the significance of such measurements is very difficult. There are published papers which do not take into account the possible experimental artifacts that can occur in the studies of whole cells and tissues, and lead to misinterpretation.

This is, e.g., the case of experiments that utilize lyophilized samples where stable free radicals may be generated by lyophilization itself. The received ESR signal then reflects only very indirectly the state of living systems. Also, rubbing, mixing, and grinding can mechanically induce the appearance of free radicals in many biopolymers and probably in tissues as well. The presence of oxygen during sample preparation may significantly affect radical formation in bacteria.

Regardless of experimental difficulties connected with nonresonance absorption of water containing biological systems, with ESR spectrometer restriction on sample size and shape, with the nature and low concentrations of paramagnetic species in cells, and with ESR power saturation, it seems that relative radical concentrations in different tissues may have some relationship to the life functions of corresponding organs (Figure 1).

The liver, kidney, and heart of a rat contain, e.g., the highest concentrations of stable free radicals, the ESR spectrum of which is a single line having the value of $g = 2.00$, which is characteristic for the signals of most organic free radicals. One should, however, realize that this spectrum may correspond to several different species which probably vary in composition among tissues. There are also other peaks at $g = 1.94, 1.97, 2.01, 2.03$, and 4 corresponding to transition metal ions and other paramagnetic ion clusters which may

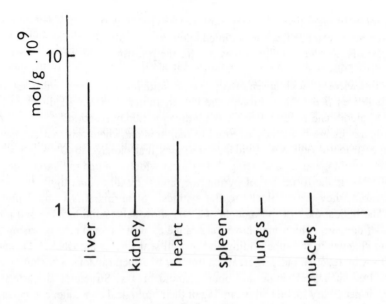

FIGURE 1.   Relative ESR signal intensities of normal rat tissues at 37°C.

sometimes be seen depending on the sensitivity of the ESR spectrometer, power saturation, and sample preparation.

ESR signals including free radicals and trace elements were also noted from different microorganisms. Examination of these signals should be conducted with extreme care since, e.g., Mn(II) hyperfine structure may disappear when the ion is a part of an aggregate or macromolecular complex. On the other hand, the shape of ESR spectra from microorganisms may prove to be the source of valuable information concerning the nature of the binding and of the function of trace elements in cells.

Stable free radical species have also been reported as an indirect result of photosynthesis on plants and seeds.[25] The usual ESR spectrum of the seed is in the first derivation from 0.5 to 2.5 mT wide and without distinctive hyperfine structure. Wetting reversibly decreases the intensity of the observable signal.

The origin and role of unpaired electrons in melanin, which is the substantial part of pigmented biological materials, is still the subject of controversy. Since melanin reacts readily with some stable free radicals, it was suggested that its free radical center accounts for some of its biological properties such as, e.g., radiation protection, etc. Provided that melanin is in contact with a solution of diphenyl picryl hydrazyl (DPPH) the concentration of DPPH radicals, however, smoothly decreases below the level of sensitivity of the ESR spectrometer, but the number and type of unpaired spins in melanin practically does not change. It seems, therefore, that reactions of added stable free radicals apparently occur with quinoid moieties of the melanin. It is of interest that unpaired electrons in melanin are very stable and can withstand 24 hr treatment with concentrated acids or bases. They react more readily with paramagnetic elements such as, e.g., copper.

On the other hand, the level of stable free radicals in melanin may be significantly increased by the effect of visible or UV light. Even though these paramagnetic particles are spectroscopically indistinguishable from those in nonirradiated samples, they are much more reactive. This may be due to the fact that they are generated on the surface of the melanin tissues. It was supposed that these light-induced species play an important role in visual processes in such a way that their subsequent reactions stimulate nerve signals.

Much research has also been devoted to the study of stable free radical changes in different states of organisms, especially in carcinogenesis and aging. The possible role of free radicals in carcinogenesis has been invoked at many opportunities, namely, when the effect of ionizing radiation and chemical carcinogens on tumor formation was studied. Experiments with several carcinogens such as 3,4-benzpyrenes indicate that they readily form complexes with suitable acceptors giving up single electrons and thereby forming free radicals.

Ionizing radiation generates a large amount of free radicals practically in all materials, so the relation between carcinogenesis and free radical reaction seems, at least in these two kinds of initiation, to be justified. From chemical carcinogens, the effect of which is satisfactorily screened, we may mention polycyclic aromatic hydrocarbons, nitroquinolines and arylnitroso compounds.

The approach of physical chemists to the carcinogenic effect of aromatic hydrocarbons is based upon the fact that a large number of hydrocarbons undergo a one electron abstraction step in oxidation to cation radicals. 3,4-Benzpyrene, a dominant carcinogen in cigarette smoke, forms, e.g., chemical linkages with DNA macromolecules, the formation of which is induced by diluted hydrogen peroxide or Fenton's reagent. It is of interest that noncarcinogenic 1,2-benzpyrene does not react with DNA under the same conditions. There is evidence that the cytotoxicity of corresponding benzpyrene is due to its transformation in the cell into 3-hydroxybenzpyrene. The pro-effect of hydrogen peroxide may thus consist in initiation of mild free radical oxidation of benzpyrene to its 3-hydroxy derivative.

The conception of free radical carcinogenesis based upon the binding of carcinogens to DNA induced by free radical initiators stimulated research on the chemical testing of potential carcinogens. The tested compounds underwent mild oxidation with suitable free radical initiating systems. The induction period of its oxidation was then put into the correlation with the results of biochemical tests. Benzaldehyde and oxygen were shown to be an especially effective initiating system for such a purpose.[26] Benzaldehyde functions here not only as the initiator but also as the substrate of oxidation; the compounds which accelerate its oxidation have a carcinogenic effect, those inhibiting oxidation were shown to be noncarcinogens. Even though each generalization in this respect is premature a good correlation with other and more expensive tests was obtained for several hundreds of organic compounds.

To complement the above idea, the different antioxidants of phenolic and amine type as inhibitors of free radical processes were reported to be effective in the treatment of malignant growths (Figure 2). Another pertinent observation is a correlation between antitumor activity and the ability of dichlorodiethyl arylamine compounds to give the observable free radicals on oxidation in weak acid solutions.[27]

What connection may this have with the stable free radical content in living tissues? It is known that one of the possible ways to form stable free radical centers may be to trap reactive radicals on a proper substrate. The indicator of the mobilization of protective functions of the organism should therefore be the increased free radical level stimulated either by increased production of such free radical traps or by a higher stationary level of reactive free radicals formed in cells.

In the rat hepatoma induced by chemical carcinogens, an ESR signal appears, having g = 2.035, several weeks before the conventional biochemical and histological indicators of tumors give positive results. Before the onset of observable malignancy this signal, however, disappears. The ESR signal apparently belongs to the unpaired spins located on a NO-Fe(II) complex with a thiol containing proteins.

In correspondance with the above idea, earliest findings have actually confirmed that malignant tissues have a lower free radical level than the standard samples, indicating, thus, that self protective functions are suppressed in the extended form of cancer.

An alternative explanation exists based upon the suggestion that the rate of cell division is controlled by free radicals; their high concentration inhibits the reactions necessary for cell division, whereas low concentration accelerates it.

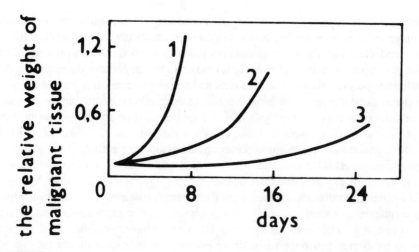

FIGURE 2.   Inhibition of the growth of the leucosis of mouse spleen by different anti-oxidants: 1 — control experiment, 2 — 2,6-di-*tert*-butyl 4-methyl phenol, 3 — 1,2-bis(diazoacetyl ethane).

The proper oscillatory regime of such reactions seems in this connection to be of significance, but considerably more evidence is necessary to verify such speculation.

Tumors generally contain sometimes quite different transition ions from normal tissues, which should be of value in the early diagnosis of cancer, but again further experimental data are needed.

Only a few experimental studies have been published on the possible correspondance of free radicals observed in vivo with the aging of organisms. From this viewpoint, the process of aging may be schematically understood as free radical cross-linking of phospholipids in cell membranes. This process irreversibly deteriorates the functional properties of membranes consisting of the transportability of ions and the activation of some enzymatic processes. Cross-linked lipids and denaturated proteins give rise to pigmented aggregates which are insoluble, and accumulate not only in the skin but also inside the organism. Oxidation and cross-linking of cell membranes proceed slowly because of the inhibiting effect of vitamin E present in membranes.

Since vitamin E is a relatively good scavenger of free radicals, the radical hypotheses of aging has found many supporters. Free radicals may be produced, e.g., in decomposition of metabolic hydrogen peroxide. As a consequence, the idea has been expressed that artificially supplied antioxidants may prolong life. Provided that 4-methyl-2,6-di-*tert*-butyl phenol was regularly added to the food of mice, the average age was increased but not the maximum lifespan. In this context, the relation between free radical content and the viability of the microorganism is of particular interest.

It seems that loss of viability of dried and lyophilized bacteria correlates well with an increase of free radical concentration in examined samples. On the other hand, if cells of bacteria were killed by heating or by chemicals, such as phenol or mercuric chloride before lyophilization, there was no detectable free radical production.

Even though it is not known whether the amount of stable free radicals observed is the cause or the result of death of an organism or whether it may be mere coincidence, the analogy with the behavior of the products of dehydration, deamination, or dehydrochlorination of synthetic organic materials is intriguing. Sugars, cellulosic materials, polyvinyl chloride, polyvinyl alcohol, etc. lose HCl and water by heat and thereby form carbonized residue giving an intense single ESR line of width 0.5 to 2 mT and g = 2.00. Apparently no reactive free radicals which could be trapped in the matrix are involved in the primary stages of the reaction; the appearance of paramagnetism should, therefore, be ascribed to a polyconjugated system of double bonds as one of its properties.[28] One may, thus, realize that the free radical hypothesis of aging is not the only way in which this unwanted process proceeds. The species giving the ESR signal may simply appear as a consequence of, e.g., dehydration of polysaccharides[29] and of deamination of proteins.

Keeping in mind that the signal of stable free radicals observed in living tissues may have its molecular origin in

1. Paramagnetic ions in macromolecular aggregates
2. Sulfur, quinoid, selenium, and other antioxidant moieties of biological systems functioning as traps of reactive free radicals
3. Conjugated systems of double bonds casually forming C-T complexes with other substrates

we may say that much should still be done and many experiments should be reinterpreted or repeated before the role of free radical species in living tissue may be satisfactorily formulated. The complexity of the living system makes experimentation difficult; moreover, the free radicals in the system are generated in small quantities over long periods of time. In spite of this, ESR investigations will be obligatory to the understanding of the roles of free radicals in biological systems, and spin trapping with more than one type of spin trap will undoubtedly play a major role as the complexities begin to be unraveled.[30]

REFERENCES

1. **Holum, J. R.**, *Fundamentals of General, Organic and Biological Chemistry*, John Wiley & Sons, New York, 1978.
2. **Mayo, S. L., Ellis, W. R. Jr., Crutchley, R. J., and Gray, H. B.**, Long-range electron transfer in heme proteins, *Science*, 233, 948, 1986.
3. **Lendzian, F., Lubitz, W., Steiner, R., Trankle, E., Plato, M., Sheer, H., and Mobius, K.**, The radical cation of bacteriochlorophyll b. A liquid-phase ENDOR and triple resonance study, *Chem. Phys. Lett.*, 126, 290, 1986.
4. **Cohen, G.**, The generation of hydroxyl radicals in biologic systems: toxicological aspects, *Photochem. Photobiol.*, 28, 669, 1978.
5. **Motohashi, N. and Mori, I.**, Thiol-induced hydroxyl radical formation and scavenger effect of thiocarbamides on hydroxyl radicals, *J. Inorg. Biochem.*, 26, 205, 1986.
6. **Witting, L. A.**, in *Free Radicals in Biology*, Vol. 4, Pryor, W. A., Ed., Academic Press, New York, 1980, 295.
7. **Yagi, K.**, Ed., *Lipid Peroxides in Biology and Medicine*, Academic Press, New York, 1982.
8. **Lippman, R. D.**, in *Review of Biological Research in Ageing*, Vol. 1, Rothstein, M., Ed., Alan R. Liss, New York, 1983, 315.
9. **Cerutti, P. A.**, Prooxidant states and tumor promotion, *Science*, 277, 375, 1985.
10. **Porter, N. A., Weber, B. A., Weenen, H., and Khan, J. A.**, Autooxidation of polysaturated lipids. Factors controlling the stereochemistry of product hydroperoxides, *J. Am. Chem. Soc.*, 102, 5597, 1980.
11. **Jindra, A., Kovacs, P., Psenak, M., and Sipal, Z.**, Biochemisty: molecular biological and pharmaceutical aspects (Slovak and Czech), Publishing House Osveta, Martin, 1985.
12. **Matsumoto, S., Matsuo, M., Iitaka, Y., and Niki, E.**, Oxidation of a vitamin E model compound, 2,2,5,7,8-pentamethylchroman-6-ol, with t-butylperoxyl radical, *J. Chem. Soc., Chem. Commun.*, 1076, 1986.
13. **Eckert-Maksic, M., Bischof, M., and Maksic, Z. B.**, Vitamin C and its radicals: tautomerism electronic structure and properties, *Croat. Chem. Acta*, 58, 407, 1985.
14. **Lolinger, J., Lambelet, P., Savoy, M. C., and Ducret, F.**, Radical exchange reactions between autooxidizing lipids, vitamin E and vitamin C in binary lipid/water systems, *Fette Seifen Anstrichm.*, 88, 584, 1986.

15. **Cort, W. M.,** in *Ascorbic Acid: Chemistry, Metabolism, and Uses,* Seib, P. A. and Tolbert, B. M., Eds., American Chemical Society, Washington, D.C., 1982, 533.

16. **Floyd, R. A., Ed.,** *Free Radicals in Cancer,* Marcel Dekker, New York, 1981.

17. **Ginter, E., Bobek, P., Babala, F., Kubec, D., Urbanova, D., and Cierna, D.,** *Advances in Physiological Science,* Vol. 12, *Nutrition, Digestion and Metabolism,* Gati, T., Szollar, L. G., and Ungvary, G., Eds., Akad. Kiado, Budapest, 1981, 79.

18. **Kolinkoeva, A. A.,** On the role of some free radicals (SH groups) in immune reactions and barrier function of spermatozoa membrane, *C. R. Acad. Bulg. Sci.,* 39, 129, 1986.

19. **Halliwell, B. and Gutteridge,** *Free Radicals in Biology and Medicine,* Clarendon Press, Oxford, 1985.

20. **Pearlman, D. A., Holbrook, S. R., Pirkle, D. H., and Kim, S. H.,** Molecular models for damage by photoreaction, *Science,* 227, 1304, 1985.

21. **Dizdaroglu, M.,** Characterization of free radical-induced damage to DNA by the combined use of enzymatic hydrolysis and gas chromatography-mass spectrometry, *J. Chromatogr.,* 367, 357, 1986.

22. **Ducret, F., Lambelet, P., Loliger, J., and Savoy, M. C.,** Antipsoriatic drug action of anthralin: oxidation reactions with peroxidizing lipids, *J. Free Rad. Biol. Med.,* 1, 301, 1985.

23. **Swartz, M., Botton, J. R., and Borg, D. C.,** *Biological Applications of Electron Spin Resonance,* Wiley Interscience, New York, 1972.

24. **Franconi, C., Ed.,** *Magnetic Resonance in Biological Research,* Gordon and Breach Science Publishers, New York, 1971.

25. **Michalov, J. and Placek, J.,** Free radicals in the root system and stem of Zea Mays, *J. Exp. Bot.,* 33, 511, 1982.

26. **Gasperik, J.,** Benzaldehyde oxidation test, a model reaction with radical mechanism, *Chem. Zvesti,* 29, 803, 1975.

27. **Emanuel, N. M.,** Chemical and biological kinetics (Russian), *Usp. Khim.,* 50, 1721, 1981.

28. **Rychly, J., Matisova-Rychla, L., and Placek, J.,** Paramagnetism of carbonization products formed from isotactic polypropylene containing synthetic zeolites, *Coll. Czechoslov. Chem. Commun.,* 44, 3665, 1979.

29. **Simkovic, I.,** Free radicals in wood chemistry, *JMS-Rev. Macromol. Chem. Phys.,* C26, 67, 1986.

30. **Harbour, J. R. and Hair, M.,** Transient radicals in heterogeneous systems: detection by spin trapping, *Adv. Coll. Interface Sci.,* 24, 103, 1986.

Chapter 11

RADICALS IN THE COSMOS AND IN THE ATMOSPHERE

I. INTRODUCTION

The high reactivity of radicals is the main reason for their generally low concentration in nature. Their large number of collisions prevent their accumulation in gaseous, liquid, or solid environments. But at specific conditions, such as low density, low temperature, and a strong radiation field, their concentration and hence their importance significantly increases in relation to the concentration of molecules with closed electronic shells. The low density of matter and low temperature prolong their lifespans and electromagnetic radiation provides energy for their continuous generation. Such convenient conditions can be encountered at higher atmospheric strata of planets or in interstellar space. Radicals are also present in the surroundings of stars and in the cometary nuclei and tails, the mass of comets being negligible relating to the mass of interstellar matter. Processes taking place in the proximity of the stars are not fully understood yet and for this reason we describe here only two previously mentioned regions of radical occurrence in the cosmos.

II. INTERSTELLAR MATTER

Even though only gas and fine dust particles are fundamental constituents of interstellar matter, they comprise a relatively high percentage of cosmos mass. In our galaxy it is, e.g., about 10% of total mass.[1] Most of it is gas which consists predominantly of hydrogen and helium, with an approximate ratio by mass of H:He:all heavier elements = 70:28:2. Dust accounts for only 1% of the mass of interstellar matter. Its predominant components are presumed to be molecular ice, silicates, and graphite with particles of size 20 to 150 nm.

The physical conditions, under which the interstellar gas and dust exist, differ from the usual conditions on Earth. The existence of microwave background radiation (which corresponds to the "big bang" theory of the evolution of the Universe) is the reason that the lower limit of the kinetic temperature of particles is 2.7 K. The dense clouds of interstellar matter may even have a kinetic gas temperature as low as 5 K. On the other hand, a tenuous gas in the interarm regions of our galaxy is hot ($T_k > 1000$ K). The density of matter can vary from 0.2 to 0.02 in hot tenuous gas to 10^5 to 10^7 particles per cubic centimeter in some black clouds. Its average value for interstellar gas in the galactic plane is about 0.5 atoms per cubic centimeter. Interstellar clouds of high density and at low temperature can, at sufficiently large mass, become gravitationally unstable and can contract to form future stars. Interstellar matter interacts significantly with radiation of different types (photons and elementary particles of different energy) originating from the stars and with the previously mentioned microwave background radiation. The UV field at wavelengths larger than 91.2 nm (the ionization threshold for a hydrogen atom) plays a dominant role in the destruction of molecules. At wavelengths in the region of 91.2 to 10 nm, photons are absorbed by the ionization of atomic hydrogen in the immediate vicinity of the stars from which they originate.

As the only source of information about structures and processes in the interstellar molecular gas we may use radiation coming from the cosmos. Its spectral analysis provides large amounts of data concerning the type of molecules and their kinetic and internal energy. On the other hand, many gaps in the spectral region (caused by absorption of photons on dust particles of interstellar clouds or in Earth atmosphere) and the limited possibility of detecting nonpolar molecules (as CH_4, CO_2, or C_2H_2) may thus distort the general chemical picture of the Universe. The most useful part of the spectrum, observed as emission or absorption lines, covers the millimeter and centimeter wavelengths and needs the use of radioastronomical techniques. Absorption or emission of radiation is connected mainly with the transition between rotational energy levels of molecules. Also, other effects such as electron and nuclear spin, nuclear quadrupole moment, and the electron orbital moment coupled with the rotational moment (leading to the splitting of rotational levels) are important for the identification of individual molecules.[2] The lowest energy levels are usually occupied according to the Boltzmann distribution law. However, there are important exceptions to this rule. Intense emissions of a maser character have been observed in some parts of space for ·OH radicals and for H_2O, SiO, and CH_3OH molecules. The origin of such giant natural masers is still not fully understood.

Chemical reactions and the chemical structure of interstellar gas are determined by physical conditions in a given part of space. In the low density interstellar clouds the main constituents are atoms with a prevailing concentration of atomic hydrogen. Dense clouds are of a molecular character because of shielding from the destructive UV radiation field by the interstellar dust. The main component here is molecular hydrogen, but also CO molecules attain relatively a large concentration. Radicals observed in such clouds are ·OH, ·CH, ·CN, ·NO, ·NS, CS·$^+$, HCC·, HCO·, ·C≡C−CN, ·C≡C−C≡CH. The most interesting molecules appearing there are cyanopolyins of general formula $H(C≡C)_nCN$ with n = 1,2,3,4,5 which are unstable at laboratory conditions.

In spite of different conditions, it is possible to consider some common features of the life cycle of interstellar molecules, the most important being their formation, relaxation, and destruction. The low kinetic temperature allows the occurrence of only those exothermic gas phase reactions which have zero activation energy. This is characteristic for reactions of cations and cation radicals with neutral molecules or radicals. Of importance here are radiative association reactions where the excess of energy is released by the emission of photons. In spite of the fact that this type of relaxation is very slow, it represents at low density conditions the only possibility of stabilization of an energized molecule. An important example is a recombination reaction (Equation 1) which is the

$$C·^+ + H_2 \rightarrow CH_2^+· + h\nu \tag{1}$$

main reaction for carbon-bearing molecules.[3] A variety of processes may then follow (Equations 2 to 7):

$$CH_2^+· + e^- \rightarrow CH· + H· \tag{2}$$

$$CH_2^+· + H_2 \rightarrow CH_3^+ + H· \tag{3}$$

$$CH_3^+ + e^- \rightarrow CH· + H_2 \tag{4}$$

$$CH_3^+ + e^- \rightarrow :CH_2 + H· \tag{5}$$

$$CH_3^+ + H_2 \rightarrow CH_5^+ + h\nu \tag{6}$$

$$CH_5^+ + e^- \rightarrow CH_4 + H· \tag{7}$$

Formation of hydrogen molecules proceeds in dense clouds through supposed recombination of hydrogen atoms on the surface of dust grains. This type of recombination can also give larger molecules but theoretical and experimental simulation of such heterogeneous processes is highly uncertain and subjective. Another example of the formation of ·OH and H_2O is the following sequence of elementary steps (Equations 8 to 15).

$$H_2 + \text{cosmic ray} \rightarrow H_2^+\cdot + e^- \tag{8}$$

$$H_2^+\cdot + H_2 \rightarrow H_3^+ + H\cdot \tag{9}$$

$$H_3^+ + \cdot O\cdot \rightarrow {:}OH^+ + H_2 \tag{10}$$

$${:}OH^+ + H_2 \rightarrow H_2O^+\cdot + H\cdot \tag{11}$$

$$H_2O^+\cdot + H_2 \rightarrow H_3O^+ + H\cdot \tag{12}$$

$$H_3O^+ + e^- \rightarrow H_2O + H\cdot \tag{13}$$

$$H_3O^+ + e^- \rightarrow \cdot OH + H_2 \tag{14}$$

$$H_3O^+ + e^- \rightarrow \cdot OH + 2\,H\cdot \tag{15}$$

Once such basic hydrides and their ions are present, many subsequent reactions leading to larger molecules are possible.[4]

Dissociation of interstellar molecules is mainly a photoinitiated process. All observed molecules can dissociate after absorption of UV photons of wavelengths $\lambda > 91$ nm. This fact decreases their lifetime and rules out the hypothesis of their formation in the atmosphere of cool stars, because they cannot survive subsequent travel to the place of their occurrence. The most stable observed molecule is CO with an exceptionally high decomposition threshold ($\lambda = 111.5$ nm). The destruction of interstellar molecules by interaction with soft X-rays and with the low-energy part of cosmic radiation is less important, because of the general decrease of the decomposition rate constants with the increasing energy of the collision partner.

III. RADICALS IN THE ATMOSPHERE

The strong ultraviolet radiation of the sun is the main factor which determines the character of chemical reactions in the atmosphere of planets. The most important absorption of solar radiation can be attributed to carbon dioxide for Mars and Venus, to molecular oxygen for the Earth, and to methane and ammonia for Jupiter and the outer planets.[5,6] Absorbed energy starts then complex cycles of chemical reactions with simple free radicals as important reactive intermediates. Because most details are known about the chemical processes in the terrestrial atmosphere we present some of them here, stressing their importance for biological processes on Earth.

Photodissociation of $\cdot O_2\cdot$ in the stratosphere (Equation 16)

$$\cdot O_2\cdot + h\nu \quad (\lambda < 244 \text{ nm}) \rightarrow 2\,\cdot O\cdot(^3P) \tag{16}$$

forms active oxygen atoms. The most important consequence of this process is the formation of a broad ozone layer according to the combination reaction (Equation 17).

$$\cdot O\cdot(^3P) + \cdot O_2\cdot + M \rightarrow O_3 + M \tag{17}$$

As third particles M are considered the nitrogen and oxygen molecules. The presence of the ozone molecules in the atmosphere is of major importance. They are able to attenuate the flux of solar photons at the surface by a factor of 10^5 at wavelengths where radiation induces significant damage to DNA in biological systems.[7] Also hypothesized from this fact is the emergence of life from the sea to land about 400 million years ago when the level of O_2 in the atmosphere reached 10% of its present value and the shielding effect of O_3 started to act. The natural destruction of ozone proceeds through two channels; as photodissociation (Equation 18) or as a reaction with $\cdot O\cdot$

$$O_3 + h\nu \rightarrow \cdot O\cdot + O_2 \tag{18}$$

atoms (Equation 19). This second reaction can proceed also through

$$O_3 + \cdot O\cdot(^3P) \rightarrow 2\cdot O_2\cdot \tag{19}$$

faster catalytical cycles with so-called minor constituents (Equations 20 and 21). Such cycles have been recognized for $X\cdot = \cdot NO$,

$$XO\cdot + \cdot O\cdot \rightarrow X\cdot + \cdot O_2\cdot \tag{20}$$

$$X\cdot + O_3 \rightarrow XO\cdot + \cdot O_2\cdot \tag{21}$$

$Cl\cdot$, $H\cdot$ and $HO\cdot$. The content of $\cdot NO$ and $\cdot NO_2$ in the atmosphere can be significantly increased by industrial fixation of N_2 leading to nitrates used as fertilizers which are biologically transformed to N_2 and N_2O. The N_2O molecules can then interact with singlet oxygen atoms (Equation 22) and form

$$N_2O + \cdot O\cdot(^1D) \rightarrow 2\cdot NO \tag{22}$$

one from components of mentioned catalytical cycles. The other danger comes from combustion motors and especially from high-flying aircraft releasing nitrogen oxides directly into the stratosphere. $Cl\cdot$ atoms are formed not only from their natural sources but also from chlorofluoromethanes (e.g., CF_2Cl_2, $CFCl_3$) by photolysis. Chlorofluoromethanes can reach the atmosphere from aerosol propellants, foam-blowing agents, sprays and refrigerators. The increasing concentration of both the nitrogen oxides and the chlorine-containing compounds can significantly disturb the equilibrium concentration of ozone in the protective gaseous layer.

Another important ecological problem is the formation of photochemical smog above large cities. Radicals such as $\cdot NO$, $\cdot NO_2$, $\cdot OH$, $\cdot OOH$, $R\cdot$, $RO\cdot$, $ROO\cdot$ are embodied in complex cycles of chemical reactions together with O_3, hydrocarbons, and many inorganic compounds. Free radicals also play an indirect role in the formation of acid rains (containing H_2SO_4 and HNO_3 as main components) above heavily industrialized regions. The burning of fossil fuels is not only the main source of anthropogenic SO_2 but also CO_2, increasing contents of which in the atmosphere can bring about the instability in energy exchange processes on Earth and cause a significant increase of temperature (greenhouse effect).[9] All the above mentioned human activities can lead to serious changes in our atmosphere which may worsen conditions for life on Earth.

The importance of free radicals in these processes may be illustrated on the case of reactive $\cdot OH$ radicals. Their sources in different levels of the atmosphere are represented by reactions 23 to 31.

$$H_2O + h\nu \rightarrow \cdot OH + H\cdot \tag{23}$$

$$O_3 + H\cdot \rightarrow \cdot OH + \cdot O_2\cdot \tag{24}$$

$$\cdot OOH + O_3 \rightarrow \cdot OH + 2\ \cdot O_2 \cdot \tag{25}$$

$$\cdot OOH + H \cdot \rightarrow 2\ \cdot OH \tag{26}$$

$$\cdot OOH + \cdot O \cdot \rightarrow \cdot OH + \cdot O_2 \cdot \tag{27}$$

$$\cdot OOH + \cdot NO \rightarrow \cdot OH + \cdot NO_2 \tag{28}$$

$$O(^1D) + H_2O \rightarrow 2\ \cdot OH \tag{29}$$

$$O(^1D) + CH_4 \rightarrow \cdot OH + \cdot CH_3 \tag{30}$$

$$O(^1D) + H_2 \rightarrow \cdot OH + H \cdot \tag{31}$$

The $\cdot OH$ radicals play an important role at mentioned catalytical cycles of ozone destruction as one from minor constituents (Equations 20 and 21). They participate also in the increase in the acidity of rains. Atmospheric degradation of natural sulfur compounds (Equation 32)

$$\cdot OH + H_2S \rightarrow \cdot SH + H_2O \tag{32}$$

is the beginning of the oxidation processes leading to SO_2. Another reaction (33) represents

$$\cdot OH + \cdot NO_2 + M \rightarrow HNO_3 + M \tag{33}$$

the decline of nitrogen oxides from the atmosphere. On the other hand, $\cdot OH$ radicals hinder the decline of part of HCl in rains by its dissociation (Equation 34) which regenerates

$$\cdot OH + HCl \rightarrow H_2O + Cl \cdot \tag{34}$$

$Cl \cdot$ atoms for further ozone-destruction cycles. The oxidation of hydrocarbons is initiated by $\cdot OH$ radicals (Equation 35) which

$$\cdot OH + RH \rightarrow H_2O + R \cdot \tag{35}$$

consequently cause smog formation during appropriate conditions. The last stage of the oxidation processes proceeds by active participation of $\cdot OH$ radicals (Equations 36 to 38) and increases the CO_2 content in the atmosphere.

$$\cdot OH + H_2CO \rightarrow H_2O + H\dot{C}O \tag{36}$$

$$H\dot{C}O + \cdot O_2 \cdot \rightarrow CO + \cdot OOH \tag{37}$$

$$\cdot OH + CO \rightarrow CO_2 + H \cdot \tag{38}$$

As follows from these examples, atmospheric chemistry can only be fully understood with the knowledge of the concentration and chemical reactions of small free radicals in different levels of the atmosphere. Different techniques of concentration measurement (e.g., Reference 10) and laboratory measurement of rate constants of such reactions (e.g., Reference 11) allow mathematical modeling of processes in the atmosphere (see also Reference 17). The predictions of these models warn us of the concentration shocks in the atmosphere and urge us to control the gaseous waste of mankind.

REFERENCES

1. **Winnewisser, G., Mezger, P. G., and Breuer, H. D.,** Interstellar molecules, *Top. Curr. Chem.*, 44, 1, 1974.
2. **Winnewisser, G.,** Molecules in astrophysics, in *Computational Techniques in Quantum Chemistry and Molecular Physics*, Diercksen, G. H. F., Sutcliffe, B. T., and Veillard, A., Eds., D. Reidel Publishing, Dordrecht, 1975.
3. **Duley, W. W. and Williams, A. A.,** *Interstellar Chemistry*, Academic Press, London, 1984.
4. **Walmsley, C. M.,** Molecular abundances in interstellar clouds, *Phys. Scripta*, T11, 27, 1985.
5. **Nicolet, M.,** Atmospheric chemistry, in *Aspects of Chemical Evolution*, Nicolis, G., Ed., John Wiley & Sons, New York, 1984, 63.
6. **McEvan, M. J. and Philips, L. F.,** *Chemistry of the Atmosphere*, E. Arnold, Christchurch, 1975.
7. **Wiesenfeld, J. R.,** Atmospheric chemistry involving electronically excited oxygen atoms, *Acc. Chem. Res.*, 15, 110, 1982.
8. **Trush, B. A.,** Laser magnetic resonance spectroscopy and its application to atmospheric chemistry, *Acc. Chem. Res.*, 14, 116, 1981.
9. **Heinloth, K.,** CO_2/O_2 Balance of the Earth's atmosphere, *Proc. 6th Gen. Conf. Eur. Phys. Soc.* Vol. 2, Prague, 1984, 661.
10. **Anderson, J. G., Hazen, N. L., McLaren, B. E., Rowe, S. P., Schiller, C. M., Schwaab, M. J., Solomon, L., Thompson, E. E., and Weinstock, E. M.,** Free radicals in the stratosphere: a new observational technique, *Science*, 228, 1309, 1985.
11. **Wine, P. H., Kreutter, N. M., Gump, C. A., and Ravishankara, A. R.,** Kinetics of OH reactions with the atmospheric sulfur compounds H_2S, CH_3SH, CH_3SCH_3, and CH_3SSCH_3, *J. Phys. Chem.*, 85, 2660, 1981.
12. **Anderson, J. G.,** Free radicals in the Earth's atmosphere: their measurement and interpretation, *Annu. Rev. Phys. Chem.*, 38, 489, 1987.

INDEX